Advances in Neurosurgery 11

Acute Non-Traumatic Intracranial Bleedings

Posterior Fossa Tumors in Infancy

Edited by
H.-P. Jensen M. Brock M. Klinger

With 186 Figures and 122 Tables

Springer-Verlag Berlin Heidelberg GmbH

Proceedings of the 33rd Annual Meeting of the
Deutsche Gesellschaft für Neurochirurgie
Kiel, May 16 – 20, 1982

Prof. Dr. Hans-Peter Jensen
Neurochirurgische Universitätsklinik Kiel
Weimarer Straße 8, D-2300 Kiel 1

Prof. Dr. Mario Brock
Abteilung für Neurochirurgie, Neurochirurgische Klinik und
Poliklinik, Universitätsklinikum Steglitz,
Freie Universität Berlin, Hindenburgdamm 30, D-1000 Berlin 45

Priv.-Doz. Dr. Margareta Klinger
Neurochirurgische Klinik der Universität Erlangen-Nürnberg
Schwabachanlage 6 (Kopfklinikum), D-8520 Erlangen

DOI 10.1007/978-3-662-05589-2

Library of Congress Cataloging in Publication Data. Deutsche Gesellschaft für Neurochirurgie. Tagung (33rd. : 1982 : Kiel, Germany) Akute non-traumatic intracranial bleedings. (Advances in neurosurgery ; 11)
"Proceedings of the 33rd Annual Meeting of the Deutsche Gesellschaft für Neurochirurgie, Kiel, May 16–20, 1982"–T. p. verso. Includes bibliographies and index.
1. Brain–Hemorrhage–Congresses. 2. Cranial fossa, Posterior–Tumors–Congresses. 3. Tumors in children–Congresses. 4. Infants–Diseases–Congresses. I. Jensen, H.-P. (Hans-Peter), 1921-. II. Brock, M. (Mario), 1938-. III. Klinger, M. (Margareta), 1943-. IV. Title: Akute non-traumatic intracranial bleedings. V. Title: Posterior fossa tumors in infancy. VI. Series. [DNLM: 1. Cerebral hemorrhage–Congresses. 2. Brain neoplasms–In infancy and childhood–Congresses. 3. Brain–Drug effects–Congresses. 4. Brain –Surgery–Congresses. W 1 AD684N v. 11 / WL 355 D4882 1982a]
RC394.H37D48 1982 616.8'1 83-10467

Originally published by Springer-Verlag Berlin Heidelberg New York in 1983

MyCopy version of the original edition 1983

2122/3140-543210

www.springer.com/mycopy

Preface

This volume of ADVANCES IN NEUROSURGERY presents the original texts of 60 papers delivered at the 33rd annual meeting of the German Neurosurgical Society held in Kiel from May 16th to 20th, 1982. These papers represent a selection from some 162 papers submitted and 96 actually given. The selection was made by the society's programme committee, of which Professor W. J. BOCK, Professor H. DIETZ and Professor W. GROTE are also members. I would like to take this opportunity to express my sincere thanks to them for their untiring cooperation.

The scientific programme dealt with three main themes:

1. Acute, non-traumatic intracranial hemorrhages, a subject that has always been of importance for neurological surgeons since the anatomist Giovanni Battista MORGAGNI in 1791 first described in detail the clinical picture and the pathological and anatomical causes of a brain hemorrhage he had observed in his servant. Indeed, at our 31st annual meeting in Erlangen in 1980 "Timing Problems in Subarachnoid Hemorrhages" was one of the main topics of discussion. For this year's meeting a cooperative study in which 27 university and hospital departments of neurosurgery participated enabled us to look into the causes and the diagnostic and therapeutic measures involved in a great number of cases of intracranial hemorrhage.

2. Tumors of the posterior cranial fossa in childhood are another topical subject for examination, especially with regard to possible X-ray therapy and cytostatic treatment. The results of a cooperative study in this field in which 31 university and hospital departments took part were also reported, thus given a comprehensive view of the operative treatment.

   In addition to their scientific value the cooperative studies carried out are also a form of "quality control" that is being called for today - and rightly so - especially by surgeons.

3. Other unrelated papers in which, on the occasion of the annual meeting, recent findings in the field of research were presented. In order to preserve the thematic unity in this volume we unfortunately must forego printing many of these papers.

Special thanks once again to all our colleagues who, with their papers, contributed of the success of the society's 33rd annual meeting. In recognition of their contribution the executive of the German Neurosurgical

Society resolved to present every speaker with the congress medal for 1982 and a certificate of participation. The congress emblem was taken from the oldest seal of the city of Kiel from the year 1365. It shows a cog with a nettle leaf on the red coat of arms of the Earl of Holstein on the stem. The sail of the heraldic boat bearing the crest was fashioned anew, receiving the emblem of the German Neurosurgical Society (Fig. 1).

To those who submitted their papers for this volume of ADVANCES in the desired form and in English, a special word of thanks.

Our sincere gratitude we also express to the Springer Verlag for its help in editing the ADVANCES IN NEUROSURGERY, Volume 11.

H.-P. Jensen

Fig. 1. Medal of the 33rd annual meeting of the German Neurosurgical Society in Kiel, 1982

Contents

Posterior Fossa Tumors in Infancy

Free Topics

List of Contributors*

* Addresses given at beginning of each contribution
1 Page on which contribution begins

The President's Opening Remarks and Address to the Delegates of the 33rd Annual Congress of the German Neurosurgical Society

H.-P. Jensen

Neurochirurgische Universitätsklinik Kiel, Weimarer Strasse 8, D-2300 Kiel 1

Mr. President of the World Federation of Neurosurgical Societies, Fellow Colleagues, Ladies and Gentlemen,

I want to extend to you all a most cordial welcome to the thirty-third annual congress of the German Neurosurgical Society here in Kiel, the most northerly university city in Germany. I welcome Professor SEITZ, the president of the German Neurological Society, Professor HAMELMANN who is representing the president of the German Association of Surgeons and Professor WAWERSIK representing the German Association of Anaesthesists. I also welcome Dr. von SCHELIHA from our state government and Councillor BARTHEL, who is representing our mayor.

I particularly extend my welcome to the many participants from other European countries and from abroad: - first to those colleagues from Austria, Italy, Spain, Switzerland, the Benelux countries, France and, last but not least, Great Britain, with whom we have been associated for years, and then to our friends from the United States and Canada. By their participation in our congress I already see us reaping the rewards of one of the early traditions of our society, that is our cultivation of long-standing personal contacts between German and American neurological surgeons as well as on the occasions of our congresses. This tradition was initiated in October 1978 when Frank MARGUTH, the president of our society at that time, invited the American Academy of Neurological Surgery to a joint meeting, an invitation that was reciprocated when the American Academy under its president Eben ALEXANDER Jr. asked us to their joint meeting in New York in October 1980. Their participation is also a symbol of the close association of neurological surgeons all over the world. This association was made very real to all of us last year on the occasion of the World Congress of Neurological Surgeons in Munich under the direction of Karl-August BUSHE, the president of that congress.

I also extend an especially warm welcome to our friends and colleagues from the Scandinavian countries. Originally we invited the Scandinavian Neurosurgical Society to a joint meeting in Kiel. But since the statutes of that venerable society forbid such joint meetings, in agreement with Jacob HUSBY, the secretary, we invited all the members of the Scandinavian society personally, just as we invited all our own members personally. I am very pleased to see that so many of our Scandinavian colleagues have accepted our invitation. You have joined the meeting, but it is not a joint meeting!

Advances in Neurosurgery, Vol. 11
Edited by H.-P. Jensen, M. Brock, and M. Klinger

I also want to send greetings to our many friends and colleagues in East Germany and other countries of Eastern Europe. Many of them have written to us but, for political reasons, are unable to take part in our congress.

Ladies and Gentlemen, after these words of welcome it is my sad duty to remember those who have passed from us. Since our last meeting the following six members of our society have died:

Mr. Walpole LEWIN, Cambridge
Professor Bent BROAGER, Copenhagen
Professor Hans-Robert MÜLLER, Hamburg
Professor Eduardo TOLOSA, Barcelona
Professor Herbert REISNER, Vienna
Professor Eric ZANDER, Lausanne

I thank you for rising in remembrance of them.

Among the president's duties are also many of an extremely pleasant nature. It is a great pleasure that I am able to inform you that on the occasion of our meeting in Tübingen on April 22, 1981 the members of our society elected two new honorary members.

They are Professor Joachim GERLACH, emeritus professor of neurosurgery and former director of the Department of Neurosurgery at the University of Würzburg and Professor Rudolf KAUTZKY, emeritus professor of neurosurgery and former director of the Department of Neurosurgery at the University of Hamburg-Eppendorf. With this honour the German Neurosurgical Society honours their services to our field of specialization and our society, their scientific research and especially their heuristic work on the ethics of medicine and the philosophical and theological problems connected with neurosurgery. I now present Professor GERLACH and Professor KAUTZKY with their honorary memberships.

In addition, I am delighted to inform you that in its membership meeting this afternoon our society decided to make Professor Karl-August BUSHE, professor of neurosurgery and director of the Department of Neurosurgery at the University of Würzburg an honorary member. It is most unusual for a member of our society to receive this honour at such a young age while still carrying out his duties in science, research and teaching. The society thus pays tribute not only to Professor BUSHE's services to neurosurgery and his scientific research but also to all the work he does in an honorary capacity. He fulfils the obligations involved in these positions in an exemplary fashion, thus promoting the standing and reputation of our society. The following are only some of his honorary positions:

In 1967 and 1968 he was president of our society and organized two congresses during this time, one in Bad Harzburg and the other in Göttingen.

In 1972 he was president of the European Society of Paediatric Neurosurgery.

In 1974 and 1975 he was president of the International Society for Paediatric Neurosurgery and as congress president was responsible for that society's congress in Würzburg in 1976.

The culmination of this honorary activities was his term as president of the 7th International Congress of the World Federation of Neurosurgical Societies in Munich in July 1981. Four long years with unbelievable personal engagement he made preparations for this congress, one in which a quarter of all the neurological surgeons in the world took part. Thanks to the excellent organization this congress was characterized not only by a most rewarding scientific programme but also by a well planned recreational programme designed to build bridges of understanding between participants of all nations. In recognition of his achievements as congress president the World Federation of Neurological Societies made Professor BUSHE an Honorary President. To honour him as a doctor and a scientist he was awarded the German Distinguished Service Medal at the beginning of his term as Dean of the Medical Faculty at the University of Würzburg. I now present Professor BUSHE with the honorary membership in our society.

The executive committee of our society has decided to present the Congress Medal and certificate of participation to all those who distinguish themselves through taking part in this year's congress. I am especially happy to have presented the first three medals to our new honorary members.

A scientific society must promote research and science in its field of specialization. Wilhelm TÖNNIS, the far sighted founder of neurosurgery in Germany, recognized this and through a generous endowment enabled the German Neurosurgical Society to award the Wilhelm Tönnis Fellowship to promising young researchers. This year's fellowship was awarded to Dr. Hans-Peter RICHTER and it enabled him to take a study trip through the United States and Canada from January 1st to February 28th, 1982. Dr. RICHTER has given us a detailed report of his study trip that clearly shows that the purpose of the fellowship, that is the promotion of research and science, has indeed been served. I congratulate Dr. RICHTER on having been awarded this fellowship and in token of this I present him with the congress medal.

Ladies and Gentlemen, I would not want to close without honouring my own teachers. First of all, I would like to mention Karl KLEIST (1879-1960) professor of neurology and psychiatry in Rostock (1916-1920) and in Frankfurt/Main (1920-1950). As professor emeritus he was the director of the Research Centre for Brain Pathology and Psychopathology in Frankfurt. His main scientific interest was the function of the brain and the localization of neurological and psychic disorders of the brain. To him I owe my fascination with the study of the human nervous system and my decision to devote my life to it. If I had a summarize in one sentence what I learned from him, it would have to be that a complete and careful medical examination of the sick individual must always be the basis of a diagnosis. Only afterwards can this diagnosis be complemented and confirmed by technical examinations and laboratory tests. Here, nothing has changed even though today we are able to examine the patient in layers with the help of computers and other equipment.

Joachim GERLACH (born in 1908) became familiar with brain research as a student in Breslau when, under his teachers Otfrid FOERSTER and Hartwig KUHLENBECK he wrote a doctoral thesis on comparative brain anatomy. He later became a general surgeon but his interest in neuroanatomical research was always uppermost. For this reason, at the urging of Wilhelm TÖNNIS he accepted a position under Hugo SPATZ at the Kaiser Wilhelm Institute for Brain Research in Berlin. The outbreak of World War II ended his dream of working exclusively in research. As a surgeon he became an assistant ot Ernst LEMKE, who was at that time the director of the military hospital for injuries of the brain, spinal cord and nerves in Berlin. When LEMKE went to a department for neurosurgery on the Eastern front in 1944 GERLACH became his successor in Berlin. At the end of the war he came to the city of Schleswig together with ROSENHAGEN, a pupil of FOERSTER's, and built up a department for neurosurgery in the hospital there. In 1948 he followed a call to Würzburg where he was able to rebuild the first separate department of neurosurgery that had once been created by Wilhelm TÖNNIS. He eventually turned it into one of the most modern departments of neurosurgery in Germany.

When I became one of GERLACH's assistants in 1950 I not only had the great fortune of being able to participate in this work but also, as his only assistant, I had private lessons in neurosurgery so to speak. It was a difficult, strenuous time - we seldom talked about the quality of life - but it was perhaps the best time of my life. The most important thing I learned from Professor GERLACH would have to be this: One must never be content with a result or success no matter how great it might seem. Rather one must constantly ponder what one might have done better. It is a very great pleasure for me to have Professor GERLACH take part in our congress and I am especially honoured to have been able to present him with an honorary membership on behalf of our society.

Werner WACHSMUTH (born in 1900), professor of surgery in Würzburg (1946-1969) with a doctorate h.c. in jurisprudence from the University of Göttingen, was always my model of a departmental head. In his many textbooks and manuals he has dealt with the techniques of surgical operations and the bases of anatomy. His books on "Practical Anatomy", published together with the anatomist Titus von LANZ, are regarded in the whole world as the fundamentals of surgery. The most important thing that I learned from him was that in a well-directed department of surgery a fracture of the radius must be treated during night duty as carefully as a heart operation - for which all the hospital's equipment is necessary - is carried out on the following day. In Professor WACHSMUTH's lecture hall hung Goethe's words:

> "What is the most difficult of all?
> What you deem to be the easiest:
> To see what is in front of you."

He loved quotations. On the door of his operating theatre were these words of Leonardo da Vinci: "Where much noise prevails, there is no clarity!".

In concluding I would also mention those to whom I owe a great deal although I worked with them for only short periods of time. They are: Ernest WORINGER, Colmar; W. McKISSOCK, London; Norman DOTT, Edinburgh; Percival BAILEY, Paul BUCY, Joseph Patrick EVANS, Louis AMADOR, Benjamin BOSHES, Chicago and Donald D. MATSON in Boston.

One can best pay tribute to one's teachers by trying to retain their best attributes and passing them on to the next generation.

It has already become a tradition for the president, on the occasion of our annual congress, to comment on basic questions involving our field of specialization, our society or our profession. Such comments are intended primarily to aid our younger colleagues to understand better their own position, their obligations and their goals. For this reason, I would like to speak this evening on

The Researcher and the Doctor in Modern Society

Both the researcher and the physician have to fulfil important obligations that must not be distorted by the ideologies of special-interest groups. The researcher's achievement is always a very personal one; he tests his ideas by using scientific methods and searches for new knowledge. Even when he works with others or when a scientific project necessitates the cooperation of many researchers working together in a team this does not affect the researcher's personal scientific responsibility; it cannot be transferred to the group.

The same holds true of the doctor in medical practice who treats the sick patient and bears complete responsibility regardless of how many others help him and consult with him. The doctor's judgment is just as much a personal one as is the researcher's and he alone is responsible. The sick person is an individual as well and he has a right to have the doctor treat not just any disease but the disease that he has.

Thus both the researcher and the doctor obviously need a great deal of personal independence and inner freedom; they cannot seek shelter in the anonymity of the crowd. But like every free individual they too are buffeted by many external factors and by inner stress, especially in times of social turbulence and political, economic and philosophical uncertainty. What will enable them to ride out these storms?

R.S. MORISON (1980) has given a clear answer to this question. He speaks of ethics, by which he means the system governing our decisions especially in difficult times. He compares ethics with the gyro-compass that helps the sailor steady his ship and keep it on course through turbulence and uncertainty. This compass-like inner force, wound up in us and nurtured by our forefathers, has been passed on as a part of our heritage. We have fostered it since early childhood. It helps us to stay on course, establish sound thinking and make decisions that represent the accumulated wisdom of our heritage. In adversity it admonishes us to take corrective action, to make our mark and to dodge neither today's hazards nor those we fear tomorrow.

Independence, inner freedom and our ethics are prerequisites for our work in research and in medical practice.

The physician and the researcher have their roots in the university, as they have both learned here the tools of their trade so to speak. For this reason I would draw your attention to Christian Albrecht's foundation charter of the University of Kiel from the year 1665. You will find this charter at the beginning of this year's programme. Here Christian Albrecht sets down important tenets of a university. He thoroughly legitimates his action by giving evidence of his own great knowledge of philosophy and politics. He quotes Xenophon in whose Kyrupaidéia he had found the ideal image of the perfect prince.

So too will each researcher examine his own knowledge and competence before beginning a research project or making a scientific pronouncement. Incidentally, the state would probably be spared a great deal of trouble and embarrassment if the same virtues were demanded of its politicians. The doctor must of necessity be conscious of his own ability in every individual case, especially in our specialty neurosurgery, where a lack of knowledge and skill puts the patient's life at risk.

In our democracies many groups without any special knowledge at all are today concerning themselves with medical practice. They criticize, give advice and recommend reforms. There is no mistaking the ideological influences at work here. Their method of re-interpreting known facts and assigning new meanings to well-defined concepts shows this. Although the doctor's role up to now has been to treat the sick, in future he is to become a mere servant in a public health system which makes health - and not the medical treatment of illness - a new, suable social and political right. The goal of every medical treatment is to help the sick patient get well again. When health is declared technically producible and the expectation is awakened that health can be manufactured just like cars the ideological paralogism becomes obvious. The doctor becomes nothing but a public health technician. It is this way of looking at things here in Germany, that has led to public health being removed from the auspices of the Ministry of Health, Welfare and the Family and being made a part of the Ministry of Work. Physicians have become "productive performers" in our public health system.

Increasingly our courts are also having to concern themselves with medical practice. They mainly have to decide whether there is evidence of malpractice or whether the sick citizen has been properly made aware of the nature of his illness and of possible complications and results of all medical measures taken. Here the doctor definitely needs legal help. Our state government here in Schleswig-Holstein commissioned lawyers to work out directives concerning patients' consent to operative measures for the protection of doctors in the medical departments of universities. When we consider that in our field of specialization the discussion between the doctor and the patient, next to the diagnosis and the indication, is one of the medical doctor's most important and most difficult tasks it is not difficult to see just how complicated the lawyers' work was. They have completed their report. These "Directives for Medical Doctors in University Medical Departments" contain 1632 words. The Lord's Prayer, in comparison, contains only 69 words and the Ten Commandments some one hundred.

I mention these two texts for a special reason. Many a patient recites the Lord's Prayer before an operation, I sometimes wonder if they don't do this in the hope that their doctor will not follow the legal directives to the letter. And the Ten Commandments could regulate many of society's problems better than all the extensive laws of the modern state have been able to.

In his foundation charter Christian Albrecht gives a further example of a virtue that must characterize a university, namely absolute honesty in research and science - naming all the sources and authors one has used in one's own work for example. He reports that in founding the university he was only carrying on the work of his father Prince Frederick, who had already obtained many rights and privileges for the university from the Emperor Ferdinand III.

Independence, inner freedom, personal ethics and absolute honesty are prerequisites for the work of the researcher and the doctor.

Christian Albrecht indicated very clearly the importance of tradition, by which means experience, knowledge and obligations are passed on from one generation to the next. Tradition is equally meaningful for the researcher and the doctor. But influences of modern society can destroy our awareness and understanding of tradition, as I will demonstrate using the upbringing of the young as an example. By the turn of the century scientists, doctors and educators alike had realized that the demand for subordination and obedience in the upbringing of a child could have damaging effects. These could be avoided if pedagogical guidance replaced the parents' and teachers' formal authority. In the child's formative years this guiding principle would aid him in his efforts at integration and in his personal development. Guidance would replace formal authority, integration and personal development based on experience would replace subordination and obedience.

This scientific knowledge was deliberately misconstrued by those who coined the expression "anti-authoritative education" - which led, in the 1960s especially, to young people being instigated to rebel against every form of authority, even authority gained from experience and ability. This led to alienation and break in tradition. Now, everyone who thinks scientifically and uses well-defined concepts will immediately realize that the expression "anti-authoritative education" is a masterpiece of ideological re-interpretation. It is also a deliberate misconstruction of findings and concepts the results of which the younger generation of doctors and researchers will have to contend with for some time to come. And whether an upbringing or education that calls for a resistance to authority violates the fourth commandment "Honour thy father and thy mother" remains to be clarified.

The University of Kiel is greatly indebted to Christian Albrecht for his services and for the prudence and wisdom he always showed. To make the university attractive and to attract as many students as possible he took great care to call the best professors from near and far and saw to their well-being so that they would stay in Kiel. In addition to their teaching and research the professors had many public functions. As a result they were held in high esteem and the citizens of Kiel accepted the privileges granted them. One such privilege was tax immunity, which meant a great financial loss for the city at that time. Although subsequently, the University of Kiel suffered many a setback, it did eventually become a typical German university. German universities have been famous in history and have been esteemed throughout the world. They have been distinguished by the accomplishments of their professors. In today's society efforts at reform count for more than tradition and chairs of medicine have been legislated away. The holders of these chairs have been exonerated from their public obligations, and their freedom to teach and to research has been curtailed. The reforms have made them mere employees of the university. Only the future will show whether this new "reformed" university of professorial employees can stand the test of time here in Germany. My judgment is that such changes in structure cannot endanger the German university with its hallowed traditions.

Even though these newly created professorial employees cannot be expected to promote our renowned institutions of higher learning in future, the university fortunately still has in the younger generation an inexhaustible reserve of innovative potential at its disposal. Here, young people who are not willing to submit meekly to certain structures of society external influences or ideologies of special-interest groups will grow and become individuals, individuals who in independence and personal freedom will achieve, individuals for whom honesty is more meaningful than demagogy and the ability to persuade, individuals who are convinced of the value of the tradition that binds the generations. These individuals will go through life, their heads held high, guided by the gyro-compass of their ethics. They will resist the temptation to seek shelter in the anonymity of the group. It will be these individuals - and not groups - who as teachers and researchers will determine the make-up of our universities. They will vouch for the high standard of the free medical profession in our country and for future developments in our specialty neurosurgery.

Having seen that the intellectual home of the doctor and the researcher is and will continue to be the university, I can look forward to the future with optimism and conclude with Christian Albrecht's words: "Many the university thrive and progress and many every effort undertaken glorify and benefit our mother country and the intellectual community".

H.-P. Jensen

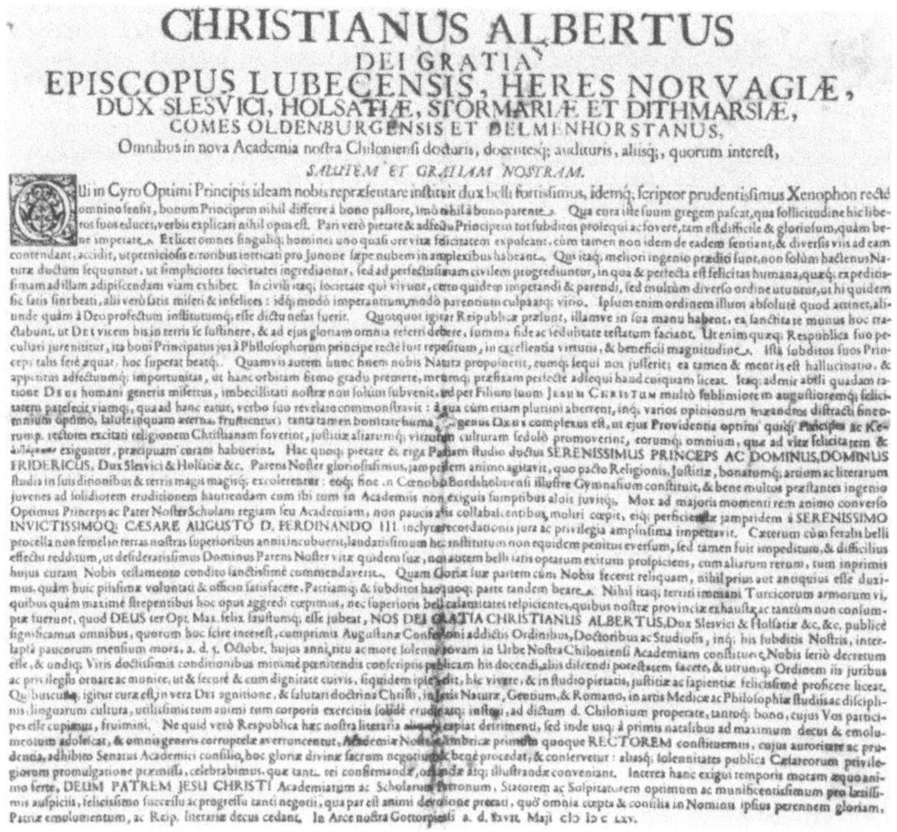

CHRISTIANUS ALBERTUS

DEI GRATIA

EPISCOPUS LUBECENSIS, HERES NORVAGIÆ, DUX SLESVICI, HOLSATIÆ, STORMARIÆ ET DITHMARSIÆ, COMES OLDENBURGENSIS ET DELMENHORSTANUS,

Omnibus in nova Academia nostra Chiloniensi docturis, docentesq; audituris, alisq;, quorum interest,

SALUTEM ET GRATIAM NOSTRAM.

Qui in Cyro Optimi Principis ideam nobis repræsentare instituit dux belli fortissimus, idemq; scriptor prudentissimus Xenophon rectè omnino sensit, bonum Principem nihil differre à bono pastore, imò nihil à bono parente. Qua cura iste suum gregem pascat, qua sollicitudine hic liberos suos educet, verbis explicari nihil opus est. Pari verò pietate & affectu Principem tot subditos prosequi ac fovere, tam est difficile & gloriosum, quàm bene imperare. Et licet omnes singuliq; homines uno quasi ore vitæ felicitatem expolcant, cùm tamen non idem de eadem sentiant, & diversis viis ad eam contendant, accidit, ut perniciosis erroribus irretiti pro Junone sæpe nubem in amplexibus habeant. Qui itaq; meliori ingenio præditi sunt, non solùm hactenus Naturæ ductum sequuntur, ut simpliciores societates ingrediantur, sed ad perfectissimam civilem progrediuntur, in qua & perfecta est felicitas humana, quæq; expeditissimam ad illam adipiscendam viam exhibet. In civili itaq; societate qui vivunt, certo quidem imperandi & parendi, sed multùm diverso ordine utuntur, ut hi quidem sic satis sint beati, alii verò satis miseri & infelices: idq; modò imperantium, modò parentium culpa atq; vitio. Ipsum enim ordinem illum absolutè quod attinet, aliunde quàm à Deo profectum institutumq; esse dictu nefas fuerit. Quotquot igitur Reipublicæ præsunt, illamve in sua manu habent, ea sanctitate munus hoc tractabunt, ut DEI vicem his in terris se sustinere, & ad ejus gloriam omnia referri debere, summa fide ac sedulitate testatum faciant. Ut enim quæq; Respublica suo peculiari jure nititur, ita boni Principatus jus à Philosophorum principe rectè fuit repositum, in excellentia virtutis, & beneficii magnitudine. Ista subditos suos Principi talis ferè æquat, hoc superat beatq;. Quamvis autem hunc finem nobis Natura proposuerit, eumq; sequi nos jusserit; ea tamen & mentis est hallucinatio, & appetitus affectuumq; importunitas, ut hanc orbitam firmo gradu premere, metamq; præfixam perfectè adsequi haud cuiquam liceat. Itaq; admirabili quadam ratione DEUS humani generis misertus, imbecillitati nostræ non solùm subvenit, sed per Filium suum JESUM CHRISTUM multò sublimiorem augustioremq; felicitatem patefecit viamq;, qua ad hanc eatur, verbo suo revelato commonstravit: à qua cùm etiam plurimi aberrent, inq; varios opinionum errandos distracti fine omnium optimo, salute inquam æterna, frustrentur; tanta tamen bonitate humanum genus DEUS complexus est, ut ejus Providentia optimi quiq; Principes ac Rerump. rectores excitati religionem Christianam foverint, justitiæ aliarumq; virtutum culturam sedulò promoverint, eorumq; omnium, quæ ad vitæ felicitatem & [illegible] exiguntur, præcipuam curam habuerint. Hac quoq; pietate & erga Patriam studio ductus SERENISSIMUS PRINCEPS AC DOMINUS, DOMINUS FRIDERICUS, Dux Slesvici & Holsatiæ &c. Parens Noster gloriosissimus, jam pridem animo agitavit, quo pacto Religionis, Justitiæ, bonarumq; artium ac literarum studia in suis ditionibus & terris magis magisq; excolerentur: eoq; fine in Cœnobio Bordsholmensi illustre Gymnasium constituit, & bene multos præstantes ingenio juvenes ad solidiorem eruditionem hauriendam cum ibi tum in Academiis non exiguis sumptibus aluit juvitq;. Mox ad majoris momenti rem animo converso Optimus Princeps ac Pater Noster Scholam regiam seu Academiam, non paucis aliis collaborantibus, moliri cœpit, eiq; perficiendæ jampridem à SERENISSIMO INVICTISSIMOQ; CÆSARE AUGUSTO D. FERDINANDO III. inclytæ recordationis jura ac privilegia amplissima impetravit. Cæterum cùm feralis belli procella non semel in terras nostras superioribus annis incubuerit, laudatissimum hoc institutum non equidem penitus eversum, sed tamen fuit impeditum, & difficilius effectu redditum, ut desideratissimus Dominus Parens Noster vitæ quidem suæ, non autem belli satis optatum exitum prospiciens, cum aliarum rerum, tum inprimis hujus curam Nobis testamento condito sanctissimè commendaverit. Quam Gloriæ suæ partem cùm Nobis fecerit reliquam, nihil prius aut antiquius esse duximus, quàm huic piissimæ voluntati & officio satisfacere, Patriamq; & subditos hac quoq; parte tandem beare. Nihil itaq; territi immani Turcicorum armorum vi, quibus quàm maximè strepentibus hoc opus aggredi cœpimus, nec superioris belli calamitates respicientes, quibus nostræ provinciæ exhaustæ ac tantùm non consumptæ fuerunt, quod DEUS ter Opt. Max. felix faustumq; esse jubeat, NOS DEI GRATIA CHRISTIANUS ALBERTUS, Dux Slesvici & Holsatiæ &c. &c. publicè significamus omnibus, quorum hoc scire interest, cumprimis Augustanæ Confessioni addictis Ordinibus, Doctoribus ac Studiosis, inq; his subditis Nostris, interlapsâ paucorum mensium mora, a. d. 5. Octobr. hujus anni, ritu ac more solenni novam in Urbe Nostra Chiloniensi Academiam constituere, Nobis seriò decretum esse, & undiq; Viris doctissimis conditionibus minimè pœnitendis conscriptis publicam his docendi, aliis discendi potestatem facere, & utrumq; Ordinem iis juribus ac privilegiis ornare ac munire, ut & securè & cum dignitate cuivis, siquidem ipse velit, hic vivere, & in studio pietatis, justitiæ ac sapientiæ felicissimè proficere liceat. Quibuscunq; igitur curæ est, in vera DEI agnitione, & salutari doctrina Christi, in Juris Naturæ, Gentium, & Romano, in artis Medicæ ac Philosophiæ studiis ac disciplinis linguarum cultura, utilissimis tum animi tum corporis exercitiis solidè erudiri atq; institui, ad dictum d. Chilonium properate, tantoq; bono, cujus Vos participes esse cupimus, fruimini. Ne quid verò Respublica hæc nostra literaria aliquid capiat detrimenti, sed inde usq; à primis natalibus ad maximum decus & emolumentum adolescat, & omnis generis corruptelæ averruncentur, Academiæ Nostræ Cimbricæ primum quoque RECTOREM constituemus, cujus autoritate ac prudentia, adhibito Senatus Academici consilio, hoc gloriæ divinæ sacrum negotium & benè procedat, & conservetur: aliasq; solennitates publica Cæsareorum privilegiorum promulgatione præmissa, celebrabimus, quæ tantæ rei confirmandæ, ornandæ atq; illustrandæ conveniant. Interea hanc exigui temporis moram æquo animo ferte, DEUM PATREM JESU CHRISTI Academiarum ac Scholarum Patronum, Statorem ac Sospitatorem optimum ac munificentissimum pro lætissimis auspiciis, felicissimo successu ac progressu tanti negotii, qua par est animi devotione precati, quò omnia cœpta & consilia in Nominis ipsius perennem gloriam, Patriæ emolumentum, ac Reip. literariæ decus cedant. In Arce nostra Gottorpiensi a. d. XXVII. Maji CIↃ IↃ C LXV.

Fig. 1. Foundation charter of the University of Kiel. Founded on October 5th, 1665 by Christian Albrecht, Bishop of Lübeck, heir apparent of Norway, Duke of Schleswig, Holstein, Stormarn and Dithmarschen, Earl of Oldenburg and Delmenhorst. Granted in Gottorf Castle on May 27th, 1665

Acute Non-Traumatic Intracranial Bleeding

Acute, Non-Traumatic Intracranial Hemorrhage – Diagnosis and Timing

F. Scheil and J. Hedderich
Neurochirurgische Universitätsklinik Kiel, Weimarer Strasse 8, D-2300 Kiel 1

Introduction

Twenty-four university or hospital departments of neurosurgery took part in the cooperative study on acute, non-traumatic intracranial hemorrhage during the years 1978, 1979 and 1980. In three instances the data from two clinics with a common group of patients were combined. With the help of an extensive questionnaire, data were gathered on the causes of the hemorrhage, when and how the diagnosis and treatment was carried out, the patient's condition on admission to hospital and to what extent his condition influenced the diagnostic procedure (computed tomography and angiography), the therapy and - where indicated - the time of the operation. A division into six grades of severity was made, using the International Cooperative Study on Timing of Aneurysm Surgery as a model:

Grade I: No symptoms
Grade II: Slight symptoms (headache, meningism, diplopia)
Grade III: Neurological focal symptoms with full retention of consciousness
Grade IV: Reduced consciousness, defensive reactions retained
Grade V: Weak defensive reactions together with stable vital signs
Grade VI: No reaction to shouting or shaking, no reaction to pain stimuli and increasing instability of vital signs.

In the diagnostic part of the questionnaire the focal points continued to be the results of CT and angiographic examinations.

Of a total of 2629 completed questionnaires 2312 could be evaluated statistically. 2253 contained information on completed initial examinations and treatment and 59 on follow-up examinations.

Results

Among the 2253 initial examinations aneurysms were the cause of the hemorrage in 1209 cases (54%) and angiomas the cause in 208 cases (9%). In 829 cases (37%) the hemorrhage had had another cause (tumors, massive hypertensive hemorrhage, hemorrhagic diathesis and others) or no cause for the hemorrhage could be found. Seven patients had an aneurysm and an angioma. Multiple aneurysms were found in 110 patients, i.e. in 9% of all cases of aneurysms reported on.

Advances in Neurosurgery, Vol. 11
Edited by H.-P. Jensen, M. Brock, and M. Klinger

For the further evaluation the instances of multiple aneurysms and the 54 aneurysms and 44 angiomas that were found during an operation or post mortem, and not through the diagnostic procedure, were not considered. The seven cases where an aneurysm and an angioma occurred together were also omitted. The final evaluation was done on 1045 single aneurysms, 164 angiomas and 829 cases of hemorrhage with another cause (miscellaneous), a total of 2038 cases.

Figure 1 shows a breakdown of the patients with aneurysms, angiomas and miscellaneous hemorrhages according to age and sex.

Figure 2 gives information on the interval of time between hemorrhage and admission to hospital and the patient's condition at admission. For greater clarity, grades I and II and IV to VI were combined. Most patients were admitted to hospital within the first twenty-four hours of the hemorrhage; this is especially true for patients whose condition was serious. The frequency of different symptoms and factors in the case history is shown in Figure 3. Headaches were observed in almost the same percentage of cases of aneurysms (29%), angiomas (31%) and miscellaneous hemorrhages (24%). Conspicuous is the high percentage of hypertensive patients in the group of miscellaneous hemorrhages (44%) and in the patients with aneurysms (30%). Only 7% of patients with angiomas were hypertensive. Previous hemorrhages were found in 18% of patients with aneurysms, in 10% of patients with angiomas and in 5% of patients with hemorrhages from miscellaneous causes. Diabetes, a bleeding tendency and hemorrhages due to anticoagulant medication occurred only rarely and then most frequently in the miscellaneous group.

A lumbar puncture as an indication of the intracranial hemorrhage was performed in 1451 or 71% of all cases. Bloody or xanthochromic cerebrospinal fluid was found in 1369 (94%) cases. In 505 cases (35% of all lumbar punctures) the cerebrospinal pressure was measured and 190 times (38% of these cases) an elevated or even greatly elevated pressure was found. In 1843 cases (90%) a CT examination was made. Of these examinations 94% were performed on patients in a serious condition (grades V or VI) within the first twenty-four hours. It was done immediately, i.e. within the first twenty-four hours, on only 53% of patients whose condition was less serious (grades I and II). Figure 4 shows the clinical findings of the CT examination. Whereas the greater percentage of the aneurysms had subarachnoid hemorrhage, in those patients with angiomas or hemorrhages from miscellaneous causes an intracerebral hematoma, a hyperdense focus or a mass effect were found. Of 914 cases examined a direct demonstration of the aneurysm by computed tomography was possible in 67 cases without contrast medium; with contrast medium a further 57 aneurysms were found. That is to say, 124 cases or 13.6% of the 914 aneurysms could be shown by computed tomography. The percentage of angiomas shown in this way was even higher - 56 of 151 cases or 37%. With computed tomography 39 angiomas could be found without contrast medium and a further 17 angiomas with.

Whereas the CT-examination was usually carried out soon after the patient's admission to hospital, angiography, as a more involved invasive method of examination, was not performed until later on. Angiograms were made in a total of 1638 patients (82% of all cases). The patient's condition was more important in determining when the arteriography was done, than was the case with the CT examination. The angiographic examination was made of 72% of patients whose condition was serious (grades V and VI) within the first twenty-four hours of admission to hospital and of only 24% of patients whose condition was good (grades I and II). Figure 5 shows a breakdown of the single aneurysms shown in angiography according to location and sex. It shows that aneurysms tended to be found in the right hemisphere and in the female sex. Aneurysms of the anterior communication artery were found most frequently and, in contrast to all other aneurysms, occurred most often in males.

In 43% of the cases reported on, the size of the aneurysm was less than 5 mm, in 49% it was between 5 mm and 2 cm and in 6% it was larger than 2 cm. In 2% of the cases no size was given.

Table 1 shows the location of the angioma in relation to the vascular zone demonstrated in the angiogram. The angiomas occurred mainly on the carotid circulation with a slight preference for the left hemisphere.

Vasospasm accompanied the aneurysm in 36% of all cases, the miscellaneous hemorrhages in 15% of all cases and the angiomas in only 5% of all cases. The frequency of vasospasm accompanying aneurysms in relation to the interval of time between hemorrhage and angiogram is shown in Table 2. Conspicuous is the frequent occurrence of vasospasm within the first forty-eight hours and, in contrast, its slight occurrence between the second and fourth day after the hemorrhage. Between the fourth and the fifteenth day it occurs again as frequently as during the first forty-eight hours and after the fifteenth day its frequency again drops. In patients who had already had intracranial hemorrhage on some previous occasion vasospasm occurred within the first forty-eight hours of the hemorrhage, more frequently than later on. Table 3 shows the occurrence of vasospasm and the location of the aneurysm.

Table 1. Distribution of location in 164 angiomas

Location	No.
Right carotid system	59
Left carotid system	67
Right and left carotid system	3
Vertebro-basilar system	19
Right carotid and vertebro-basilar system	8
Left carotid and vertebro-basilar system	7
Right and left carotid and vertebro-basilar system	1

Table 2. Time of angiography and incidence of vasospasm in 967 cases

		Days after onset of the last SAH before admission			
		0 - 2	2 - 4	4 - 15	15 →
One attack before admission	n=793	56 (142) 39%	36 (134) 27%	147 (363) 41%	49 (154) 32%
More than one attack before admission	n=174	13 (23) 57%	11 (29) 38%	26 (69) 38%	14 (53) 26%
Total	n=976	69 (165) 42%	47 (163) 29%	173 (432) 40%	63 (207) 30%

Table 3. Location of aneurysm and vasospasm in 1040 single aneurysms

Site	No.	Spasm	%
Anterior cerebral artery	61	26	43
Middle cerebral artery	220	91	42
Basilar artery	32	13	41
Anterior comm. artery	415	158	38
Posterior comm. artery	51	16	31
Internal carotid artery	243	71	29
Vertebral artery	18	2	11

Table 4. Timing of surgery in 848 aneurysms after hemorrhage

a) Operations within 48 hours

Grade at admission and operation

Admission Grade \ Operation	I	II	III	IV	V	VI	Admission
I							0
II	1	3	1	1			6
III			2	1			3
IV				14	5		19
V					14	1	15
VI						3	3
Operation	1	3	3	16	19	4	46

Preoperative condition and operative results on discharge

Grade	Excellent	Good	Fair	Poor	Exitus
I (n= 1)	1				
II (n= 3)	1	1			1
III (n= 3)		2	1		
IV-VI (n=39)	2	7	3	7	20

b) Operations within 2 to 4 days

Grade at admission and operation

Admission Grade \ Operation	I	II	III	IV	V	VI	Admission
I	3						3
II		13		3			16
III		2	3		2		7
IV		2	3	7		1	13
V				1	2		3
VI				1		1	2
Operation	3	17	6	12	4	2	44

Preoperative condition and operative results on discharge

Grade	Excellent	Good	Fair	Poor	Exitus
I (n= 3)	1				2
II (n=17)	6	5	2		4
III (n= 6)	1		2		3
IV-VI (n=18)	2		5	1	10

c) Operations within 4 to 15 days

Grade at admission and operation

Grade (Admission) \ Operation	I	II	III	IV	V	VI	
I	20	2	2	1			25
II	39	83	10	5	1		138
III	2	14	26	4	1	1	48
IV	1	19	12	20	1		53
V			1	7	5	1	14
VI				1			1
	62	118	51	38	8	2	279

Preoperative condition and operative results on discharge

Grade (Admission)	Excellent	Good	Fair	Poor	Exitus
I (n=62)	45	6	4		7
II (n=118)	64	21	11	3	19
III (n=51)	19	14	3	3	12
IV-VI (n=48)	4	6	8	6	24

d) Operation after 15 days and later

Grade at admission and operation

Grade (Admission) \ Operation	I	II	III	IV	V	VI	
I	52	7	1		1		61
II	48	83	11	6			148
III	4	14	40	5	1	2	66
IV	12	17	25	15	3		72
V		4	7	6	1		18
VI			1				1
	116	125	85	32	6	2	366

Preoperative condition and operative results on discharge

Grade (Admission)	Excellent	Good	Fair	Poor	Exitus
I (n=116)	95	11	2	1	7
II (n=125)	79	26	6	6	8
III (n=85)	17	43	13	7	5
IV-VI (n=40)	3	4	7	10	16

e) Operation within 10 to 21 days

Grade at admission and operation

Grade (Admission) \ Operation	I	II	III	IV	V	VI	
I	21		1				22
II	14	32	3	2			51
III		4	22		1		27
IV		5	3	4	1		13
V							0
IV							0
	35	41	29	6	2	0	113

Preoperative condition and operative results on discharge

Grao › (Admission)	Excellent	Good	Fair	Poor	Exitus
I (n=35)	30	3	2		
II (n=41)	27	7	2		5
III (n=29)	11	9	4	1	4
IV-VI (n= 8)			1		7

Table 4a - e show the time of the operation and the patient's condition on discharge, in 848 cases of single aneurysms. After a hemorrhage the operation was performed within the first forty-eight hours, between the second and fourth day, between the fourth and fifteenth day and after the fifteenth day. Since in the questionnaire the intervals of time between the hemorrhage and the admission to hospital and between the admission to hospital and diagnosis and operation were different categories there were a number of patients whose time of operation could not be worked into this scheme. These were cases operated on between the tenth and twenty-first day. These results are shown in Fig. 9e.

The charts show the patient's condition on admission to hospital and at the time of operation and the result on discharge compared to the condition at the time of operation. Here one can see how many patients improved between the time of admission to hospital and the time of the operation (below the diagonal line) and how many got worse (above the diagonal line). The patient's condition at discharge was judged according to his/her ability to work. "Excellent" means fully able to work, "good" partially able to work, "fair" unable to work, "poor" unable to work and dependent on help from another.

In only 5.4% of all cases was an operation performed within forty-eight hours of the hemorrhage. Most of these patients were in a serious condition (grades IV to VI) and died after the operation. Of the seven patients who were in better condition (grades I to III) when operated on, only one died after the operation. The mortality rate of patients operated on in this period was 46%. Although the number of patients in a serious condition (grades IV to VI) among those operated on between the second and fourth day was markedly fewer than in the above-mentioned group the mortality rate here was almost as high, namely 43%. The reason for this is the much higher mortality rate in grades I to III. The mortality rate in patients operated on between the fourth and fifteenth day was 22%, in patients operated on after the fifteenth day 10%. In patients operated on between the tenth and twenty-first day the mortality rate was 14%.

Of those patients operated on within the first forty-eight hours 30% were fully or partially able to work. For the following periods, i.e. the second to the fourth day, the fourth to the fifteenth day and after the fifteenth day, the percentages were 34, 64, and 76 respectively. In the group of patients operated on between the tenth and twenty-first day the percentage was 77.

Patients in whom vasospasm was found before the operation show poorer postoperative results than those in whom pre-operatively no vascular constrictions were found.

Discussion

The main causes of the intracranial, non-traumatic hemorrhages in this cooperative study were aneurysms and angiomas and, in the large miscellaneous group, massive hypertensive bleeding. In comparison with the Cooperative Study of Intracranial Aneurysms and Subarachnoid Hemorrhage (13, 19) there is an almost identical percentage of aneurysms (54% and 51%) and angiomas (9% and 8%). The age and sex breakdown in cases of aneurysms, with a prevalence of female patients and an onset of the disorder in the sixth decade of life also corresponds to the results in LOCKSLEY's study.

Extensive Japanese studies (24, 27) have, however, shown a predominance of male patients in a similar age group. While PERRET (19) found no great differences in the number of male and female patients with angiomas, ALBERT (1) had more male patients and an earlier onset of the disorder as did this cooperative study.

This study confirms the rarity of vasospasm in angiomas. In this study vasospasm accompanied the angioma in only 5% of the cases; NISHIMORA (17) found vasospasm in only 6 of 52 cases. In SASAKI's study vasospasm was found only in one of 12 cases (21). Vasospasm accompanies an aneurysm much more often. Other authors have found it in from 21 to 62% of all cases (2-6, 12, 15, 18, 23, 26). In this study vasospasm could be shown in 36% of the single aneurysms that were angiographed. The frequent occurrence of vasospasm shortly after the hemorrhage is conspicuous. A frequent occurrence of vasospasm within the first two days of the hemorrhage was also found by ALLOCK and OHTA (2, 18), while KWAK, HAMER (5, 12) and others found only a slight incidence during this period of time. Hemorrhages previous to the one leading to admission influence to a great extent the possibility of vasospasm occurring. When an angiogram is made within the first forty-eight hours or the first four days of a hemorrhage there is a higher incidence of vasospasm in patients with, than in patients without, previous hemorrhages. When the angiogram is made at a later date, the incidence is lower.

The question of the most opportune time for an operation after rupture of an aneurysm has been often discussed in the last few years (7-11, 14, 16, 20, 22, 25). To avoid possible complications from recurring hemorrhages or vasospasm, an early operation on patients in good condition - i.e. within the first three days or even the first forty-eight hours of the hemorrhage - has often been recommended. Since in this study only a small number of patients was operated on within the first forty-eight hours, a definitive judgement cannot be made on this, especially since most of these patients were in a serious condition. In the following interval of time, i.e. from the second to the fourth day, the high mortality rate of patients operated on in a good condition (grades I to III) is conspicuous. Other authors also give unsatisfactory results of operations performed between the third and seventh day of the hemorrhage (8, 9, 22, 25). As far as the patient's ability to work and the mortality rate are concerned the best results are found when operations are performed after the fifteenth day, especially on patients in good condition. It is worth mentioning that of the patients who were operated on within the first forty-eight hours of the hemorrhage and whose pre-operative condition was serious (grades IV to VI) 23% were able to work again, at least partially, afterwards. When operated on between the second and fourth day 11% were able to work afterwards, between the fourth and fifteenth day it was 18% and after the fifteenth day 21% (patients with grades IV to VI).

Summary

In this cooperative study in which 24 university and hospital departments of neurosurgery in Germany and Austria took part, data could for the first time be compiled from a great number of cases of acute, nontraumatic intracranial hemorrhage. The results given here contain information on the diagnostic procedures, the findings and the subsequent treatment. The results are comparable with those of similar studies published earlier.

References

1. Albert, P.: Personal experience in the treatment of 178 arteriovenous malformations of the brain. Act. Neurochirurg. 61, 207-226 (1982)
2. Allcock, J.M., Drake, C.G.: Ruptured intracranial aneurysms. The role of arterial spasm. J. Neurosurg. 22, 21-29 (1965)
3. Ecker, A., Riemenschneider, P.A.: Arteriographic demonstration of spasm of the intracranial arteries with special reference to saccular arterial aneurysms. J. Neurosurg. 8, 660-667 (1951)
4. Fletcher, T.M., Taveras, J.M., Pool, J.L.: Cerebral vasospasm in angiography for intracranial aneurysms. Incidence and significance in 100 consecutive angiograms. Arch. Neurol. 1, 38-47 (1959)
5. Hamer, J., Penzholz, H., Götte, B.: Time course and clinical significance of cerebral vasospasm after aneurysmal subarachnoid hemorrhage. In: Advances in Neurosurgery, Vol 9. Schiefer, W., Klinger, M., Brock, M. (eds.), pp. 197-200. Berlin, Heidelberg, New York: Springer 1981
6. Heilbrun, M.P.: The relationship of neurological status and the angiographical evidence of spasm to prognosis in patients with ruptured intracranial saccular aneurysms. Stroke 4, 973-979 (1973)
7. Hey, O., Schürmann, K.: Timing of aneurysms of cerebral vessels - Documentation of 200 patients in a retrospective study. In: Advances in Neurosurgery, Vol. 9. Schiefer, W., Klinger, M., Brock, M. (eds.), pp. 233-237. Berlin, Heidelberg, New York: Springer 1981
8. Hori, S., Suzuki, J.: Early and late results of intracranial direct surgery of anterior communicating artery aneurysms. J. Neurosurg. 50, 433-440 (1979)
9. Hugenholtz, H., Elgie, R.G.: Considerations in early surgery on good risk patients with ruptured intracranial aneurysms. J. Neurosurg. 56, 180_185 (1982)
10. Hunt, W.E., Hess, R.M.: Surgical risk as related to time of intervention in the repair of intracranial aneurysms. J. Neurosurg. 28, 14-20 (1968)
11. Jensen, H.P., Hansmann, H., Muhtaroglu, U.: Timing in subarachnoid hemorrhages. 31st Annual Meeting Scandinavian Neurosurgical Society. 1979. Ouli. Act. Neurochirurg. 51, 257-262 (1979)
12. Kwak, R., Niizuma, H., Ohi, T., Suzuki, J.: Angiographic study of cerebral vasospasm following rupture of intracranial aneurysms: Part I. Time of appearance. Surg. Neurol. 11, 257-262 (1979)
13. Locksley, H.B.: Report on the cooperative study of intracranial aneurysms and subarachnoid hemorrhage. Section V, Part I: Natural history of subarachnoid hemorrhage, intracranial aneurysms and arteriovenous malformations based on 6368 cases in the cooperative study. J. Neurosurg. 25, 219-239 (1966)
14. Mehdorn, H.M., Grote, W., Tenfelde, B.: Timing of aneurysm surgery after spontaneous subarachnoid hemorrhage. In: Advances in Neurosurgery, Vol. 9. Schiefer, W., Klinger, M., Brock, M. (eds.), pp. 243-249. Berlin, Heidelberg, New York: Springer 1981
15. Millikan, C.H.: Cerebral vasospasm and ruptured intracranial aneurysm. Arch. Neurol. 32, 433-449 (1975)
16. Muhtaroglu, U., Klinge, H., Rautenberg, M.: Treatment of acute subarachnoid hemorrhage. In: Advances in Neurosurgery, Vol. 9

Schiefer, W., Klinger, M., Brock, M. (eds.), pp. 255-260. Berlin, Heidelberg, New York: Springer 1981

17. Nishimura, K., Hawkins, T.D.: Cerebral vasospasm with subarachnoid hemorrhage from arteriovenous malformations of the brain. Neuroradiology 8, 201-207 (1975)

18. Ohta, T., Kawamura, J., Osaka, K., Kajikawa, H., Handa, H.: Angiographic classification of so-called cerebral vasospasm. Correlation between existence of vasospasm and postoperative prognosis in subarachnoid hemorrhage. Brain Nerve (Jap.) 21, 1019-1027 (1969)

19. Perret, G., Nishioka, H.: Report on the cooperative study of intracranial aneurysms and subarachnoid hemorrhage. Section VI: Arteriovenous malformations. J. Neurosurg. 25, 467-490 (1966)

20. Pia, H.W.: Grading and timing of operation of cerebral aneurysms. In: Advances in Neurosurgery, Vol. 9. Schiefer, W., Klinger, M., Brock, M. (eds.), pp. 217-225. Berlin, Heidelberg, New York: Springer 1981

21. Sadaki, T., Mayanagi, Y., Yano, H., Kim, S.: Cerebral vasospasm with subarachnoid hemorrhage from cerebral arteriovenous malformations. Surg. Neurol. 16, 183-187 (1981)

22. Sano, K., Saito, I.: Timing and indication of surgery for ruptured intracranial aneurysms with regard to cerebral vasospasm. Act. Neurochirurg. 41, 49-60 (1978)

23. Schneck, S.A., Kricheff, I.I.: Intracranial aneurysm rupture, vasospasm and infarction. Arch. Neurol. 11, 668-681 (1964)

24. Suzuki, J., Hori, S., Sakurai, Y.: Intracranial aneurysms in neurosurgical clinics in Japan. J. Neurosurg. 35, 34-39 (1971)

25. Suzuki, J., Onuma, T., Yoshimoto, T.: Results of early operations on cerebral aneurysms. Surg. Neurol. 11, 407-412 (1979)

26. Wilkins, R.H., Alexander, J.A., Odom, G.L.: Intracranial arterial spasm: A clinical analysis. J. Neurosurg. 29, 121-134 (1968)

27. Yoshimoto, T., Kayama, T., Kodama, N., Suzuki, J.: Distribution of intracranial aneurysm. In: Cerebral aneurysm. Suzuki, J., (ed.), pp. 14-19. Tokyo: Neuron 1979

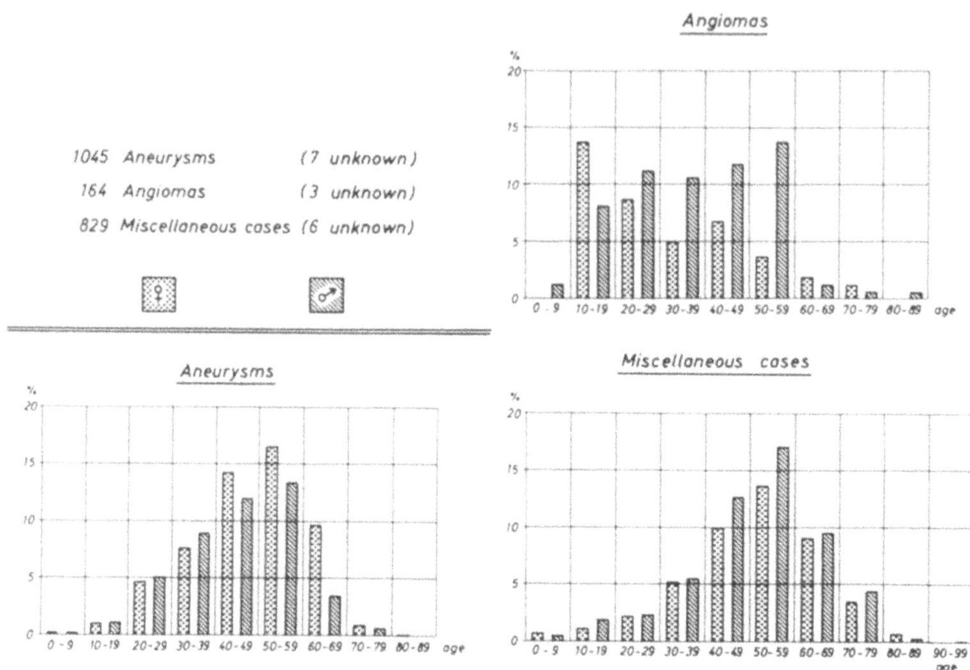

Fig. 1. Distribution of age and sex in 2038 cases of acute, non-traumatic intracranial hemorrhages

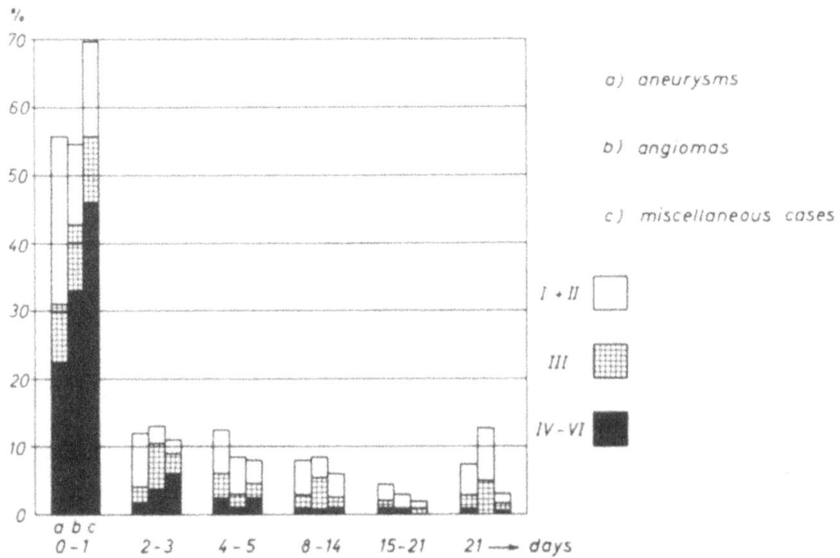

Fig. 2. Time between hemorrhage and admission of patients with aneurysms, angiomas or miscellaneous cases

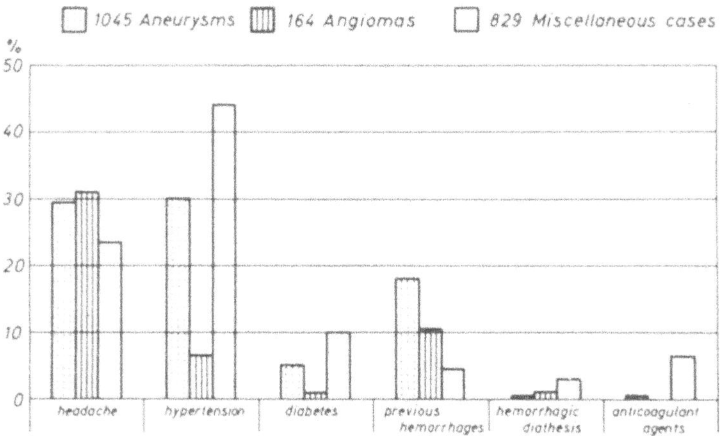

Fig. 3. Case histories in 2038 cases of acute, non-traumatic intracranial hemorrhages

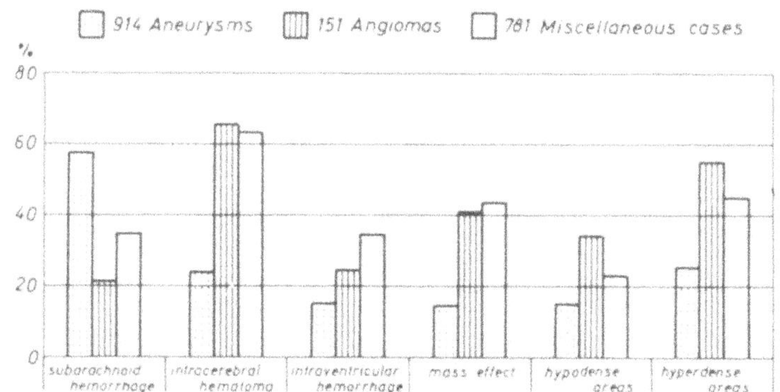

Aneurysms detectable: 124 (13,6 %), without enhancement 67, with enhancement 57

Angiomas detectable: 56 (37 %), without enhancement 39, with enhancement 17

Fig. 4. Results of CT findings

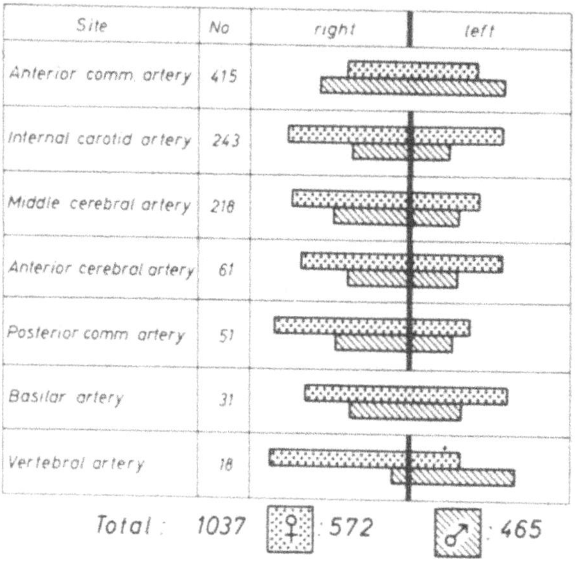

Site	No
Anterior comm. artery	415
Internal carotid artery	243
Middle cerebral artery	218
Anterior cerebral artery	61
Posterior comm. artery	51
Basilar artery	31
Vertebral artery	18

right | left

Total: 1037 ♀: 572 ♂: 465

Fig. 5. Location and sex distribution in 1037 ruptured single aneurysms

Acute, Non-Traumatic Intracranial Hemorrhage – Treatment and Prognosis

E. Kraus and J. Hedderich

Neurochirurgische Universitätsklinik Kiel, Weimarer Strasse 8, D-2300 Kiel 1

Introduction

The questionnaire worked out for the cooperative study of acute, non-traumatic intracranial hemorrhage dealt with, in their relation to the treatment and prognosis, the pre-operative treatment (measures, length of time they were carried out and their results) and the operative treatment. Of special importance was when the operation was performed and the patient's neurological condition at that time. The division into grades of severity given in Part 1 is in keeping with the International Cooperative Study.

Further aspects of the series of questions in the questionnaire were the postoperative condition 10 and 20 days after the operation, the results of the postoperative angiography and the postoperative complications.

To facilitate the judging of the patient's condition at the time of discharge and during the follow-up period a socially and medically weighted division into grades of severity was chosen. The objective criterion here was the extent of the patient's ability to work:

Excellent: No or only slight physical, neurological or psychic sequelae following the hemorrhage, without a reduction in the patient's ability to work

Good: Physical, neurological or psychic consecutive symptoms which reduced the patient's ability to work

Fair: Physical, neurological or psychic deficiency symptoms, unable to work

Poor: Dependent on help from another person

Dead:

When devising the questionnaire we purposely did not allow for all eventualities, so as to permit a certain measure of flexibility in the judgement of individual cases. Our intention here produced not only positive results since as a result some of the participants failed to supply valuable information. To this situation must also be ascribed the fact that in a number of cases no clear distinction was made between conservative and pre-operative treatment, meaning - in the case of a death after an aneurysmal hemorrhage before the final operation, for example - that an exact calculation of the entire management mortality was not possible.

Advances in Neurosurgery, Vol. 11
Edited by H.-P. Jensen, M. Brock, and M. Klinger

From the wealth of information gathered only the most important aspects can be discussed here.

Results

Of the 2038 intracranial hemorrhages that corresponded to the criteria of selection and were therefore considered more closely, 62.7% were operated on (Table 1) viz. 891 aneurysms, 132 angiomas and 254 of the heterogeneous group of miscellaneous hemorrhages.

Conservative Treatment

The conservatively treated cases are made up mainly of those subarachnoid hemorrhages where no bleeding-source could be found and where there was no indication for an operation, and of the cases in which the patients refused to be operated on or where an operation did not seem possible. Here it is worthy of note that there was little difference in the neurological condition of the conservatively treated patients at the time of admission to the hospital (if the causes of the hemorrhage are considered separately). In each group approximately one half of the cases had disturbances of consciousness, one sixth focal symptoms and one third only slight or no symptoms (Table 2). But the results of the treatment differ greatly; a mortality rate of 60% was observed in cases of aneurysms not operated on and of 19% in cases of angiomas not operated on. The number of patients who were in excellent condition when discharged and who had only slight or no sequelae after the hemorrhage was the reverse; there were three and a half times as many angioma as aneurysm patients.

The conservative or pre-operative methods of treatment, including preparatory operations, are listed in Table 3. Cortisone, dehydration, sedatives, antihypertensive and antifibrinolytic agents were used most frequently. The cause of the hemorrhage determined how often each of these was used. Ventricular drainage was performed in 101 cases. Of these, 71 were hemorrhages from the ventricle (445 cases) and in 30 cases there was another indication for the operation. The results of the conservative or pre-operative treatment are shown in Table 4.

Recurrent hemorrhages were observed most often with aneurysms, namely in 10.6% of all cases. The more serious the patient's condition on admission to hospital, the more likely a recurrence of the hemorrhage was. With patients of grade I there was a recurrence in 5.4% of the cases, with patients of grade II in 5.9%, of grade III in 13.2%, of grade IV in 18.8%, of grade in 13.0% and of grade VI in 25.0%.

This also holds true for the angiomas where in grades I to III there was a recurrence in 3% of the cases and in grades IV to VI in 8.9%,

Table 1. Acute, non-traumatic intracranial hemorrage: 2038 cases

Causes	Number	Operations
Aneurysms	1045 (51,3%)	891 (85,3%)
Angiomas	164 (8,0%)	132 (80,5%)
Miscellaneous cases	829 (40,7%)	254 (30,6%)
Total	2038 (100 %)	1277 (62,7%)

Table 2. Results of conservative treatment

Admission Grade	Aneurysms No. 154	Angiomas No. 32	Miscellaneous cases No. 578
I	3	22	6
II	27	9	28
III	24	16	17
IV	30	28	23
V	9	16	14
VI	7	9	12

Discharge Grade	Aneurysms No. 154	Angiomas No. 32	Miscellaneous cases No. 578
Excellent	17	59	39
Good	13	13	15
Fair	5	6	8
Poor	5	3	6
Dead	60	19	32

Table 3. Conservative or preoperative treatment (in %)

Methods	Aneurysms n = 1045	Angiomas n = 164	Misc. cases n = 829
Ventricular drainage	3.7	3.1	7.1
Lumbar drainage	2.2	0	0.5
Intracranial pressure monitoring	1.4	0.6	2.7
Shunt operation	3.5	2.5	3.6
Dexamethasone	68.8	73.6	63.7
Dehydrating hypertonic agents	18.3	19.0	25.8
Fluid restriction	10.5	9.8	16.4
Diuretics	8.1	11.7	17.6
Sedatives	53.3	36.8	39.8
Antihypertensives	26.5	9.2	34.9
Antifibrinolytics	37.5	16.0	21.2

Table 4. Clinical course under conservative or preoperative treatment (in %)

Condition	Aneurysms n = 1045	Angiomas n = 164	Miscell. cases n = 829	Total n = 2038
Improved	30.2	32.1	26.7	29.0
Unchanged	50.9	59.3	44.4	48.9
Deteriorated	18.9	8.6	28.9	22.1
Recurrent hemorrhage	10.6	5.1	7.3	

and for the miscellaneous hemorrhages where the figures were 1.8% and 4.5% respectively.

A recurrent hemorrhage was observed more frequently when vasospasm was also present. With the aneurysms there was a recurrence in 12.7% of cases with vasospasm, but only in 9.6% of cases where no spasm was found. The corresponding figures for the miscellaneous hemorrhages are 4.2% and 1.9%. The angiomas are an exception; no recurrence of hemorrhage was observed in cases where spasm had followed the hemorrhage from the angioma.

The probability of a recurrent hemorrhage increases with the size of the aneurysm. With aneurysms of up to 5 mm there was a recurrence in 10% of the cases whereas the percentage was 11.4 in aneurysms of up to 2 cm and 15.3 in aneurysms of over 2 cm. There was, however, no relationship between a recurrence of the hemorrhage and a previous hemorrhage. The results of this study do not allow any judgement on the effects and success of antifibrinolytic therapy.

Operative Treatment

Miscellaneous cases. With the heterogeneous group of miscellaneous cases the operative measure was usually the removal of a space-occupying intracerebral hematoma. At the time of admission to hospital 70% of all cases were suffering from a serious cerebral functional disturbance and a disturbance of consciousness. At the time of the operation this percentage was even somewhat higher, namely 74% (Table 5). The operative mortality rate was 39% as against a mortality rate of 32% in cases of miscellaneous hemorrhages not operated on. In cases with a ventricular hemorrhage the mortality rate rose to 61%. This makes the fact all the more surprising that almost 10% of patients with ventricular hemorrhages were later fully able to work and 11% were partially able to work. The results of operations on patients receiving anticoagulant therapy were somewhat less favourable than for the group of miscellaneous cases as a whole. The cases showing a hemorrhagic diathesis had even worse results.

Angiomas. Of the 164 patients with hemorrhage due to arteriovenous malformation 132 were operated on and 32 received conservative therapy. In 130 cases an intracranial operation was performed; these operations can be broken down into one embolisation, nine partial resections and 120 extirpations of the angioma.

As Table 6 shows, the condition of patients on admission to hospital was the worst; by the time of the operation it had improved remarkedly

Table 5. Surgical results in 248 miscellaneous cases. Classification of grading see text

Grade	Admission		Operation		Grade	Discharge	
	No.	%	No.	%		No.	%
I	2	1	3	1	Excellent	44	18
II	13	5	16	6	Good	50	20
III	59	24	46	19	Fair	20	8
IV	89	35	88	36	Poor	37	15
V	65	26	67	27	Dead	97	39
VI	23	9	28	11			

Table 6. Surgical results in 132 angiomas. Classification of grading see text

Grade	Admission No	%	Operation No.	%	Grade	Discharge No.	%
I	15	11	28	21	Excellent	66	50
II	28	21	31	23	Good	42	32
III	42	32	39	30	Fair	7	5
IV	32	24	18	14	Poor	1	1
V	14	11	13	10	Dead	16	12
VI	1	1	3	2			

and at the time of discharge 50% of the angioma patients operated on showed only slight or no sequelae and were fully able to work. The operative mortality rate was 12% here compared with a mortality rate of 19% in the group not operated on and 24% in the group with ventricular hemorrhage of whom one third or 32% nevertheless were in excellent condition on discharge. Of the seven cases with pre-operative spasm four were in excellent and three in good condition when discharged; all seven were able to work.

It is interesting to note that between the time of admission to hospital and operation the neurological status of only eight patients had deteriorated and only in three cases was this deterioration greater than one grade of severity. This fact is to be explained by the relatively rare recurrence of hemorrhage and vasospasm and the fact that hemorrhages due to arteriovenous malformation were operated on relatively soon after admission to hospital. Of the angioma operations 17% were performed within forty-eight hours of the hemorrhage and 24% within the first ninety-six hours.

Aneurysms. Of the 1045 ruptured single aneurysms 891 were operated on, and of these the typical clip procedure was used in 851 cases. In the remaining 40 cases carotid ligation, catheter or other operative methods were employed. At the time of operation the condition of over 60% (61.4%) of the patients was grades I and II; this is an improvement of 6.5% over the condition on admission to hospital. Similarly, the percentage of patients with a disturbance of consciousness decreased by 8.4%, so that by the time of operation 18% of the patients still had a disturbance of consciousness.

It became obvious that some departments of neurosurgery participating were more, some less generous in their indications for operation. This study also includes cases in which an operation was performed only as a last resort, with only a very limited chances of success. In this way this study differs from many other publications in which the cases included were carefully chosen. This must be kept in mind when the average results of all operations are considered. The results of the operations were excellent in 48% of all cases, good in 20%, fair in 9% and poor in 5% (Table 7).

In the cases where clipping was performed the operative mortality rate was 18%, i.e. less than a third of the mortality rate in the cases not operated on. For each grade of severity at the time of admission and operation the mortality rate here was lower than with the conservatively treated patients. The operative mortality rate in the 40 cases in which other operative techniques were used was 30% (12 cases).

Table 7. Preoperative grade and result on discharge in 851 aneurysms. Classification of grading see text

		Excellent		Good		Fair		Poor		Dead	
Grade	No.	No.	%	No.	%	No.	%	No.	%	No.	%
I	217	172	79	20	9	8	4	1	0	16	7
II	306	178	58	60	19	22	7	9	3	37	12
III	175	48	28	69	39	23	13	11	6	24	14
IV	104	11	11	15	14	20	19	14	14	44	42
V	39	-	-	1	3	4	10	7	18	27	69
VI	10	-	-	2	20	-	-	2	20	6	60
Total	851	409	48	167	20	77	9	44	5	154	18

The extent of the cerebral functional disturbance as indicated by the grade of severity very definitely determines the outcome of operation. The differences here are very obvious (Table 7). According to the neurological grade of severity the excellent and good operative results range from 90% to 3% and the mortality rate from 7% to almost 70%.

During the period in which the study was carried out (1978 - 1980) immediate and early operations were performed only rarely. A mere 46 patients, i.e. 5.4%, were operated on within the first forty-eight hours of the hemorrhage and a further 44 or 5.2% within the first ninety-six hours.

In the instances where an operation was performed immediately, i.e. within the first forty-eight hours of the hemorrhage, the condition of only 9% of the patients was grades I and II whereas 45% of the patients operated on after three to four days were from grades I and II, 65% from five to fifteen days and 66% when the operation was performed even later. Since only 10% of the operations discussed here were immediate or early operations and two thirds of these were doubtless emergency operations on patients from grades IV to VI, any judgement on the influence of the timing of operation on its outcome, does not seem appropriate.

We must emphasize that for the period of this study (1978 - 1980) the operations performed immediately were primarily emergency operations. The conclusions of the detailed discussion on the question of timing at the 31st annual meeting of the German Neurological Society at Erlangen in 1980 had no consequences for this study.

Our statistics show that the outcome of the operation bears a direct relationship to the patient's pre-operative condition. In spite of the extent of the study no binding claim can be made about the relevance of the timing of the operation, i.e. at what time after the hemorrhage the operation is performed, for the final outcome.

How other factors influenced the outcome of operation was also investigated, constitutional factors such as age and sex, risk factors such as high blood pressure and arteriosclerosis and especially the factors that are characteristic of this disorder, namely pre- and post-operative vasospasm, recurrent hemorrhage, blood in the subarachnoid space, intracerebral hematoma, ventricular hemorrhage and mass effects and finally the size and location of the aneurysm. The surgical results for each of these factors in relation to the preoperative condition of the patient will not be set out in detail here. It has already been discussed elsewhere (5).

As was to be expected, the worst results were from patients with a recurrent hemorrhage and pre-operative spasm, high blood pressure and patients in advanced age. This is also true of cases where there was a large intracerebral hematoma and mass effect, ventricular hemorrhage, serious arteriosclerosis and especially the occurrence of post-operative generalized spasm.

Post-Operative Angiography and Complications

Angiography was carried out post-operatively in 275 cases; in 202 aneurysms, 60 angiomas and 15 of the miscellaneous cases. In 23% of the aneurysms a local vasospasm was found, in 9% a generalized one. In approxiamtely one half of the cases (46%) with a postoperative spasm, spasm had also been found in the pre-operative angiogram. With the angiomas and miscellaneous cases the number with spasm, 5 and 3 respectively, was very low.

Further findings from the angiography in relation to elimination of the source of bleeding and arterial occlusion are to be found in Table 8.

The complications after operations for aneurysm and angioma listed in Table 9 differ in many respects.

Follow-up

The follow-up results of the patients operated on show an interesting difference when compared to the follow-up results of patients not operated on. Where the period of observation was between one month and four years the mortality rate in cases of aneurysmal hemorrhage not operated on was 53%. The surgical cases had an operative mortality rate of 18%; the mortality rate of a three months period of observation was

Table 8. Postoperative angiography of 202 aneurysms and 60 angiomas

Findings		Aneurysms n = 202	Angiomas n = 60
Source of bleeding	eliminated		
	total	185 (91.6%)	52 (86.7%)
	partial	12 (5.9%)	7 (11.7%)
	no	5 (2.5%)	1 (1.7%)
Arterial occlusion		18 (8.9%)	4 (6.7%)

Table 9. Postoperative complications

Complications	Aneurysms	Angiomas
Hemorrhage	4.7	6.2
Brain edema	25.0	14.6
Ischemic lesion	23.6	6.9
Seizures	3.5	6.9
Meningitis	1.3	0.8
Wound infection	4.8	5.3
Pulmonary	11.3	8.5
Cardiovascular	9.5	6.2

13%, from four to twelve months 3% and after one year 1%. In all, the mortality rate for the period of observation was 4%. Accordingly, in this period the percentage of patients with full ability to work who had been operated on rose from 66 to 75 while in the cases not operated on it was 21.

The late mortality rate of patients with angiomas not operated on was 17%, in those operated on 3%. Of patients with angiomas not operated on 58% regained full ability to work as against 74% of the cases where an operation was performed.

Discussion

After the diagnosis has been made and the cause of the hemorrhage identified, the treatment of acute, non-traumatic intracranial hemorrhage varies. In keeping with the clinical picture and the possible sequelae and risks of hemorrhages due to ruptured aneurysms and angiomas and other hemorrhages of hypertensive and vascular origin, the conservative or pre-operative methods of treatment and their success vary.

The frequency and occurrence of rebleeding and the clinical course of conservative or pre-operative treatment are corroborated by other authors (1, 7, 10).

Due to the heterogeneous make-up of our group of miscellaneous cases it is difficult to compare the results of operations here with results described in other publications; as far as numbers are concerned they would seem to be comparable with the results of other investigators (2, 11).

In our study the results of operations on angiomas were excellent in 50% of the cases reported on, good in 32%, fair in 5% and poor in 1%. The mortality rate was 12%. Results of angioma operations described elsewhere vary greatly, showing mortality rates of between 4% (6) and 14% (9).

In cases of aneurysms the outcome of the operation is definitely determined by the patient's neurological condition (3). In patients from grade I the mortality rate was 7%, a rate which rose in grades V and VI to 69% and 60% respectively. The percentage of "excellent" results fell correspondingly from 79 to 0.

The operative mortality rate in our study was 18%, while in the Cooperative Aneurysm Study it was 30% (7). The mortality rate in the Cooperative Aneurysm Study three months after the hemorrhage was 36% (4).

Summary

The purpose of this study was to provide information on the diagnosis and treatment of acute, non-traumatic intracranial hemorrhage in Germany and Austria. Although many very different clinics took part and the data were compiled without pre-selection, the results can be judged to be thoroughly satisfactory.

References

1. Adams, H.P., Kassell, N.F., Torner, J.C., Nibbelink, D.W., Sahs, A.L.: Early management of aneurysmal subarachnoid hemorrhage. A report of the Cooperative Aneurysm Study. J. Neurosurg. 54, 141-145 (1981)
2. Guidetti, B., Gagliardi, F.: Experiences with operation in 149 cases of spontaneous ICH (1966-1977). In: Spontaneous intracerebral haematomas. Pia, H.W., Langmaid, C., Zierski, J. (eds.), pp. 247-251. Berlin, Heidelberg, New York: Springer 1980
3. Hunt, W.E., Hess, R.M.: Surgical risk as related to time of intervention in the repair of intracranial aneurysms. J. Neurosurg. 28, 14-20 (1968)
4. Kassell, N.F., Boarini, D.J., Adams, H.P., Sahs, A.L., Graf, C.J., Torner, J.C., Gerk, M.K.: Overall management of ruptured aneurysm: Comparison of early and late operation. Neurosurgery, Vol. 9, 2, 120-127 (1981)
5. Kraus, E.: Inauguraldissertation der Christian-Albrecht-Universität Kiel, 1983 (in press)
6. Mingrino, S.: Supratentorial arteriovenous malformations of the brain. In: Advances and technical standards in neurosurgery, Vol. 5. Krayenbühl, H. et al. (eds.), pp. 93-123. Wien, New York: Springer 1978
7. Post, K.D., Flamm, E.S., Goodgold, A., Ransohoff, J.: Ruptured intracranial aneurysms. Case morbidity and mortality. J. Neurosurg. 46, 290-295 (1977)
8. Sahs, A.L., Perret, G., Locksley, H.B., et al. (eds.): Intracranial aneurysms and subarachnoid hemorrhage. A Cooperative Study. Philadelphia: J. B. Lippincott Co. 1969
9. Sano, K., Aiba, T., Jimbo, M.: Surgical treatment of cerebral aneurysms and arteriovenous malformations. Neurol. Medicochir. 7, 128-131 (1965)
10. Sundt, T.M., Kobayashi, S., Fode, N.C., Whisnant, J.P.: Results and complications of surgical management of 809 intracranial aneurysms in 722 cases. Related and unrelated to grade of patient, type of aneurysm and timing of surgery. J. Neurosurg. 56, 753-765 (1982)
11. Walder, H.A.D.: Surgery in spontaneous ICH. In: Spontaneous intracerebral haematomas. Pia, H.W., Langmaid, C., Zierski, J. (eds.), pp. 260-262. Berlin, Heidelberg, New York: Springer 1980
12. Yoshimoto, T., Uchida, K., Kaneko, U., Kayama, T., Suzuki, J.: An analysis of follow-up results of 1000 intracranial saccular aneurysms with definitive surgical treatment. J. Neurosurg. 50, 152-157 (1979)

Acute, Non-Traumatic Intracranial Hemorrhage – Critical Remarks on the Results of the Cooperative Study

H.-P. Jensen

Neurochirurgische Universitätsklinik Kiel, Weimarer Strasse 8, D-2300 Kiel 1

Patients with acute, non-traumatic intracranial hemorrhages usually make up a large percentage of all patients in departments of neurosurgery. Not until the cause, type and extent of the hemorrhage has been clarified can a decision be made as to whether an operation is possible and indicated or whether a conservative method of treatment is preferable. For this very important decision the introduction of computed tomography has meant great progress.

This study was carried out over a period of three years (1978 - 1980) during which time most major university and hospital departments of neurosurgery had computed tomography at their disposal.

Our aim in this cooperative study was to find out how soon after the hemorrhage and in what condition the patients come to the hospital; which diagnostic procedures are carried out, when and in what form; and at what intervals of time and with what results further treatment is undertaken.

Therefore, the results can really not be compared with publications dealing primarily or exclusively with the treatment of aneurysms or angiomas even though these were the source of the hemorrhage in a great number of the cases reported on. And, against the usual practice of other cooperative studies, there was no selection as to which university and hospital departments would participate. On the contrary, in May 1981 all departments of neurosurgery, neurology and neuro-radiology in German speaking countries in continental Europe were invited to take part.

A very detailed questionnaire was worked out. Twenty-four clinical centers participated, returning 2629 completed questionnaires, 2312 of these were evaluated statistically. The number of cases reported on by the individual centres varied from 10 to 265. Eleven departments had had less than 60 cases and eight had had more than 150 cases (Fig. 1). As the period of the study was three years, 11 of the departments had thus treated fewer than 20 cases a year and 8 more than 50.

This indicates that the individual values of the statistical evaluation of the study cannot be seen to be representative either for the individual departments taking part or for neurosurgery in Germany and Austria.

Advances in Neurosurgery, Vol. 11
Edited by H.-P. Jensen, M. Brock, and M. Klinger

It is obvious that in a cooperative study of such a nature many important results - and especially experiences made by individual surgeons - hardly appear in the results which are only average values. For this reason we owe a vote of thanks to all those colleagues who took the time and participated in this study.

The value of such an investigation lies in the great number of cases reported on. My co-workers SCHEIL and KRAUS have brought together a number of interesting facts, a few of which I would like to emphasize. With such a great number of cases it is doubtless of great importance that the cause of the hemorrhage in 54% of them was aneurysms, in 9% angiomatous malformations and in 37% other factors. These percentages differ greatly from clinic to clinic. Seven patients had both an aneurysm and an angioma. In 9% of all cases of aneurysms there were multiple aneurysms.

The hemorrhages from aneurysms and those from other causes occurred mainly in patients between 50 and 70 years of age. There is a relatively even distribution of occurrence of the angiomas between the second and the seventh decade of life. Women, with 55%, had more aneurysms; men, with 59%, suffered more angiomas. In addition, men were more often represented in the group of hemorrhages from other causes where the percentage was 54.

The interval of time between the hemorrhage and admission to the hospital is obviously related to the degree of severity of the clinical picture. Patients in a serious condition generally go to a neurosurgical department right away - something that is especially true for vascular causes and hypertensive hemorrhages. Doubtless there has been a great change in Germany in this area in the last few years. Today many more patients who have suffered a stroke or a cerebral hemorrhage from an unknown cause initially go to a department of neurosurgery. Diagnosis and treatment in such cases used to be carried out more often in departments of non-operative medicine. This new development is due to the superior diagnostic facilities of neurosurgical departments which permit a rapid clarification of the cause.

Of special interest in our study was the importance of computed tomography. As was to be expected the CT examination was performed either immediately following or within twenty-four hours of admission, especially with patients in serious condition (grades five and six) - in our study in 94% of the cases reported on. With patients in a less serious condition (grades one and two) it was made in only 53% of the cases and in the remaining cases it was not made until later. The results of the CT scanning show subarachnoid hemorrhages most often accompanying aneurysms, whereas intracerebral hematomas, intraventricular hemorrhages and mass effects occur most frequently with angiomas and other bleedings. Where hyperdense areas are described relatively often after intracranial hemorrhages, we are dealing with calcifications or, as in cases of angiomas and larger aneurysms, a direct demonstration of the malformation. In the cases of the miscellaneous hemorrhages we are dealing with tumors as the cause of the hemorrhage or with diffuse bleeding into the parenchyma where tissue density does not yet permit identification as a hematoma. With the hypodense zones we are, in general, dealing with circulatory disturbances which manifest themselves as collateral edema following the hemorrhage or with ischemic areas resulting from the vascular spasm. Whether or not the aneurysm or angioma can be demonstrated directly - with or without enhancement - depends on both the size of the malformation and the scanner's power of resolution.

Many of the examinations reported on in this study were undoubtedly made with first or second generation scanners. We are surely justified in our expectation that in future improved computed tomography will also enable us to recognize angiomas and smaller aneurysms more often.

In connection with the results of the CT diagnosis, it is interesting to note that in our study a lumbar puncture, which up to now has been considered the classical indication of a subarachnoid hemorrhage, was performed as an acute diagnostic measure in only 71% of the cases.

It is very obvious that computed tomography has become an instrument of great value in the diagnosis of acute intracranial hemorrhages. It cannot, however, replace cerebral angiography, especially in the diagnosis of aneurysms and angiomas. When indicated, cerebral angiography was carried out as an invasive method, generally after the CT examination. Whereas a CT examination was performed in 1843 cases in our study, that is in 90% of the cases here evaluated, cerebral angiography was carried out as a second diagnostic measure in only 1683 cases or in only 82%. The patient's condition also determined when the angiography was done. The angiography was done within the first 24 hours on 72% of the patients showing grades five and six but on only 24% of patients showing grades one and two.

Single aneurysms shown with angiography tended to be found more often in the right hemisphere and in females. Aneurysms of the anterior communicating artery were found most frequently and, in contrast to all other aneurysms, occurred most often in males. In 43% of the cases reported on, the size of the aneurysm was less than 5 mm, in 49% it was between 5 mm and 20 mm and in 6% it was larger than 20 mm. In 2% of the cases no size was given.

In our study special attention was paid to vasospasm because of its importance especially in aneurysmal hemorrhages. Of the 976 aneurysms vasospasm was shown angiographically in 36% of them. The interval of time between hemorrhage and angiography reveals important differences. Vasospasm was found most often in angiograms made within the first 48 hours of the hemorrhage (42%). Angiograms made within two to four days showed vasospasm in only 29% of the cases; those made between the fourth and the fifteenth day showed more vasospasm - namely 40% - and in angiograms made after the fifteenth day only 30% of all cases showed vasospasm. They occurred most frequently in patients who had already had a subarachnoid hemorrhage, that is to say in 57% of the cases where an angiogram was made within the first 48 hours, in 38% when angiography was carried out between the second and the fifteenth day and in only 26% when an angiogram was not done until later. Interesting as these statistics are, they say little about the pathophysiology of cerebral vascular constriction. In the last analysis they only confirm what we all know from experience, namely that vascular spasm is frequently found longer than two weeks after a subarachnoid hemorrhage.

The question of the most opportune time for operation after a subarachnoid hemorrhage has been much discussed in the past few years and often enough the extremely "early operation" has been recommended. During the period of time covered by our study operations were only rarely performed within two or four days of the hemorrhage and then they were generally emergency operations to remove a space-occupying intracerebral hematoma. The degree of severity in these cases was from grades four to six. Nevertheless, relatively speaking, many operations were performed between the fifth and the fifteenth day, even on patients where the degree of severity was between grades one and three.

This can probably be attributed to the danger of a recurrence of the hemorrhage. More than half of the patients in our study were operated on more than two weeks after the hemorrhage.

The results of operation are especially interesting in the cases of aneurysms. Once again we must point out that the results of our study are average values from departments where conditions and the number of cases treated vary greatly. In our tabular summary (Table 1 and 2) we have broken down the results at the time of discharge according to the patient's condition before operation. We have worked out the percentage of all patients who, after the operation, were able to work (excellent and good) and those that were unable to work but could take care of themselves (fair) and those that were dependent on help from a third party (poor).

Naturally, those patients whose pre-operative condition was good, that is grades one and two, showed the best results. More meaningful is nonetheless the fact that a significant number of patients from grades four to six could be discharged after operation in good condition.

In a further tabular summary (Table 2) we have compiled the results of operation showing the pre-operative degree of severity and the different sites of the aneurysms. We found the best results with aneurysms of the middle cerebral artery. The results for internal carotid aneurysms were somewhat poorer. Due to the small number of cases the results of operations on the remaining aneurysms are not really comparable. Six cases of aneurysms operated on could not be included in this list since their location was not indicated clearly enough in the questionnaire.

In conclusion it must be pointed out that our co-operative study did not bring any fundamentally new scientific information to light, but this was not to be expected - given the nature of the study and the fact that case data from very different departments of neurosurgery were compiled without pre-selection. Nevertheless, we did produce a comprehensive view of the prevalent treatment of acute, non-traumatic intracranial hemorrhages today. From this review emerges a number of interesting facts and it can also serve as a point of reference for the "quality control" of medical practice constantly being called for today.

Against older, comparable studies the following points are worth bearing in mind:

Table 1. 851 operations on single aneurysms: preoperative grade and results on discharge

Grade	No.	Excell. No.	% (Excell.+Good)	Good No.	Fair/Poor		Dead	
					No.	No.	No.	%
I	217	172	89	20	9	4	16	7
II	306	178	78	60	31	10	37	12
III	175	48	67	69	34	19	24	14
IV	104	11	25	15	34	33	44	42
V	39	-	3	1	11	28	27	69
VI	10	-	20	2	2	20	6	60
Total	851	409	68	167	121	14	154	18

Table 2. Location of 1037 aneurysms of which 845 were operated on: pre-operative grade and results on discharge

Site	No.	Operation No.	Grade	No.	Excell. No.	Excell./Good %	Good No.	Fair/Poor No.	Fair/Poor %	Dead No.	Dead %
Ant. comm. art.	415	335	I	95	80	95	10	1	1	4	4
			II	136	72	78	35	13	10	16	12
			III	45	13	71	19	4	9	9	20
			IV-VI	59	4	14	4	21	35	30	51
Int. carot. art.	243	210	I	48	35	79	3	4	8	6	13
			II	68	40	79	14	6	9	8	12
			III	54	13	65	22	10	19	9	18
			IV-VI	40	2	18	5	14	35	19	47
Middle cer. art.	218	181	I	44	35	86	3	3	7	3	7
			II	48	32	77	5	4	8	7	15
			III	51	17	73	20	11	21	3	6
			IV-VI	38	3	26	7	8	21	20	53
Ant. cer. art.	61	53	I	9	9	100	-	-	-	-	-
			II	26	17	73	2	4	15	3	12
			III	9	-	44	4	1	33	2	23
			IV-VI	9	1	22	1	3	33	4	45
Post. comm. art.	51	41	I	9	6	78	1	1	11	1	11
			II	18	14	83	1	2	11	1	6
			III	11	4	64	3	4	36	-	-
			IV-VI	3	1	33	-	1	33	1	33
Basilar. art.	31	13	I	5	3	-	1	-	-	1	-
			II	2	1	-	-	-	-	1	-
			III	3	-	-	1	1	-	1	-
			IV-VI	3	-	-	-	-	-	3	-
Vertebr. art.	18	12	I	4	2	-	1	-	-	1	-
			II	6	1	-	2	2	-	1	-
			III	2	-	-	-	1	-	1	-
			IV-VI	-	-	-	-	-	-	-	-
Total	1037	845									

1. Today patients with acute intracranial hemorrhages come more frequently either immediately or at least much earlier than before to a department of neurosurgery and are not treated first for longer periods of time in general departments or departments of neurology or internal medicine.
2. Computed tomography was performed in almost all cases immediately or shortly after, admission, giving essential information on the cause of the hemorrhage.
3. Cerebral angiography is used more purposefully after the CT examination and is used earlier to detect the presence of aneurysms and angiomas.
4. As in all other similar studies the results of operation in cases of aneurysms depend on the patient's condition and on the location of the aneurysm.
5. Since our study is based on average values of 24 university or hospital departments of neurosurgery the operative results can - compared with other studies - be judged thoroughly satisfactory.

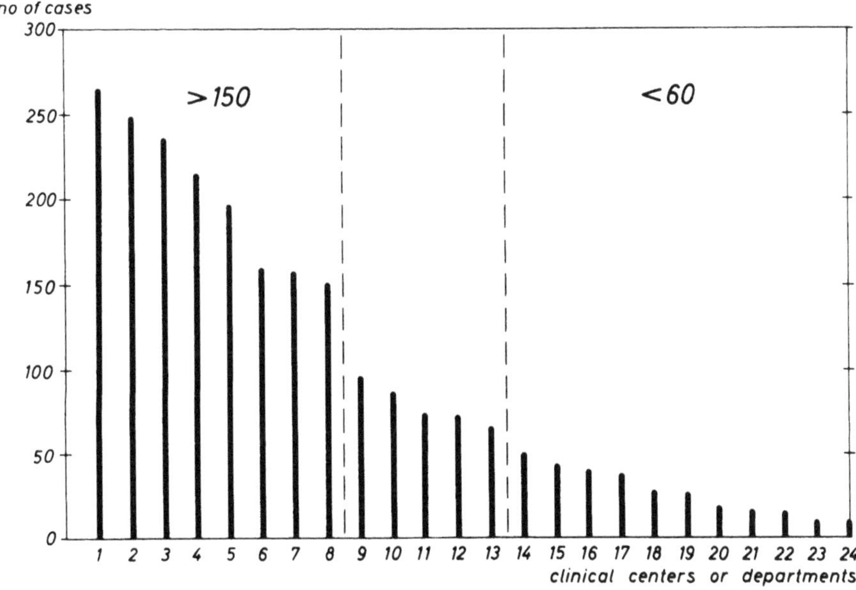

Fig. 1. Number of case registrations in the clinical centers or departments which took part in the cooperative study on acute, non-traumatic intracranial hemorrhage during the years of 1978, 1979 and 1980

Dynamic Multi-Plane CT in the Diagnosis of Intracranial Aneurysms

D. Vonofakos, H. Hacker, and H. Grau

Abteilung für Neuroradiologie, Zentrum der Radiologie, Klinikum der Johann Wolfgang Goethe-Universität, Schleusenweg 2–16, D-6000 Frankfurt

The detection of subarachnoid or intracerebral bleeding by plain Computer Tomography in the first days after rupture of a vascular malformation is easy and is based on the high density of blood clots. Therefore one can attempt to predict the probable site of the ruptured malformation by analysing the location of blood clots in the subarachnoid spaces (5, 6, 8, 12, 13).

However the direct visualization of intracranial aneurysms after the use of contrast medium is possible only in a limited number of cases and depends on the size and location of the aneurysm and seems questionable or impossible in cases of massive subarachnoid bleeding (1-3, 10, 12, 13).

The role of multi-plane dynamic CT for the direct visualization of vascular structures is emphasized in recent papers (4, 7, 9, 11).

Technique

A pre-contrast scan is always performed in any case of suspected subarachnoid bleeding, because it might give important diagnostic information about the location of blood clots and the probable site of a ruptured aneurysm. In this way we can choose the area which must be examined by Dynamic Multi-plane CT (Table 1).

A total of 100 - 150 ml 66% contrast medium (Telebrix 300[1]) is injected mechanically into an antecubital vein with a flow of 1 - 1.5 ml/sec. After the injection of about 30 ml contrast medium, a series of 2 mm

Table 1. Location of intracranial aneurysms directly visualized by dynamic multi-plane CT

Anterior communicating artery	3
Internal carotid artery	1
Posterior communicating artery	3
Middle cerebral artery	4
Vertebral/basilar artery	2
Negative	6
Total	19

[1] Byk Gulden

Advances in Neurosurgery, Vol. 11
Edited by H.-P. Jensen, M. Brock, and M. Klinger

cuts with table incrementation of 2 mm is obtained at the level of the circle of Willis. With our Unit (Somatom 2N[2]) we can perform six scans/min maximally and thus a series of 6 - 10 scans can be obtained during the injection time.

The high blood iodine concentration during the injection allows a direct visualization of all vascular structures because they are intensively opacified. The density of the basal vessels predominates over the density of subarachnoid or intracerebral bleeding. Therefore the following "High Lighting" of all the cuts allows the marking of blood vessels, which can be distinguished from blood clots.

Finally with multilevel summation of the marked images the whole vascular anatomy of the examined area appears in a single image, while brain structures disappear. Consequently arteries and veins are equally opacified and cannot be differentiated from each other, but the anatomic correlation allows the recognition of the circle of Willis, anterior and middle cerebral arteries, basilar artery, posterior and middle cerebral arteries, sometimes posterior communicating arteries, as well as temporal and tentorial veins and the sinuses. Therefore an aneurysm down to a few millimeters in size can be directly visualized, as a saccular structure adjacent to an artery.

Results

We have examined 19 cases, of suspected intracranial aneurysm. The location of the detected aneurysms is presented in the Table 1. The smallest one had a size of 3 mm (Fig. 2). In all cases the diagnosis was confirmed angiographically and in the positive cases also surgically. The most difficult problem was the presence of massive subarachnoid bleeding in the basal cisterns and in these cases we had to increase the blood iodine concentration further more, i.e. the injection flow had to be higher (about 1.5 ml/sec) and thus densities of 120 - 160 Hounsfield Units could be achieved in the main blood vessels; densities, which lay far above the density of blood clot. Representative cases in correlation to the angiogram are demonstrated in the Fig. 1 - 4.

Discussion

Multi-plane dynamic CT cannot be used for evaluation of the time course of enhancement patterns. It only enables a better approach to the intracranial morphology and especially to the vascular system, while functional studies are only possible by single-plane dynamic CT (11).

The role of multi-plane dynamic CT in the detection of small intracranial aneurysms in our series of 19 cases is obvious, with a diagnostic accuracy which approaches 100%. Giant intracranial aneurysms on the contrary can be easily detected by conventional Computer Tomography, but in this case the single-plane dynamic CT is of great value for the specific diagnosis, as we will be reporting in a future paper.

Noteworthy is the reliability of the method in negative cases, whereby angiographically no aneurysm could be detected either.

---

[2] Siemens

Concerning the clinical value of early localization of a ruptured aneurysm by computer tomography, before angiography, we believe that this is important diagnostic information and can facilitate better timing of the subsequent angiography; for example, early angiogram of one carotid artery by direct puncture for the evaluation of spasm and four-vessel angiography at a later stage.

References

1. Bryan, R.N., Shah, C.P., Hilal, S.: Evaluation of subarachnoid hemorrhage and cerebral vasospasm by Computed Tomography. CT 3: 144-152 (1979)
2. Findlay, G.F.G.: Computer-Assisted (Axial) Tomography in the management of subarachnoid hemorrhage. Surg. Neurol. 13: 125-128 (1980)
3. Inoue, Y., Saiwai, S., Miyamoto, T., Ban, S., Yamamoto, T., Takemoto, K., Taniguchi, S., Sato, S., Namba, K., Ogata, M.: Postcontrast Computed Tomography in subarachnoid hemorrhage from ruptured aneurysms. J. Comput. Assist. Tomogr. 5: 341-344 (1981)
4. Koehler, P.R., Anderson, R.E.: Computed angiotomography. Radiology 137: 843-845 (1980)
5. Liliequist, B., Lindqvist, M., Valdimarsson, E.: Computed Tomography and subarachnoid hemorrhage. Neuroradiology 14: 21-26
6. Liliequist, B., Lindqvist, M.: Computer Tomography in the evaluation of subarachnoid hemorrhage. Acta Radiol. (Diagn.) 21: 327-331 (1980)
7. Reese, D.F., McCullough, E.C., Baker, H.L., Jr.: Dynamic sequential scanning with table incrementation. Radiology 140: 719-722 (1981)
8. Scotti, G., Ethier, R., Melancon, D., Terbrugge, K., Tchang, S.: Computed Tomography in the evaluation of intracranial aneurysms and subarachnoid hemorrhage. Radiology 123: 85-90 (1977)
9. Weinstein, M.A., Duchesneau, P.M., Weinstein, C.E.: Computed angiotomography. AJR 129: 699-701 (1977)
10. Weisberg, L.: Direct visualization of intracranial aneurysms by Computed Tomography. Comput. Tomogr. 5: 191-199 (1981)
11. Wing, S.D., Anderson, R.E., Osborn, A.G.: Cranial computed angiotomography. Radiology 143: 103-107 (1982)
12. Yock, D.H., Jr., Larson, D.A.: Computed Tomography of hemorrhage from anterior communicating artery aneurysms, with angiographic correlation. Radiology 134: 399-407 (1980)
13. Zuccarello, M., Iavicoli, R., Pardatscher, K., Fiore, D.L., Mingrino, S.: Die diagnostische Bedeutung der CT bei subarachnoidaler Blutung aus Hirnarterienaneurysmen. Neurochirurgia 24: 47-50 (1981)

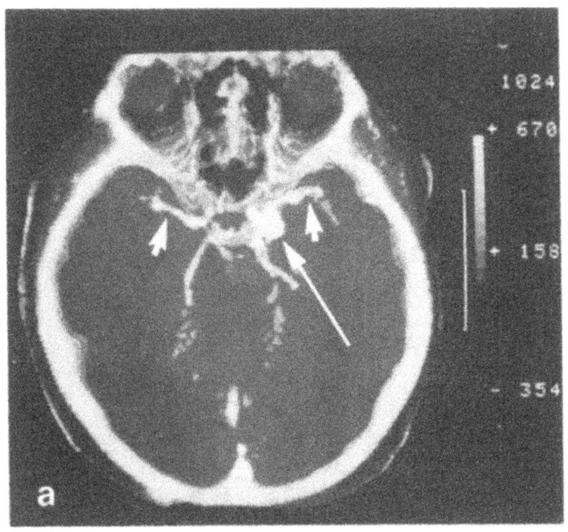

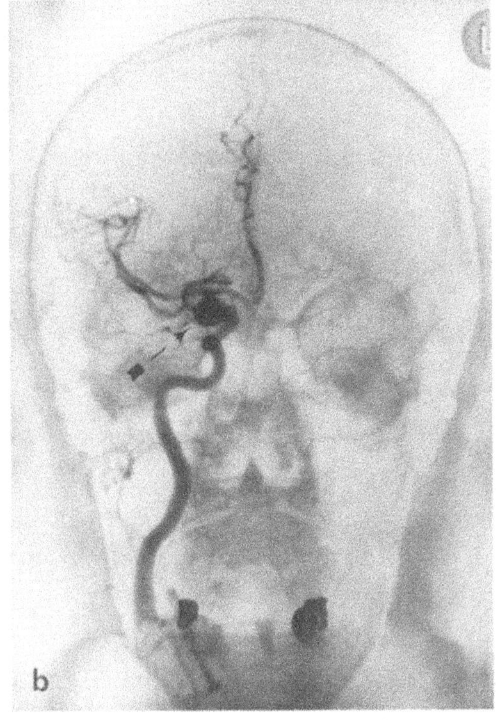

Fig. 1. a Dynamic multi-plane CT image after high-lighting and multi-level image summation. Note the middle cerebral arteries *(arrow heads)* and the aneurysm on the right carotid artery *(arrow)*; b The same patient as in a. Angiographic demonstration of the carotid artery aneurysm *(arrow)*

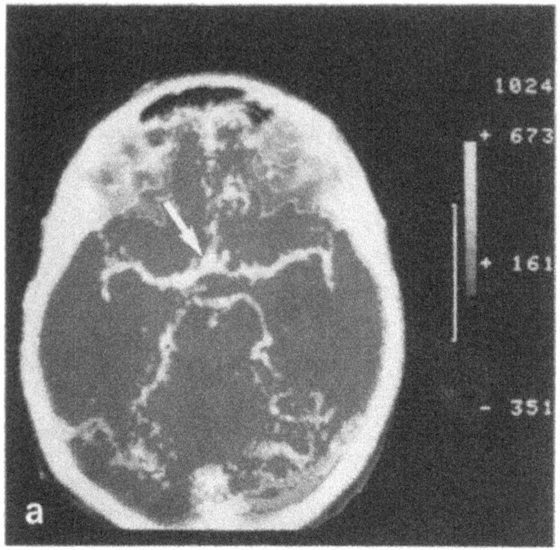

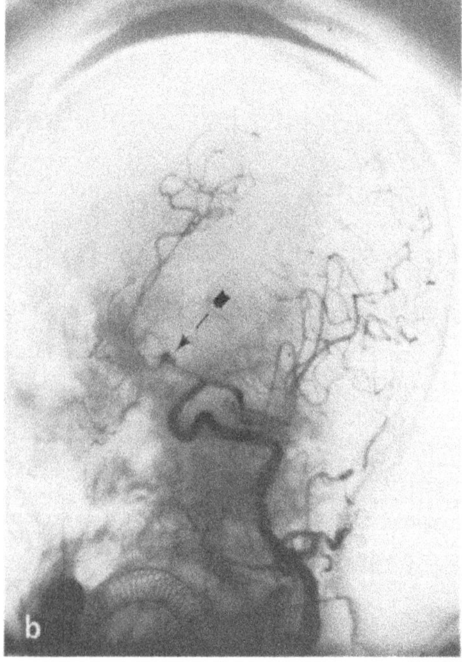

Fig. 2. a Dynamic multi-plane CT image after high-lighting and multi-level image summation. Small aneurysm on the anterior communicating artery *(arrow)*. The circle of Willis and the middle cerebral arteries are well recognizable; b The same patient as in a. Small aneurysm on the anterior communicating artery. Diameter: 3 mm

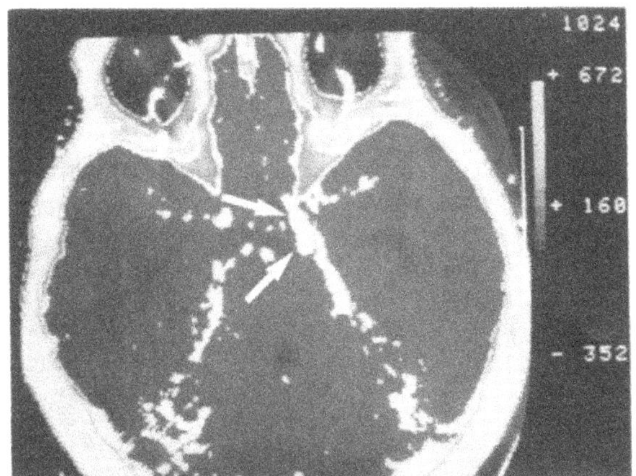

Fig. 3. Dynamic multi-plane CT image after high-lighting and multi-level image summation. Long aneurysm on the right posterior communicating artery *(arrows)*

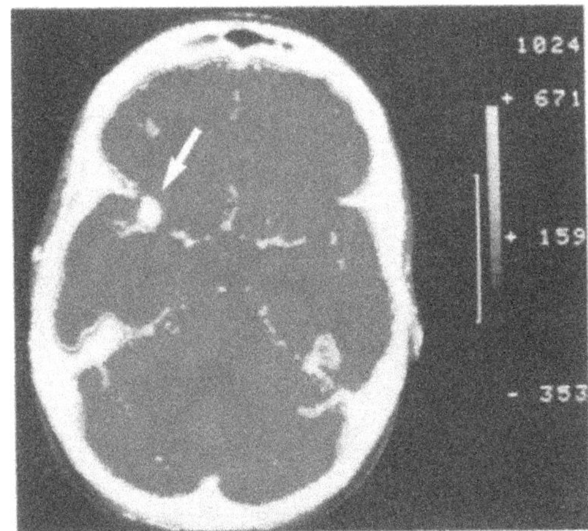

Fig. 4. Dynamic multi-plane CT image after high-lighting and multi-level image summation. Note the aneurysm on the bifurcation of the left middle cerebral artery *(arrow)*

Resolving Spontaneous Intracerebral Hematoma – Diagnostic Problems in Enhanced CT

A. Thron

Abteilung für Neuroradiologie, Institut für Radiologie der Universität Tübingen, D-7400 Tübingen

The CT appearance of acute intracerebral hemorrhage is well known and has virtually diagnostic value. The course of resolving intracerebral hematoma has also been well described (1, 3, 4, 8, 9-11). In some reports attention has been drawn to the fact, that in contrast-enhanced studies during resorption of hematoma a peripheral ring blush is frequently observed (5-7, 9, 12). Without a previous scan for comparison this finding may pose various differential diagnostic problems with respect to the stage of resorption (2, 4).

Patients Material

Between 1979 and 1980 in a series of 95 patients with spontaneous intracerebral hematoma contrast-enhanced CT-controls have been performed in 34 cases. Table 1 summarizes the time-related number of scans following the ictus. Only hematomas more than 2 x 2 cm in size have been considered in this study.

Results

During the first two weeks usually no ring of contrast enhancement can be seen (Table 1). In one case a ring was demonstrable ten days after the onset of symptoms. Beginning with the third week, a ring blush is an almost constant finding, which, however, may be of variable development and demarcation. Ring formation does not depend upon the age of the patient or the extent of the hematoma, but there is a slight tendency to more pronounced ring formation in younger patients (Table 2)

Table 1. CT-control studies in resolving intracerebral hemorrhage (>2 x 2 cm)

Time	No. of patients	No. of contrast enhanced scans	No. of rings
1 - 6 days	95	-	-
7 - 13 days	11	5	1
14 - 27 days	26	22	19
4 - 5 weeks	11	10	9
6 - 9 weeks	12	11	9
>9 weeks	1	1	1

 Advances in Neurosurgery, Vol. 11
Edited by H.-P. Jensen, M. Brock, and M. Klinger

Table 2

Patient age (years)	Ring-enhancement +	(+)	Ø	Total
< 40	5	2	-	7
40 - 60	8	3	1	12
> 60	8	3	4	15
Total	21	8	5	34

Table 3

Size of hematoma (diameter)	Ring-enhancement +	(+)	Ø	Total
2 - 3 cm	3	2	2	7
3 - 4 cm	11	5	1	17
4 - 5 cm	7	1	2	10
Total	21	8	5	34

Table 4

Location	Ring-enhancement +	(+)	Ø	Total
Lobar	11	4	3	18
Basal ganglia (+ lobar)	10	4	-	14
cerebellum	-	-	2	2
Total	21	8	5	34

and in larger hemorrhages (Table 3). The location of the hematoma is also not decisive. In two infratentorial lesions we had negative results (Table 4).

Compared to small hematomas with an extent of less than 2 x 2 cm, the decrease in size and density of large intracerebral hematomas is delayed (3). But it is of greater importance that the resorbing clot does not show homogeneous diminution of attenuation values. On the contrary, there is a centripetally progressing decrease of density. Fig. 1 shows the typical appearance of enhanced control studies in different stages of resorption. The acute intracerebral hematoma (Fig. 1a) of increased attenuation is surrounded by a small hypodense rim, thought to represent edema. After ten days (Fig. 1b) the peripheral parts of the lesion have changed to low attenuation values, due to the increase in water content and the reduced hemoglobin molecule concentration (11). No ring enhancement can be found at that time and the surrounding edema is even more pronounced. After three weeks the hematoma resolution has proceeded to a stage, where only a central part remains of increased density (Fig. 1c). This central part is surrounded by an area of hypodensity, representing that part of the clot in which hemoglobin and other breakdown products have already been resorbed.

Enhanced CT now demonstrates a ring blush which demarcates the zone of hypodensity peripherally and corresponds to the original size of the hematoma. Mass effect and edema have only slightly diminished.

In the next stage after 3 - 5 weeks the whole of the hematoma has developed into a homogeneously lucent area. As a ring blush is still noted, the lesion now appears as a cystic formation. In the following days the mass effect and the volume of the hematoma diminishes rapidly. Some months later the healed hematoma normally demonstrates as a small lucent defect of brain tissue with ipsilateral enlargement of cerebrospinal fluid (CSF) spaces (Fig. 1d).

Table 5 shows the characteristics of four stages of hematoma resorption in enhanced CT-controls of large intracerebral hemorrhages. The acute hemorrhage (Stage I) is followed by a pattern of resorption resembling a cockade (Stage II). Four to six weeks after the ictus the resolving hematoma is observed as a ring-shaped or cystic lesion with preserved but decreasing mass effect (Stage III). Post-hematoma residua, consisting of a defect of brain tissue and enlargement of surrounding CSF-spaces, characterize the stage of healing (IV).

Although the ring blush as a transient phenomenon of stages II and III has no specific characteristics, it can frequently be observed as a narrow, irregularly shaped and not closed ring formation (Fig. 6).

Discussion

Our investigations suggest, that there is no isodense phase of resolving intracerebral hemorrhages of more than 2 x 2 cm size as is assumed in some reports (1, 4, 8). The decrease in density by an average of 0.7 EMI units per day (3) is a centripetally progressing process. The time-dependent development of ring-enhancement, beginning with the third week of resorption is of variable distinctness. It is not yet clear, which are the main determining factors for a more or less pronounced ring formation. Fig. 2 shows a very clear ring blush in a hematoma of the right caudate nucleus in the third week of resorption. In another hematoma of comparable size, site, and age (Fig. 2b) there is only slight peripheral enhancement. According to LASTER et al. (6) the early enhancement is modified by steroids and might therefore solely be due to breakdown of the blood-brain barrier. Beginning with the third week vascular proliferation and formation of an increasing capsule of granulation tissue is observed pathologically (6, 9, 12). The blood-brain barrier function of these new-formed vessels is incomplete and extravasation of contrast medium besides increased vascularity are probably the underlying mechanisms of ring formation in the late stage II and stage III.

Table 5. Stages of hematoma resorption in enhanced CT (hematoma size > 2 x 2 cm)

I	Hematoma of homogeneously increased density. Progressive surrounding edema. No ring-enhancement (first two weeks).
II	Hematoma with centripetally progressing decrease of density. Ring enhancement corresponding to the original size of the hematoma. ("Cockade-Form"). No reduction of volume of edema (3rd - 5th week).
III	Hematoma homogeneously lucent with marked ring blush ("Cystic Form"). Gradual reduction of volume and edema (4th - 8th week).
IV	Healed hematoma. Small lucent defect of brain tissue (>6th week).

In stage I of hematoma resolution diagnostic confusion only arises in cases with unusually extensive perifocal edema (Fig. 3), which may be caused by compression of veins or a primary venous hemorrhage. Of course, an underlying neoplasm, masked by the dense hematoma cannot be excluded (12), but is a rare cause of intracerebral hemorrhage.

The "cockade-pattern" of stage II in an enhanced CT-control may erroneously be taken for neoplasm, a misinterpretation, which we observed several times (Fig. 4). On the other side the peripheral enhancement allows differentiation of liquefied clot from edematous brain. Thus it demonstrates, that despite the decrease in size of the dense portion of the hematoma the volume of the whole clot remains relatively unchanged until stage III.

The differential diagnosis of stage III in the absence of an initial scan of the acute hematoma includes an array of ring-shaped or cystic lesions especially primary or secondary brain tumors and abscesses (Fig. 5) (5, 9).

In most cases a definitive diagnosis may be made by control-scans over an adequate interval of time together with the patients history and clinical findings. To avoid aggressive procedures for diagnosis or treatment the demonstrated CT-pattern of hematoma resolution in enhanced control studies at different times should be known.

Ring formation is identical in non-traumatic and traumatic intracerebral hemorrhage (Figs. 5 and 6).

Fig. 6 shows abscess and resolving intracerebral hemorrhage simultaneously in one patient and gives an impression of the somewhat different character of both ring lesions.

Summary

A ring of contrast enhancement is common in resolving intracerebral hematoma more than 2 x 2 cm in size. In a CT-study of 34 patients the time of appearance and other characteristics of this ring blush were determined. In general, ring enhancement occurred between the third and the sixth week of hematoma resorption. Ring formation may be distinct or faint but it depends neither upon the age of the patient nor on the extent, location or etiology of the hematoma. The underlying pathological mechanisms are briefly discussed.

To separate other lesions of ring-shaped appearance the typical course of resolving intracerebral hemorrhages and their distinguishing features in enhanced CT are demonstrated.

References

1. Bergstroem, M., Ericson, K., Levander, B., Svendson, P., Larsson, S.: Variation with the time of the attenuation values of intracranial hematomas. J. Comp. Assist. Tomogr. 1: 57-68 (1977)
2. Coulam, C.M., Seshul, M., Donaldson, J.: Intracranial ring lesions: can we differentiate by computed tomography. Invest. Radiol. 15: 103-112 &1980)
3. Dolinskas, C.A., Bilaniuk, L.T., Zimmermann, R.A.: Computed tomography of intracerebral hematomas: I. Transmission CT observations on hematoma resolution. Am. J. Roentgenol. Radium Ther. Nucl. Med. 129: 681-688 (1977)

4. Grumme, Th., Lanksch, W., Wende, S.: Diagnosis of spontaneous intracerebral hemorrhage by computerized tomography. In: Lanksch, W., Kazner, E. (eds.), Cranial computerized tomography, pp. 284-287. Berlin, Heidelberg, New York: Springer 1976

5. Kazner, E., Steinhoff, H., Wende, S., Mauersberger, W.: Ring-shaped lesions in the CT scan - Differential diagnostic considerations. In: Wüllenweber, A., Wenker, H., Brock, M., Klinger, M. (eds.), Advances in neurosurgery, Vol. 6, pp. 80-83. Berlin, Heidelberg, New York: Springer 1978

6. Laster, D.W., Moody, D.M., Ball, M.R.: Resolving intracerebral hematoma: alteration of the "ring sign" with steroids. Amer. J. Roentgenol. Radium Ther. Nucl. Med. 130: 935-939 (1978)

7. Messina, A.V.: Computed tomography: contrast enhancement in resolving intracerebral hemorrhage. Am. J. Roentgenol. 127: 1050-1052 (1976)

8. Messina, A.V., Chernik, N.L.: Computed tomography: the "resolving" intracerebral hemorrhage. Radiology 118: 609-613 (1975)

9. Nasher, H.C., Nau, H.-E., Reinhardt, V., Löhr, E.: Computertomographische Befunde und Verlaufskontrollen bei konservativ und operativ behandelten intracerebralen Blutungen nicht traumatischer und nicht aneurysmatischer Genese. Radiologe 20: 122-129 (1980)

10. Schumacher, M., Rossberg, Ch., Stoeter, P.: CT-Differentialdiagnose und Verlaufsuntersuchungen bei intracerebralen Blutungen. Arch. Psychiatr. Nervenkr. 231: 171-185 (1982)

11. Weigel, K., Ostertag, C.B., Mundinger, F.: CT follow-up control of traumatic intracerebral hemorrhage. In: Frowein, R.A., Wilcke, O., Karimi-Nejad, A., Brock, M., Klinger, M. (eds.), Advances in neurosurgery, Vol. 5, pp. 62-67. Berlin, Heidelberg, New York: Springer 1978

12. Zimmermann, R.A., Leeds, N.D., Naidich, T.P.: Ring blush associated with intracerebral hematoma. Radiology 122: 707-711 (1977)

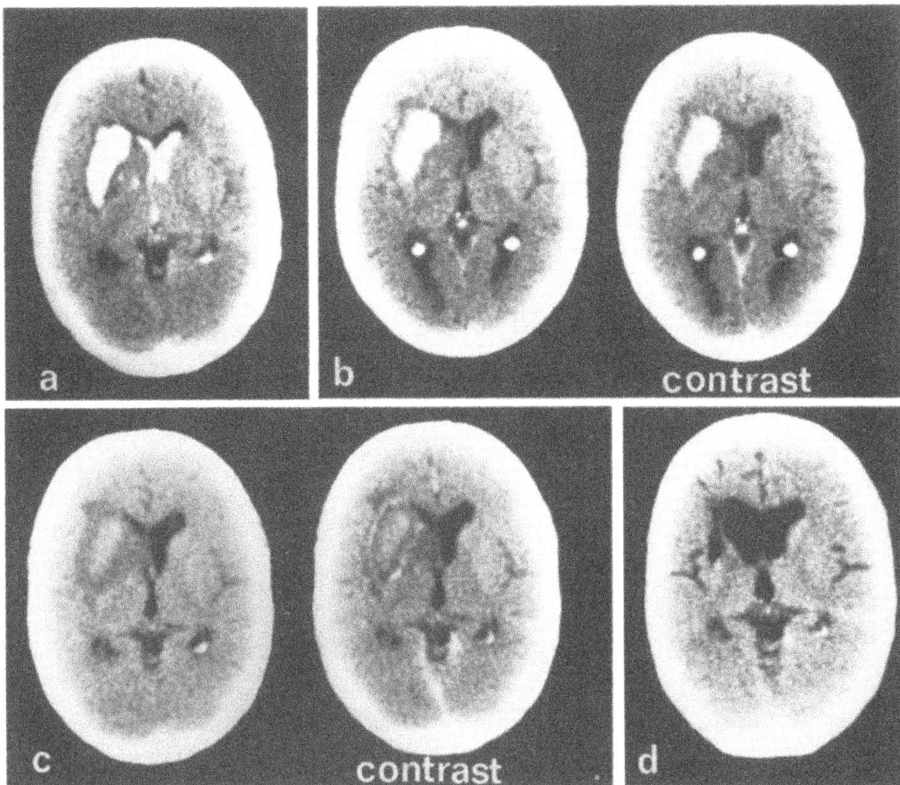

Fig. 1 a-d. CT-studies in different stages of hematoma resorption. a Acute hematoma of the left basal ganglia with ventricular penetration; b CT control after 10 days. No ring is seen after contrast medium enhancement; c Stage II. Centripetally progressing diminution of attenuation values. Ring of enhancement around the periphery of values. Ring of enhancement around the periphery of the hematoma after three weeks. d Stage IV. Small area of decreased density at the site of the original hematoma and focal ventricular dilatation

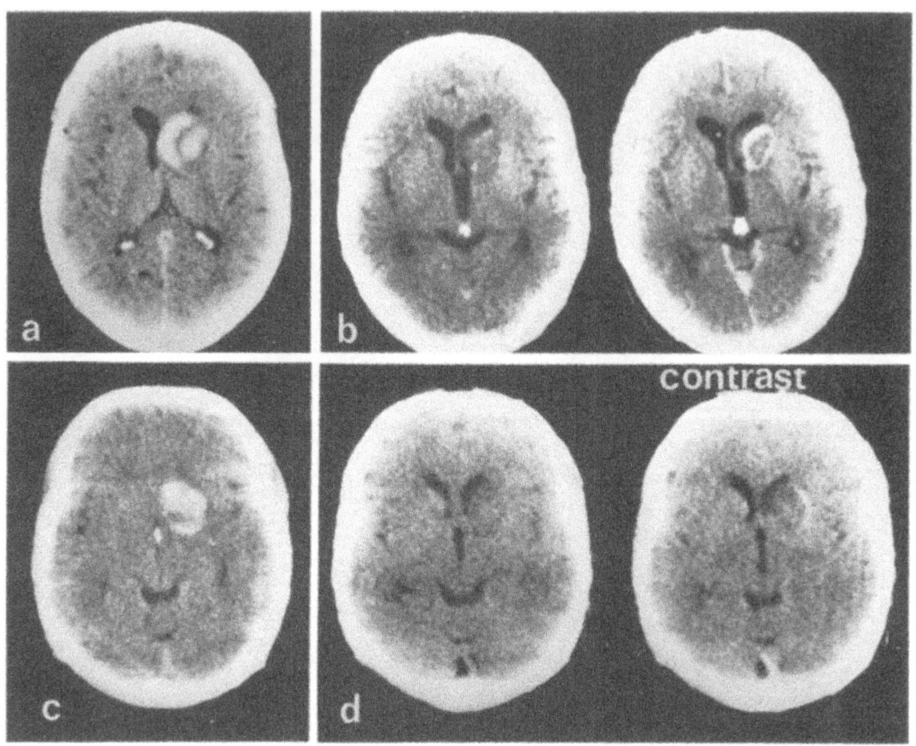

Fig. 2 a-d. Distinct ring formation (a, b) and faint ring blush (c, d) in two patients of equal ages and hematomas of similar location and extent in the fourth week of hematoma-resorption (Stage III)

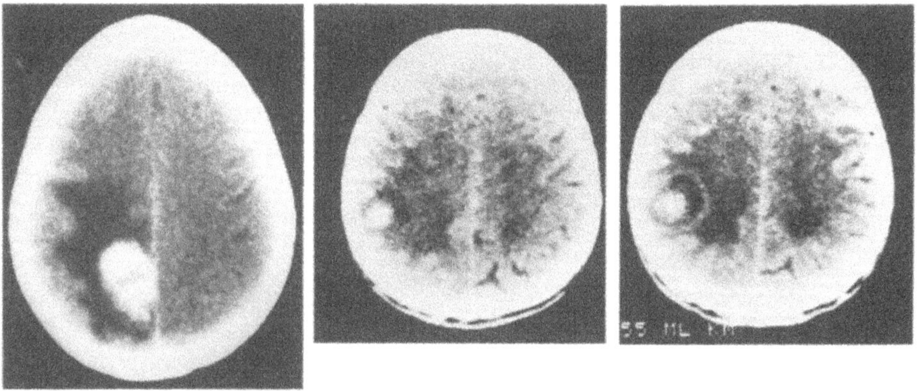

Fig. 3. *Left:* Unusually extensive edema surrounding a parietal hematoma

Fig. 4. *Right:* The "Cockade-pattern" of stage II resembling a neoplastic lesion

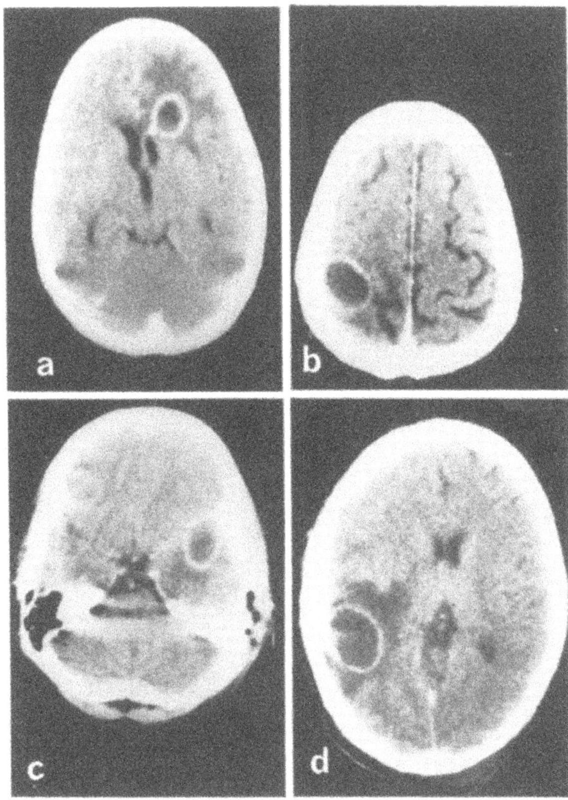

Fig. 5 a-d. Ring lesions with central hypodensity in cases of abscess (a and c), neoplasm (b) and resolving hemorrhage in stage III (d). In the case of d an abscess was suspected four weeks after traumatic hemorrhage. Operation revealed a so-called "chocolate cyst"

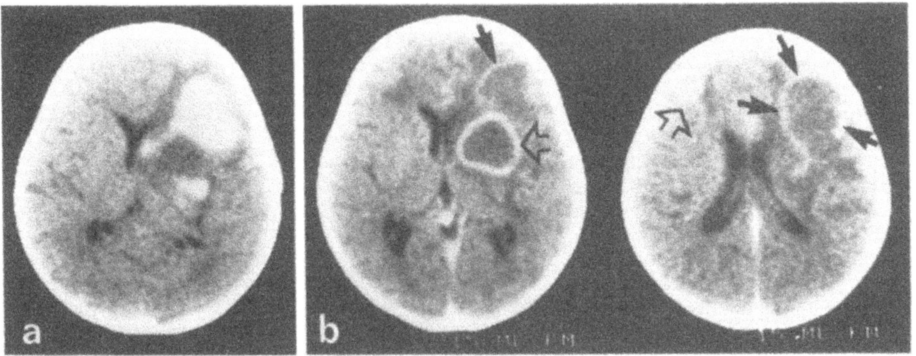

Fig. 6 a,b. Ring lesions caused by abscesses *(open arrows)* and resolving hematoma *(closed arrows)* in the same patient (b). The hemorrhage resulted from abscess puncture (a)

Differential Diagnosis and Resorption Behavior of Spontaneous Cerebral Hemorrhages in CT

M. Schumacher and Ch. Rossberg

Abteilung für Neurologie, Neurologisch-Psychiatrische Klinik Christophsbad, D-7320 Göppingen

Since the introduction of computed tomography (CT) views on the treatment of intracerebral hemorrhages have changed. The decision about medical or surgical treatment of large, space-occupying hematomas does not anymore depend exclusively on clinical criteria but also on CT criteria, such as site, size and space-occupying character of the hemorrhage (GRUMME et al. 1976; MAYWARD and O'REILLY 1977; LÖHR et al. 1977; PINEALA 1977). CT follow-up examinations have decisively influenced the therapeutic procedure through the information acquired about the spontaneous course and processes of resorption in spontaneous and traumatic hemorrhages (LEGRE et al. 1978; WEIGEL et al. 1978).

In the present study on 44 spontaneous hemorrhages an attempt will be made to establish CT criteria that contribute to the differential diagnosis of the bleeding source and to determine what additional diagnostic procedure is needed. At the same time, the relationship between location, size and rate of resorption in conservatively treated hemorrhages would be shown.

Material and Methods

The CT examination were conducted in part, with a SIRETOM 2000 and in part with SOMATOM 2 on 44 patients aged from 14 months to 81 years. The slices were carried out with standard positioning, with a slice thickness of 8 mm and in special cases, for the purpose of reconstruction, of 2 mm. All patients were examined for the first time within the first two days after the acute clinical event. The follow-up examinations were performed at weekly intervals until total resorption of the hemorrhage. In two-thirds of the patients supplementary CT with contrast medium (1.5 ml/kg) were performed. According to the size, the hemorrhages were classified as small (maximum 1 cm in diameter), moderate (1 - 3 cm in diameter) and large (larger than 3 cm in diameter). An additional differentiation was made between cortical and subcortical hemorrhages, and hematomas of the basal ganglia. The speed of resorption of hemorrhages with a diameter greater than 2 cm, was compared with those lesser than 2 cm.

Results

The commonest sources of the bleeding were hemorrhagic infarctions or hypertensive hemorrhages (in 36.4% of the cases). In none of these cases could a stenosis or an occlusion of the extracranial arteries

Advances in Neurosurgery, Vol. 11
Edited by H.-P. Jensen, M. Brock, and M. Klinger

be demonstrated by means of the doppler sonography. An additional clarification by means of angiography was not conducted in this group. In almost one third of the spontaneous hemorrhages no cause could be found, in spite of an intensive search. In cases of vascular dysplasia (11.4%), bleeding from tumors (4.5%) and thrombosis of the brain sinuses (4.5%) the source of the bleeding was verified angiographically. The causes of hemorrhages associated with coagulation disorders were from anticoagulant therapy, severe hepatopathy and leukemic diseases.

In two-thirds of the patients the hematomas measured more than 2 cm and were most commonly found in the white matter and basal ganglia (Fig. 1). In 6.8% (3 patients) of the cases rupture into the ventricular system occurred (Figs. 2, 3). In 4.5% (2 patients) a secondary hemorrhage appeared two days later in a primary ischemic infarct (Fig. 4). In all the cases surrounding edema had developed that was always detectable at the first examination and in the course of the first week increased considerably. The size of the hematomas and the extent of the surrounding edema showed no clear correlation.

Differential Diagnosis

Table 1. Etiology and CT characteristics in 44 spontaneous intracerebral hemorrhages

Etiology	N	CT characteristics
Stroke	16	Edema +++, relationship to regions of arterial blood supply, luxury perfusion
Hypertensive hemorrhage		Space-occupying, white matter and basal ganglia
Coagulation disorder	6	Small, cortical and subcortical extending, gyri-outlined SAH
Vascular malformation	5	SAH with penetrating effect, structure of the vessels after blood resorption seen CT with enhancement
Sinus thrombosis	2	Small, multiple, bilateral; edema (compression of the ventricle), empty torcular Herophili
Tumor	2	Discrepancy of hematoma size, edema (+++) and space-occupying character, three layers, follow-up CT examination (blood resorption without decrease of the space-occupying)
Unknown	13	

From CT results (Table 1) it was only possible in exceptional cases to obtain any reliable hint as to the source of the bleeding. In most of the cases (only a suspicion existed) and a correct and adequate diagnosis was possible only by considering the clinical history.

Hypertensive bleedings (Fig. 1) were mostly characterized by their size and preferred location in the basal ganglia. Hemorrhagic infarctions were classified as such, when they were limited to the vascular region of the anterior, middle or posterior cerebral artery or to isolated vascular branches. As a rule the area of hemorrhagic infarction was surrounded by intense edema. After injection of contrast material the total infarct region was clearly shown and marked as a luxury perfusion or a ring enhancement (Fig. 1).

Hemorrhages after embolic vascular occlusion generally showed small rounded areas of hemorrhage situated in the cortex and subcortical areas (Fig. 5). Any reliable differentiation in relation to smaller post-traumatic hemorrhages or in cases of reduced coagulability of the blood was nevertheless not possible.

Spontaneous bleeding in vascular dysplasia generally overlays the malformation in the initial stages. The control examination and especially the follow-up CT with contrast medium after complete resorption of the hematoma were decisive for the etiological classification. Extensive subarachnoid hemorrhage with rupture into the neighbouring parenchyma suggested an aneurysm as bleeding source. A rare complication in one patient with pronounced vascular spasm caused by a subarachnoid hemorrhage, was a secondary ischemic infarction (Fig. 6). Compared with subarachnoid hemorrhages caused by aneurysms, bleedings by coagulopathies show a slight increase in density of the subarachnoidal fluid. The combination of hemorrhages near to the cortex, small and at times outlined by the gyri was found mostly in cases of coagulopathy (Fig. 7).

Finger-shaped, and sometimes multiple bilateral hemorrhages in the grey and white matter, especially in the centrum semiovale indicates spontaneous bleeding in thrombosis of the venous sinuses (Fig. 8). The combination of bilateral brain edema, compression of the ventricular system and the failure to demonstrate the torcular Herophili confirms the diagnosis of an occlusion of the superior sagittal sinus.

The discrepancy of a small sized hematoma with considerable surrounding edema and also evidence of a considerable shift were characteristic of hemorrhage into a necrotic tumor (Fig. 9). Three layers from the center to the periphery were visible: a central hemorrhagic component surrounded by a zone of higher density consisting of tumor tissue and continuing into an irregularly shaped zone of lower density corresponding to the edema. A valuable criterion for the differential diagnosis proved to be the increasing formation of edema and the space-occupying character, in spite of the advanced blood resorption.

Resorption

During the first five days there were no detectable changes in density in the hemorrhagic region and there were no changes in the amount of the bleeding. The surrounding edema, that was already visible a few hours after the bleeding, increased considerably to the end of the first week, remained constant up to the second week or increased slightly in the second week after the hemorrhage. From the third week on there was a constant decrease of the edema observed (Fig. 3). All hemorrhages showed a logarithmic resorption velocity rate (Fig. 10), whereby *circa* 40% were totally reabsorbed by the fourth week. After seven weeks larger hemorrhages were totally reabsorbed. The resorption always advanced from the periphery to the center. The center of the hemorrhage degenerates in most cases into a cystic area, whereby the peripheral zone in part was resorbed free of any residuum. There was no difference in the resorption velocity between large and moderate hemorrhages, but in small hemorrhages up to 2 cm diameter regression was seen within the first week.

Conclusions

With the highly sensitive CT apparatus of the last two generations, it is possible to detect intracerebral hemorrhages of 2 - 3 mm size as well as bleedings in the subarachnoid space. Compared with the high quota of detection, the etiology in many cases remains obscure. The differential diagnosis frequently depends only on morphological and topographical criteria (Table 1), although correlating the clinical history with the clinical symptoms permits a relative reliable interpretation. Consequently there remain in this study, the same as seen by other authors (GRUMME et al. 1976), approximately 30% of the spontaneous hemorrhages of unknown etiology.

Additional diagnostic problems occur in delayed bleedings as for example is seen in secondary hemorrhagic infarcts, (approximately 5.5% of our patients), that occur two to three days after the onset of symptoms. False negative results caused by a masking effect in the isodense stage of resorption is uncommon (LANKSCH et al. 1978).

Ring structures in cases of partial resorption of hematomas have to be differentiated from inflammatory and tumor lesions. Excluding cases of hypertensive massive hemorrhages, hemorrhagic strokes and hemorrhages caused by other medical diseases e.g. coagulation disorders, more diagnostic investigations are needed, especially when according to the CT criteria and the clinical findings suspicion is aroused that tumor bleeding, a vascular malformation or a sinus thrombosis exist. This is important for the treatment.

Only in a few space-occupying hematomas with a midline shift of less than 5 mm (KARIMI-NEJAD et al. 1979) can a restrained attitude be taken by close follow-up CT and clinical controls of a spontaneous resorption. In our experience even larger hematomas were resorbed up to the eighth week and the patients survived without surgical measures (Fig. 2).

Contrary to the results reported by DOLINSKAS et al. (1977), the resorption velocity had a more logarithmic course, which is more in agreement with the results made by LEGRE et al. 1978. The rapid resorption that occurs after the fourth week is probably due to the capillary growth in the peripheral margins of the hematoma. Smaller hemorrhages under 2 mm probably resorb without the development of capillary vessels and therefore sooner. In these small hematomas ring enhancement was never seen.

The high velocity of resorption in hemorrhages in the white matter and basal ganglia could not be fully clarified in our prevailing study. It is possible that the matrix of the central white matter and the basal ganglia possess the capacity to stimulate capillary formation which is visible in some particular cases (Fig. 1) as a reinforced ring enhancement in CT.

References

1. Dolinskas, C.A., Bilaniuk, L.T., Zimmermann, R.A.: Computed tomography of intracerebral hematomas: I. Transmission CT observations on hematoma resolution. Am. J. Roentgenol. Radium. Ther. Nucl. Med. 129: 681-688 (1977)
2. Grumme, Th., Lanksch, W., Wende, S.: Diagnosis of spontaneous intracerebral hemorrhage by computerized tomography. In: Lanksch, W., Kazner, E. (eds.), Cranial computerized tomography, pp. 284-287. Berlin, Heidelberg, New York: Springer 1976

3. Hayward, R.D., O'Reilly, G.V.: Computerized tomography and intracerebral hemorrhage. Am. Heart. J. 93: 126-133 (1977)

4. Karimi-Nejad, A., Hamel, E., Frowein, R.A.: Verlauf der traumatischen intracerebralen Hämatome, Nervenarzt 50: 432-435

5. Lanksch, W., Grumme, Th., Kazner, E.: Schädelhirnverletzungen im Computertomogramm. Berlin, Heidelberg, New York: Springer 1978

6. Legre, J., Debaene, A., Lamoureux, J.M., Tapias, P.L.: Interest of control by CT of intracerebral spontaneous and traumatic hemorrhage at different stages of evolution. Neuroradiology 15: 51 (1978)

7. Löhr, E., Grote, W., Weichert, H.C., Fiebach, D., Bock, W.: Die Computertomographie bei intrakraniellen Blutungen. Radiologe 17: 177-184 (1977)

8. Messina, A.V.: Computed tomography: contrast enhancement in resolving intracerebral hemorrhage. Am. J. Radiol. 127: 1050-1056 (1976)

9. Nahser, H.C., Nau, H.-E., Reinhardt, V., Löhr, E.: Computertomographische Befunde und Verlaufskontrollen bei konservativ und operativ behandelten intracerebralen Blutungen nicht traumatischer und nicht aneurysmatischer Genese. Radiologe 20: 122-129 (1980)

10. Pineala, A.: Computer tomography in intracerebral hemorrhage. Surg. Neurol. 8: 55-62 (1977)

11. Weigel, K., Ostertag, C.B., Mundinger, F.: CT follow-up control of traumatic intracerebral hemorrhage. In: Frowein R.A., Wilcke O., Karimi-Nejad, A., Brock, M., Klinger, M. (eds.), Advances in neurosurgery, Vol. 5, pp. 62-67. Berlin, Heidelberg, New York: Springer 1978

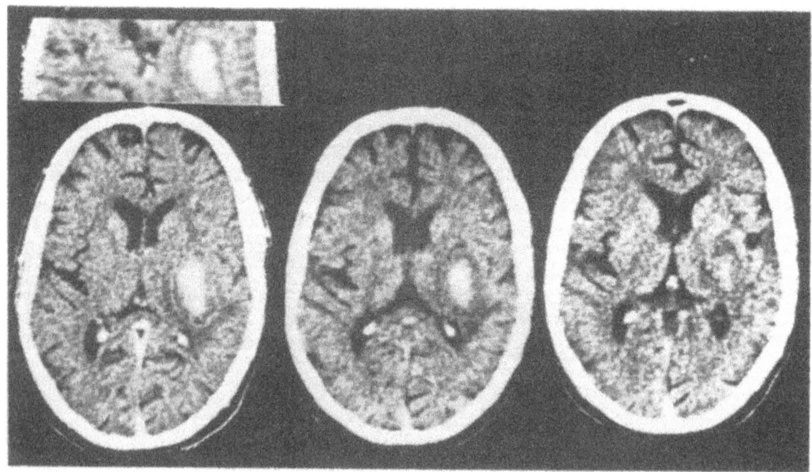

Fig. 1. Massive hypertensive hemorrhage into the right temporal lobe, course of resorption over a period of four weeks. Note, the more strongly developed ring enhancement medially (coronal reconstruction left picture)

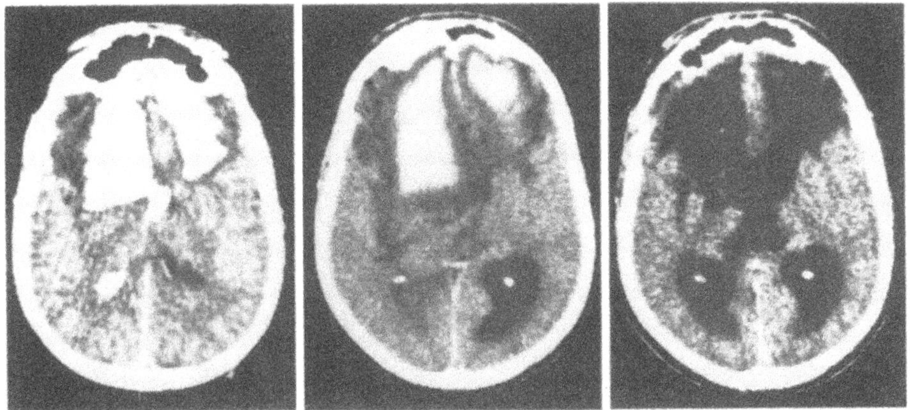

Fig. 2. Massive bifrontal bleeding with ventricular rupture in alcoholic hepatopathy, total resorption of the hemorrhage within seven weeks, development of a cystic cavity (no trauma)

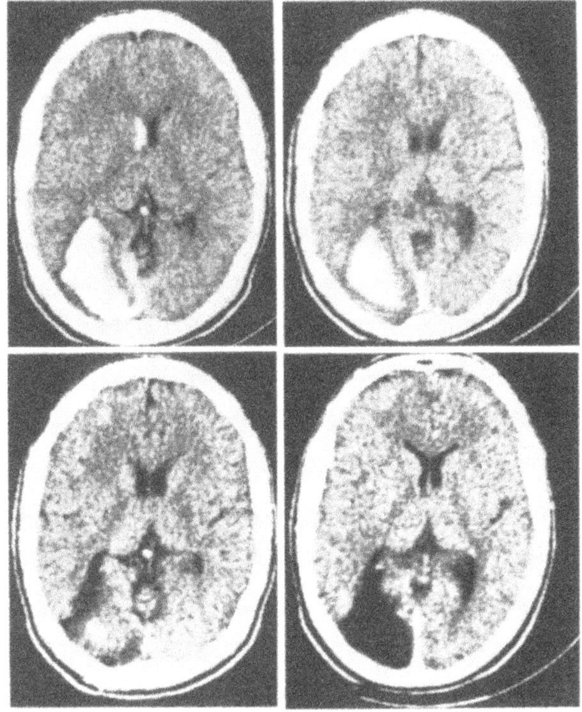

Fig. 3. Left temporo-occipital spontaneous hemorrhage with ventricular rupture, complete resorption in seven weeks with cystic parenchymal defect. Normal angiogram, probably spontaneous healing of a vascular malformation as a result of the hemorrhage. (19-year-old patient)

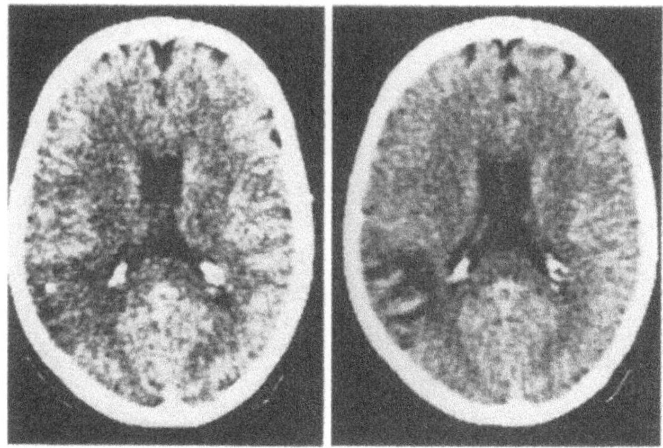

Fig. 4. Ischemic lesion of the left temporal lobe, secondary bleeding after three days with more pronounced outline of the ischemic area

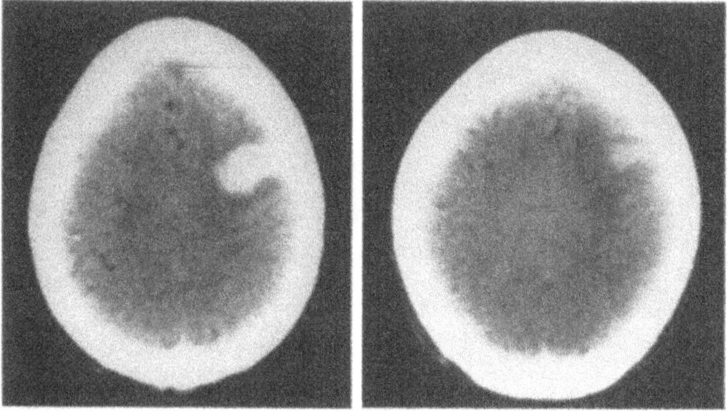

Fig. 5. Embolic hemorrhagic infarct in the course of endocarditis lenta. Remaining hematoma after three weeks (right picture) with demarcation of the ischemic area

►

Fig. 6 a-b. Spasmogenic hemorrhagic infarct in the left front-temporal lobe from aneurysm of the middle cerebral artery, combination of subarachnoid and intracerebral bleeding with a lesser extent in the infarcted area (a). Pronounced arterial spasm of the right carotid siphon, the anterior and middle cerebral arteries (b upper pictures), as well as of the left carotid siphon, the anterior and posterior cerebral arteries. (b lower pictures) with subtotal stenosis of the middle cerebral artery

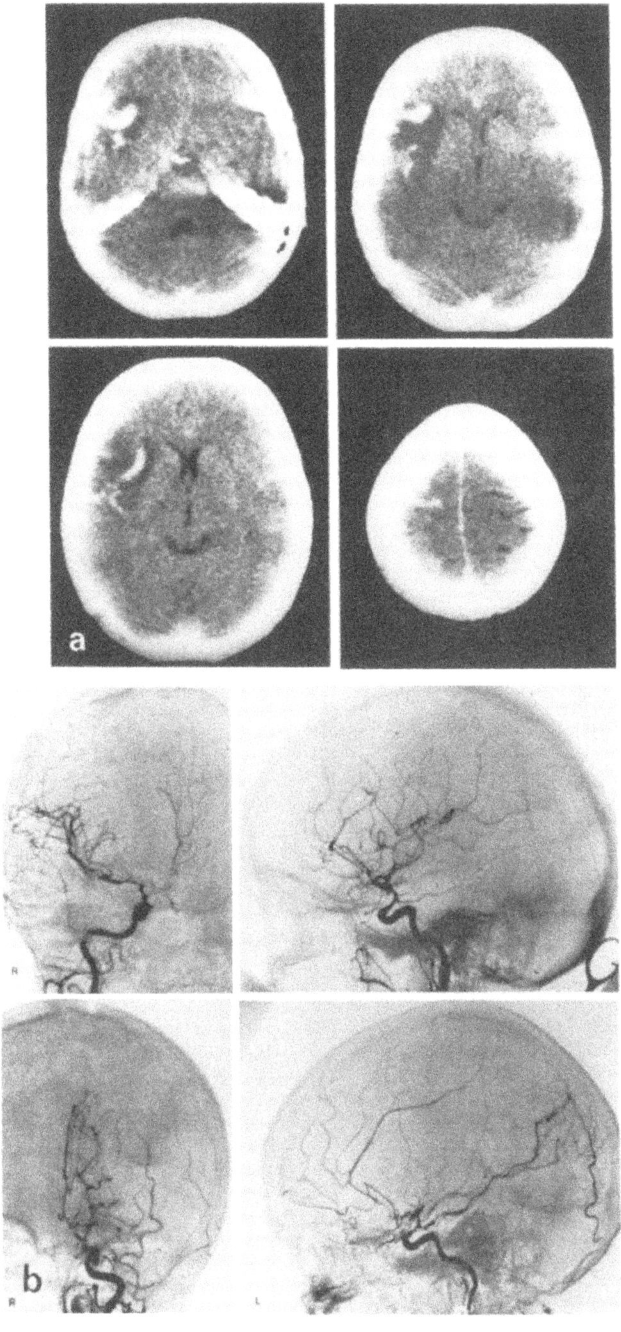

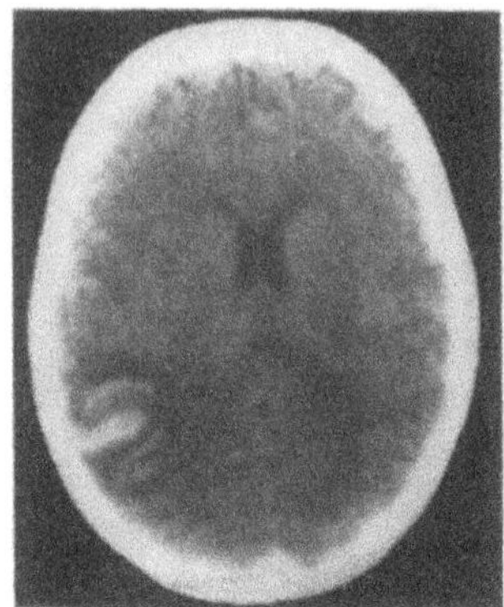

Fig. 7. Spontaneous bleeding in the left temporal lobe in chronic lymphatic leukemia, combination of cerebral and subarachnoid hemorrhage

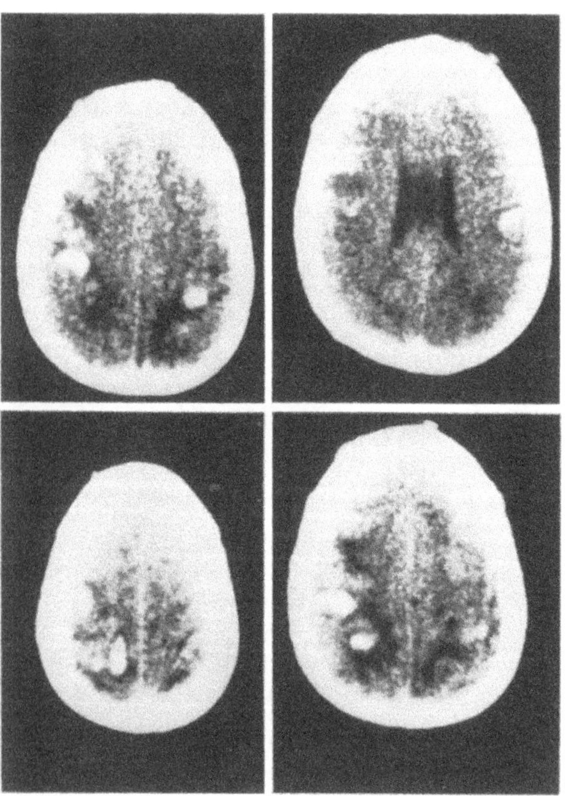

Fig. 8. Bilateral spontaneous hemorrhages in superior sagittal sinus thrombosis, cortical and subcortical location of the hemorrhage with flattening of the ventricles caused by bilateral brain edema. (I owe this picture to the courtesy of the Department of Neuroradiology, University of Tübingen)

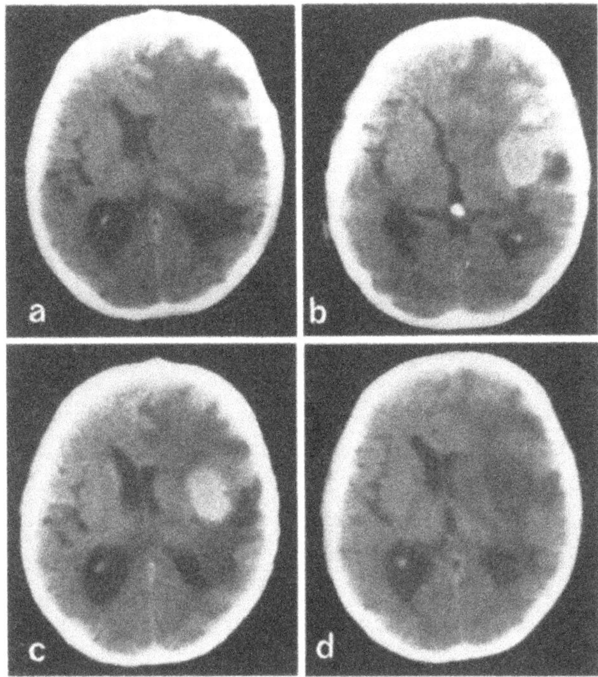

Fig. 9 a-d. Tumor hemorrhage of an oligodendroglioma of the right temporal lobe, disproportion between the size of the hematoma, edema and degree of space-occupation. Three layers with a central bleeding, middle tumor margin and peripheral edema. Stage before (a) and after bleeding (b, c); total resorption and formation of a cystic cavity after three weeks (d)

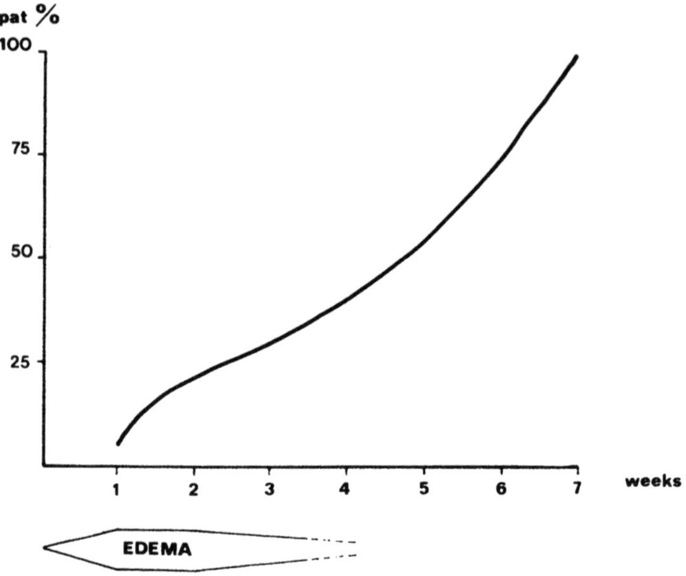

Fig. 10. Course of resorption and development of brain edema in spontaneous hematomas. Resorption and edema of 44 spontaneous cerebral haematomas. (Curve = completed resorption)

Familial Occurrence of Intracerebral Hematomas

W. Luyendijk and G. Th. A. M. Bots
Departments of Neurosurgery and Neuropathology, University Medical Center, NL-2333 Leiden

A short report is given on two large families, with (quasi) spontaneous intracerebral hematomas, living in Katwikijk, a fisherman's village on the North Sea.

During the age period between 40 - 60 years a considerable number of family members met with an acute (quasi) spontaneous cerebral bleeding. As all relatives are familiar with this threatening danger, they feel relieved after having reached the age of 60 years.

Altogether 14 cases have been operated on because of such an intracerebral hematoma and in addition several other non-surgical cases can be still added to this material. Anyhow both families together show a rather high incidence of these bleedings as is clearly illustrated by the two pedigrees (Figs. 1 and 2). Among the 91 cases studied at the age of 40 to 60 years 21 hematomas have been confirmed at operation or autopsy so far (23.1%).

Angiography never demonstrated any underlying vascular anomalies. No hematological abnormalities were found in any of the patients.

In only one case a slight arterial hypertension (170/100) was found. All other patients had a normal pre-existing viz. postoperative blood pressure.

The prognosis appears to be bad; half of the patients died sooner or later after operation.

Fourteen cases were examined neuropathologically. In all of them a cerebral arteriosclerosis was established, and it was characterized by a hyaline thickening of the arteriolar walls which contained amyloid deposits. This amyloid arteriosclerosis appeared to be restricted to the gray substance of the brain and presented itself predominantly in the cerebral cortex. Sometimes the same was also found in the leptomeningeal vessels.

In Holland eleven similar other cases have been studied by Wartendorf (The Hague).

His patients belonged to one family from Scheveningen another fisherman's village on the North Sea coast, lying directly south of Katwijk.

Outside of Holland familial amyloid-arteriosclerotic hematomas are described by GUDMUNDSSON et al. from Iceland.

Advances in Neurosurgery, Vol. 11
Edited by H.-P. Jensen, M. Brock, and M. Klinger

It is not known yet how these three families are interconnected - legally or illegally. Genealogical investigations have been negative up to now.

As Dutch family names are found in old graveyards in Iceland it has to be assumed that mutual contacts via the North Sea must have occurred in the past.

References

1. Gudmundsson, G., Hallgrinsson, J., Jonasson, T.A., Bjornason, O.: Hereditary cerebral haemorrhage with amyloidosis. Brain 95, 387-404 (1972)
2. Jellinger, K.: Cerebrovascular amyloidosis with cerebral haemorrhage. Journal of Neurology 214, 195-206 (1977)
3. Luyendijk, W., Bots, G.Th.A.M.: Familial Incidence of ICH. In: Spontaneous intracerebral haematomas. Pia, H.W., Langmaid, C., Zierski, J. (eds.), pp. 50-56. Berlin, Heidelberg, New York: Springer 1980
4. Wattendorff, A.R., Bots, G.Th.A.M., Went, L.N., Endtz, L.J.: Familial cerebral amyloid angiopathy presenting as recurrent cerebral haemorrhage. Journal of Neurol. Sciences (1982), in press

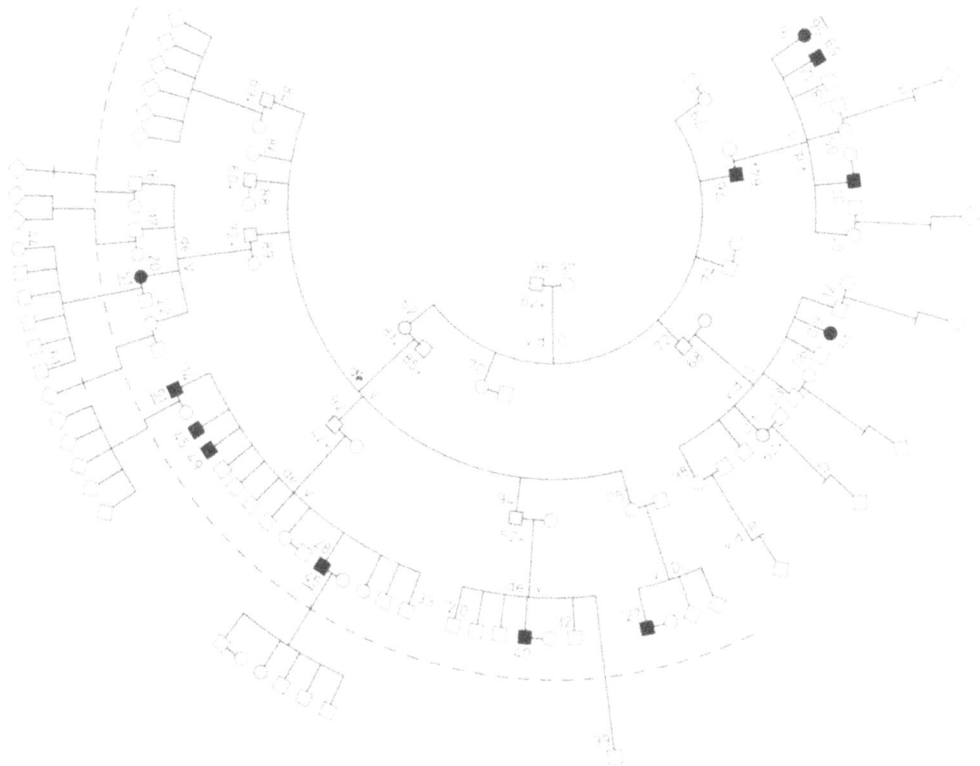

Fig. 1. Pedigrees of the two Katwijk families showing the incidence of intracerebral bleeding. Black: cases of hematoma at operation or autopsy (total 26); grey: hematomas or hemorrhages not surgically established

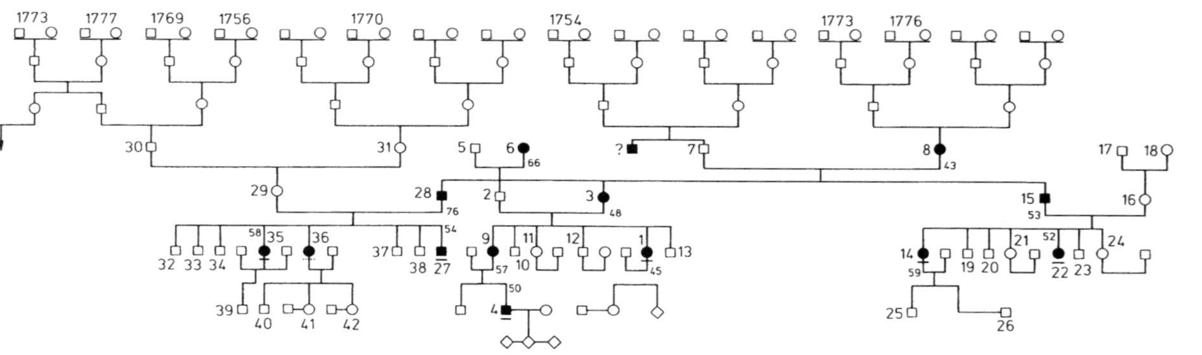

Fig. 2. See legend of Fig. 1

nrCBF for Timing of Angiography and Operation in Subarachnoid Hemorrhage

G. Meinig, P. Ulrich, R. Pesch, and K. Schürmann

Abteilung für Neurochirurgie, Klinikum der Johannes Gutenberg-Universität Mainz, Langenbeckstrasse 1, D-6500 Mainz

Introduction

Early operation on cerebral aneurysms is demanded not least because of the threat of spasm of the cerebral vessels. As is generally recognized, this is to be expected in a very high percentage within a few days after the first aneurysm hemorrhage and frequently cannot be diagnosed clinically (4, 5). Many neurosurgeons have rated the risk of operating in the presence of vasospasm so high that, like C.G. DRAKE in 1975 (2), they demanded control angiography with its own risk, especially in existing vasospasm in early angiography and the aneurysm operation itself could not take place rapidly.

We now attempt with the aid of traumatic measurement of cerebral blood flow (xenon inhalation) to detect the development, course and influence of treatment on cerebral vasospasm and establish the consequences for the timing of the aneurysm operation including angiography. In the present paper, preliminary results on single observations will be presented because of the small number of cases.

Methods

The nrCBF measurements were performed under common resting conditions in the supine position with a ^{133}Xe inhalation system (NOVO diagnostic systems, Hadsund) simultaneously monitoring 16 homologous regions of both hemispheres. Calculations were performed with a Hewlett Packard (Model 9825/M) desk-top computer using the two compartmental analysis of OBRIST et al. 1975 (8) and RISBERG et al. 1975 (9). F_1 (the flow of rapidly perfused grey matter compartment) and ISI (the initial slope index calculated between the initial 30 to 90 seconds after the start of the measurement) were the CBF observations used in this study. The arterial pCO_2 was determined in blood samples from the radial artery by an AVL gas check and controlled recordings of the end-tidal CO_2 concentrations (NOVO capnograph).

Calcium antagonist was generally administered for treatment of cerebral vasospasm. In this study we only gave nimodipine (test preparation from the company BAYER) in a dose of 1 - 2 mg/h. The influence on cerebral blood flow was evaluated by nrCBF follow-up studies.

Advances in Neurosurgery, Vol. 11
Edited by H.-P. Jensen, M. Brock, and M. Klinger

Results

The correlation between cerebral blood flow and neurological state described by various authors (here after HUNT and HESS, 6) appears to be confirmed in 16 patients with subarachnoid hemorrhage, when the measurement is performed within one week, as is to be seen in Figure 1a. However, five patients in whom we found normal cerebral blood flow values in clinical stages IV - V immediately after the subarachnoid hemorrhage are not included in Figure 1a. On the other hand in a few cases indication of spasm by CBF did not correlate with the neurological state of the patient. The relation is less unequivocal in 18 patients (Fig. 1b), in whom the cerebral blood flow was measured 2 - 4 weeks after subarachnoid hemorrhage. Despite normal brain blood flow in the majority of patients, grade I is only occasionally reached. In all patients in whom we had diagnosed a cerebral vasospasm angiographically, the result was confirmed by measurement of cerebral blood flow.

In a series of patients in whom the angiogram was not yet available, we were able to demonstrate the indication of vasospasm. In the case example described below (Fig. 2), nimodipine (test preparation from the company Bayer) was administered. It was a 47-year-old female patient in whom angiography was performed on the sixth day after subarachnoid hemorrhage in stage I according to HUNT and HESS. A high-grade global bilateral vasospasm, which could not be suspected on the basis of clinical findings, was found. Over the following days, the patient deteriorated clinically into stage IV (almost V). The measurement of cerebral blood flow finally performed showed a dramatically reduced perfusion in the blood supply on both sides. The ISI showed a value of 24 ml/100 g/min. After intravenous administration of 1 mg nimodipine within one hour, the cerebral blood flow was raised to 30 ml/100 g/min; two days later, there was a rise to 39 ml/100 g/min with the continued infusion of nimodipine (1 - 2 mg were administered intravenously per hour). In the course of a further three weeks, an almost normal cerebral blood flow was re-established. However, the severe neurological symptoms only regressed slowly and incompletely.

In the majority of our cases, there was an improvement of cerebral blood flow with nimodipine. However, one will by no means be able to expect the dramatic improvement of blood flow in every case, especially when an improvement in the blood flow is impossible from the start because of raised intracranial pressure. This applies irrespective of whether this results from hydrocephalus, brain edema or hematoma, cardiopulmonary disorders, hypoxia, disorders of autoregulation etc.

Figure 3 shows the case example of a 59-year-old female patient in whom administration of nimodipine resulted in an improvement of cerebral blood flow. After a secondary fall, a return to normal of the blood flow occurred with delay after withdrawal of the medication (cardiac problems). However, the discrepancy between cerebral blood flow and clinical symptoms is striking. By the third day after the subarachnoid hemorrhage, a continuous improvement of clinical symptoms occurred, although the brain blood flow had fallen globally after the withdrawal of nimodipine.

Figure 4: Without exception, all patients showed indications for disorders of blood flow of the cerebral cortex (corresponding to the F_1 value) after the subarachnoid hemorrhage. The regional disorders are likewise to be demonstrated practically always in the distribution pattern, although regional reductions of the blood flow by 10% below the hemispheric average show a trend but cannot be regarded as significant. Mostly, the blood flow disorders persist over weeks (not uncommonly over months, as in Fig. 4). As we interpret our observations, they are also accompanied by slight organic brain disorders which cannot be detected with the HUNT and HESS scale.

Conclusions

We carried out measurements of cerebral blood flow by means of xenon inhalation. In some cases only sporadic measurements, and in some cases progress investigations were carried out in about 30 patients. We were able to observe that angiographically demonstrated vasospasm could always be confirmed by measurement of cerebral blood flow. Moreover, the atraumatic measurement of cerebral blood flow (especially the progress observation) has shown that cerebral vascular spasm can be detected even when it is not suspected from the neurological findings. We believe that serious damage to the patients in angiography or operation can be avoided by early diagnosis of vasospasm. In addition, we believe that control angiographies for exclusion of vasospasm immediately before the operation can be obviated by means of xenon inhalation measurement. Beyond this, measurement of nrCBF gives us an important factor for evaluating the alterations in cerebral blood flow during treatment of cerebral vasospasm. We have ourselves been able to observe a marked improvement of blood flow in more than half of the cases in a small group of patients in whom we administered the calcium antagonist nimodipine intravenously. A further result of the cerebral blood flow measurement, the depression of blood flow in the region of the cerebral cortex which frequently persists over weeks in patients who have apparently become normal clinically requires further observation, possibly together with psychological monitoring.

Summary

In order to avoid additional damage by angiography or operation for a patient with SAH in the period of vasospasm, the ^{133}Xe inhalation method for measurement of rCBF proved to be of some help. It may also provide indications of vasospasm in cases without clearly corresponding neurological signs. Manifest vasospasm in angiography was confirmed by nrCBF in all cases. Alterations of cortical flow persisting for several months are detectable in many cases, correlating well with a longer lasting psycho-organic syndrome (not detectable with the HUNT and HESS scale). For observation of the efficiency of medical treatment nrCBF measurements seem to be suitable and reliable.

References

1. Drake, C.G.: Surgical risk as related to time of intervention in the repair of intracranial aneurysms. J. Neurosurg. 28, 19-20, 1968
2. Drake, C.G.: Intracranial aneurysms. In: Tower, D. (ed.), The Nervous System. Vol. 2. The clinical neurosciences. Raven Press, New York, 287-295, 1975
3. Grubb, R.L., Raichle, M.E., Eichling, J.O., Gado, M.H.: Effects of subarachnoid hemorrhage on cerebral blood volume, blood flow, and oxygen utilization in humans. J. Neurosurg. 46, 446-453, 1977
4. Hamer, J.: Häufigkeit und klinische Bedeutung des cerebralen Vasospasmus nach aneurysmatischer Subarachnoidalblutung. Nervenarzt 52, 108-112, 1981

5. Hartmann, A., Menzel, J., Buttinger, C., Lange, D., Albert, E.: Die regionale Gehirndurchblutung des Pavians beim ischämischen Hirninfarkt unter Dexamethasonbehandlung. Fortschr. Neurol. Psychiat. 49, 380-392, 1981

6. Hunt, W.E., Hess, R.M.: Surgical risk as related to time of intervention in the repair of intracranial aneurysms. J. Neurosurg. 28, 14-20, 1968

7. Ishii, R.: Regional cerebral blood flow in patients with ruptured intracranial aneurysms. J. Neurosurg. 50, 587-594, 1979

8. Obrist, W.D., Thompson, H.K., Wang, H.S.: Regional cerebral blood flow estimated by [133] Xenon inhalation. Stroke 6, 246-256, 1975

9. Risberg, J., Ali, Z., Wilson, E.M.: Regional cerebral blood flow by [133] Xenon inhalation. Stroke 6, 142-148, 1975

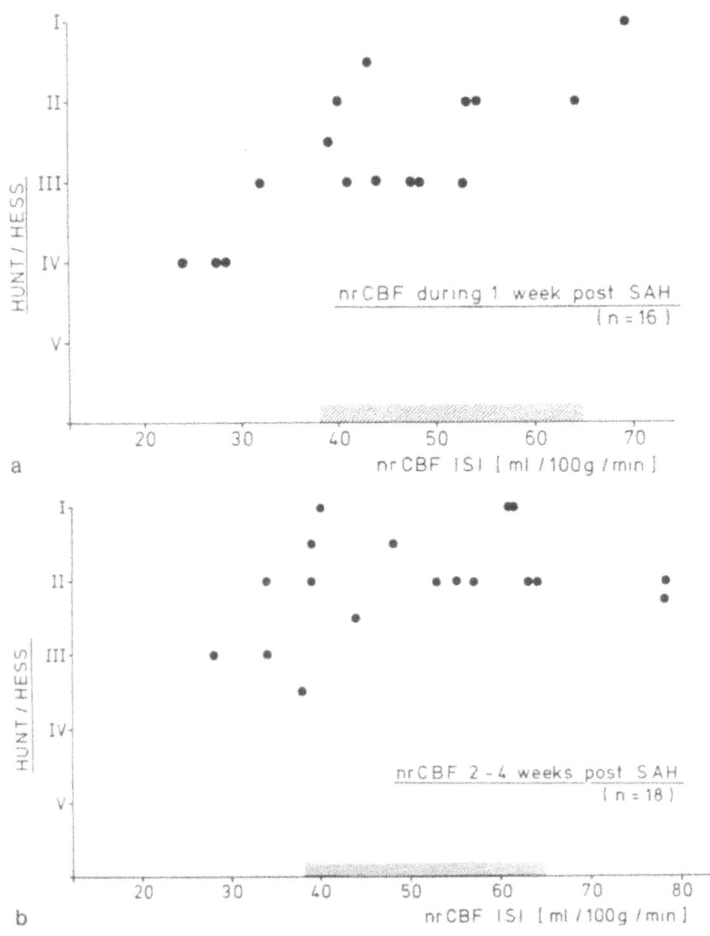

Fig. 1 a. Correlation of nrCBF (ISI) and grading after HUNT and HESS in 16 patients with SAH (measurement during the first week after the SAH)b.Correlation of nrCBF (ISI) and grading after HUNT and HESS in 18 patients with SAH (measurements 2 - 4 weeks after SAH)

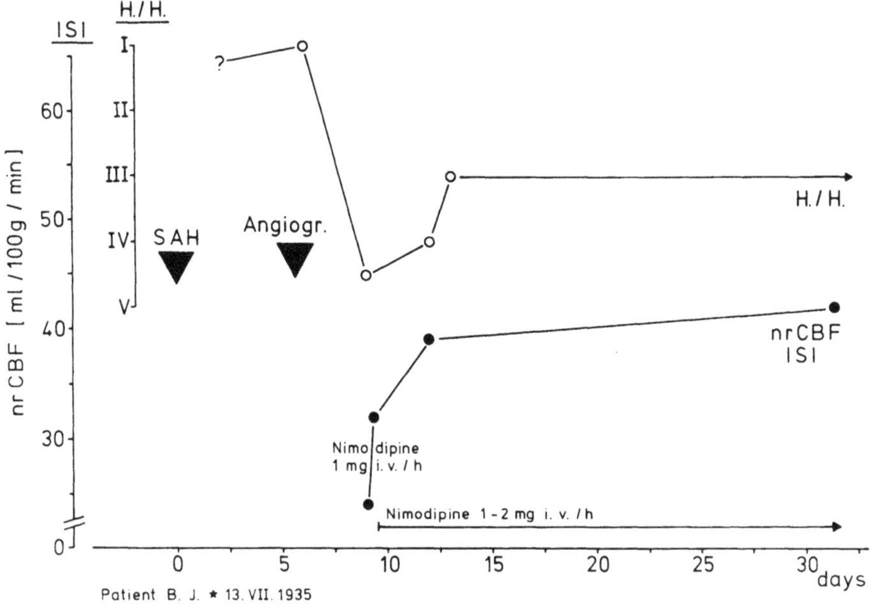

Fig. 2. 47 year-old female (B.J.) with SAH. Clinical course after HUNT and HESS (white circles) and nrCBF before and during administration of Nimodipine

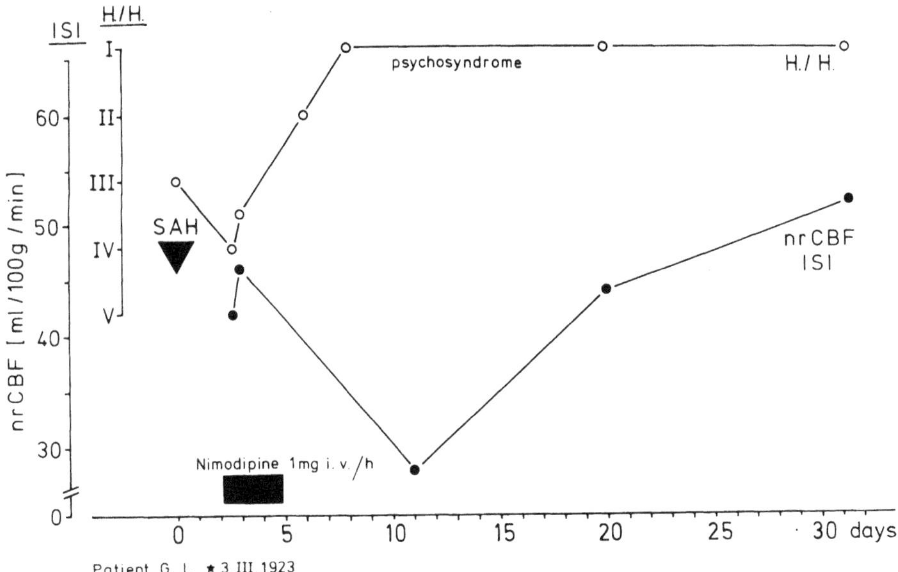

Fig. 3. 59-year-old female (G.L.) with SAH. Relation of clinical course after HUNT and HESS (white circles) and nrCBF (black points)

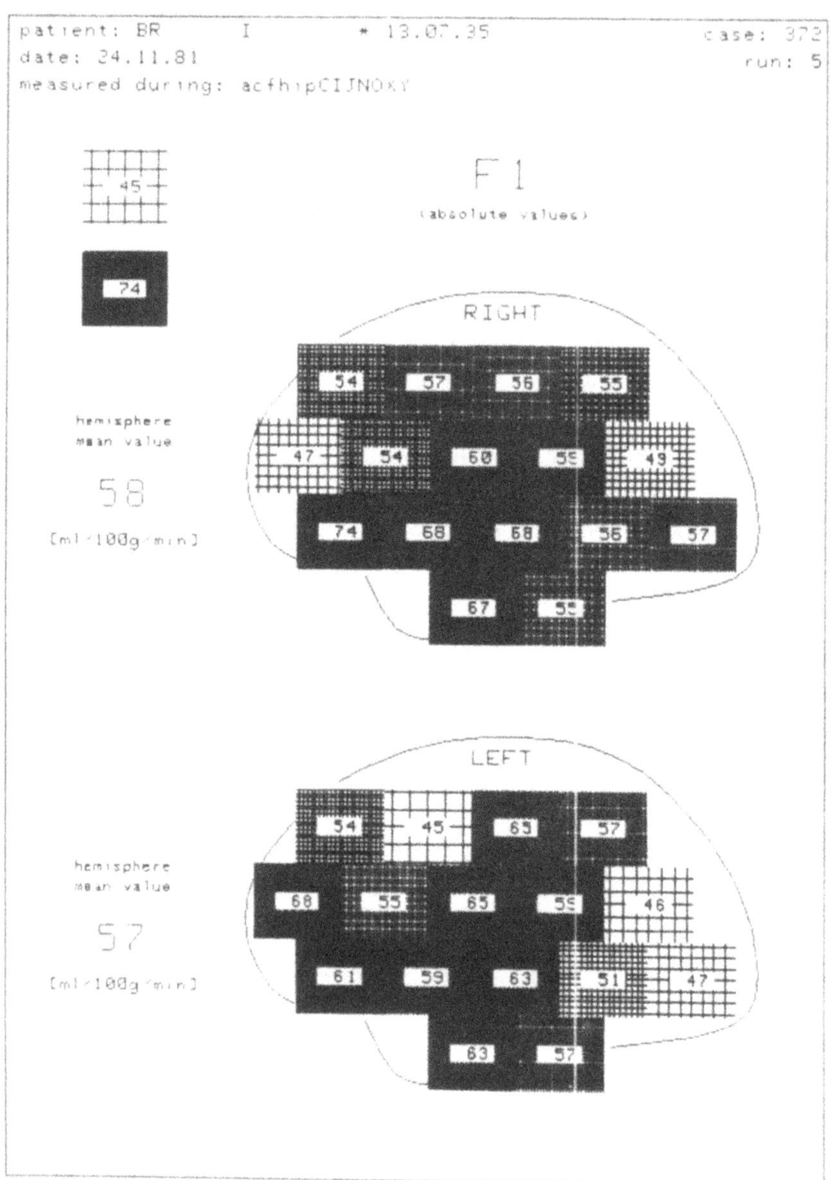

Fig. 4. Altered cortical flow pattern of the right and left hemisphere in a 47-year-old female (B.J., see Fig. 2) five weeks after SAH

An Angiographic Study of the Carotid Arterial System and CT Scan in the Cat After Experimentally Induced Subarachnoid Hemorrhage

F. Ulrich[1], D. Otto[1], W. J. Bock[1], H. Seibert[1], H.-J. Freund[2], and A. Aulich[2]*

[1] Abteilung für Neurochirurgie, Medizinische Einrichtungen der Universität Düsseldorf, Moorenstrasse 5, D-4000 Düsseldorf

[2] Abteilung für Neurologie, Medizinische Einrichtungen der Universität Düsseldorf, Moorenstrasse 5, D-4000 Düsseldorf

Summary

The following investigations were performed to induce experimental cerebral vasospasms in cats. Three days after production of SAH 17 CT scans and angiograms were performed. Seven CT scans of cats showed a hypodense area in the right hemisphere especially in those animals in which the cerebral angiograms revealed constricted cerebral vessels possibly indicating a recent ischemic infarct or cerebral edema.

Today it is almost no problem to clip cerebral aneurysms two days after SAH in grade I - II (III patients (HUNT and HESS). But grade IV - V patients are still in a very dangerous situation, often dying before the aneurysm can be successfully dealt with.

Autopsies after death due to SAH showed that ischemia is undoubtedly the principal cause of brain damage in the acute stage of SAH.

Cerebral vasospasm is one of the main factors causing neurological deficits in man (1-5).

The following investigations in animals were performed to induce cerebral vasospasm experimentally in cats and to study the physiological pathology of severe SAH.

Materials and Methods

Twenty-eight adult cats weighing between 1500 g and 3100 g were anesthetised by ketanest and rompun (0.1 ml rompun/kg and 5 - 10 mg ketamine/kg).

The normal level of blood gases was maintained by mechanical ventilation.

A burr-hole 4 mm in diameter was made subfrontally on the right side of the skull and the dura mater and arachnoid were opened. 2 ml blood was taken from the femoral artery. The SAH was produced by injection of 1 - 2 ml of autologous blood through a catheter introduced fronto-

*Technical assistance was provided by Mrs. M. Wegener, Mrs. G. Friedrich and Mrs. Lein-Filbrich. The author thank Mrs. S. Brinkmann and Mrs. F. Siemers for their assistance in preparation of the manuscript

Edited by H.-P. Jensen, M. Brock, and M. Klinger

basally. EEG, CMP, ECG and systemic blood pressure were monitored continuously (Statham transducer P 23 cb 500 48 10 Volt, P 23 Db 40079 10 Volt; Beckmann dynograph recorder, type R 411).

In nearly all cases there was a transitory abrupt rise in ICP up to the level of arterial diastolic pressure immediately after injection, accompanied by a suppression of EEG activity, as described by Asano in 1978 (6).

Results

After production of SAH seven cats were beyond help and died within the next 24 hours, three cats were somnolent in the following three days, six cats showed extremely small pupils or varying pupil size. In four cats an incomplete right oculomotor paresis was seen. After experimentally induced SAH three cats were in a relatively good condition without any neurological deficit.

Three days after induction of SAH 21 intracranial angiograms (by injection of 3 ml telebrix in approximately 2 seconds) were performed after introduction of a catheter into the right common carotid (B-D Longdwel 6746 F 50-9.2062) Rapid serial films were taken with CGR film changer at a rate of 2 films/sec for 3 seconds (titanus p 100 S, 55 KV, 40 mAs, magnification 1).

In 16 cats the angiographic study showed poor visualization of all intracranial cerebral vessels in contrast to normal non-treated cats.

In three cases the small anterior cerebral arterial branches and in one case the middle cerebral artery could not be identified.

In nine cases neither anterior cerebral, nor middle cerebral artery were seen in the angiograms.

Three angiograms showed generalised cerebral vasoconstriction.

Three days after SAH 17 CT scans were performed. Seven of the CT scans showed a hypodense area in the right hemisphere especially in those animals in which the cerebral angiograms revealed constricted cerebral vessels - possibly indicating a recent frontotemporal ischemic infarct or cerebral edema.

These experiments on cats augment other studies on ICP, EEG, ECG and systemic blood pressure in SAH by intracranial panangiograms and CT scans.

In contrast to all *in vitro* experiments of isolated cerebral vessels this *in vivo* investigation of cats made it possible to study late vasospasm including cerebral edema in an experimental animal, as well as factors such as ICP, EEG, ECG and systemic blood pressure in severe SAH.

In future we will test the influence of vasodilator drugs - first of all calcium antagonists - in these animals.

References

1. Kagström, E., Greitz, T., Hansen, J., Galera, R.: Excerpta Med. Int. Cong. Series, 110, 629-633

2. Symon, V.: In: Smith, E. (ed.), Background to Migraine, pp. 78-95. London: Heinemann Publishing Co. (1967)

3. Bergvall, U., Galera, R.: Acta Radiol. (Diagn.), 9, 229-237 (1969)

4. Hashi, K., Nishimura, S.: Excerpta Medica, pp. 152-156, Amsterdam (1977)

5. Weir, B., Grace, M., Hansen, J., Rothberg, C.: J. Neurosurg. 48, 173-178 (1978)

6. Asano, T., Basugi, N., Saito, I., Sano, K.: In: Intracranial pressure III. J.W.V. Beks, D.A. Bosch, M. Brock (eds.), pp. 14-17. Berlin, Heidelberg, New York: Springer 1976

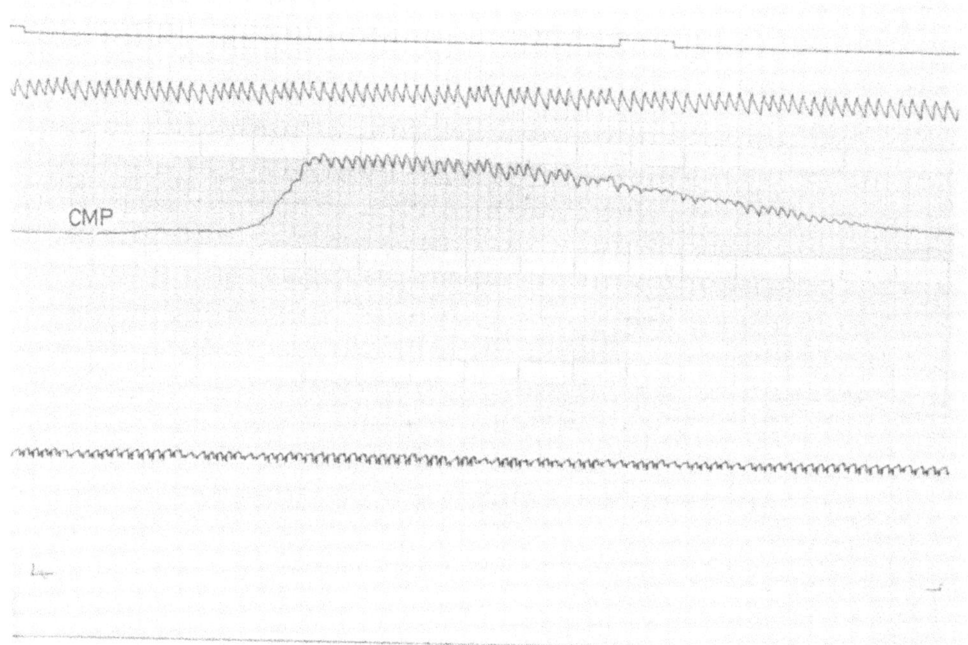

Fig. 1

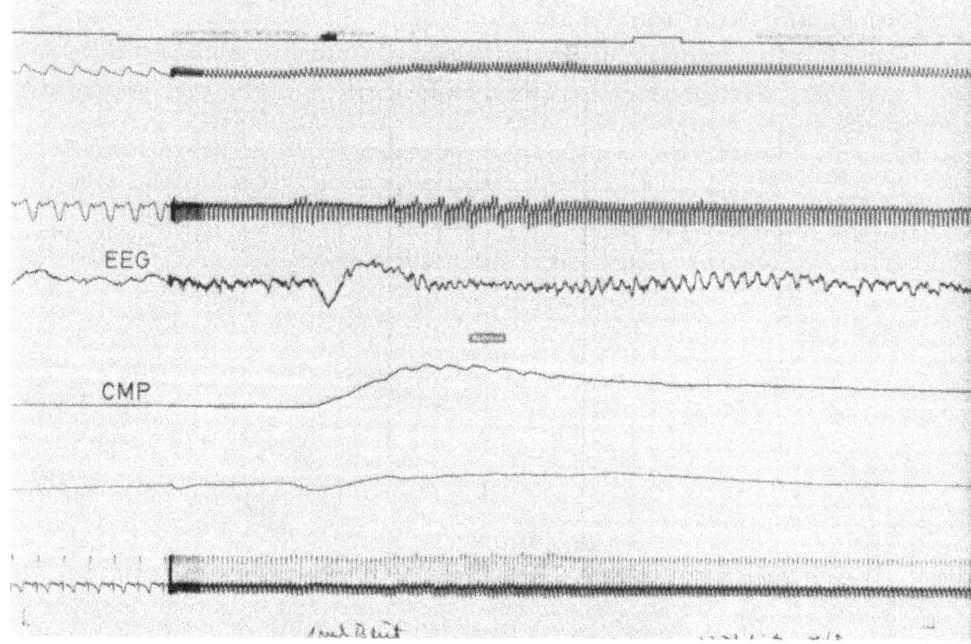

Fig. 2

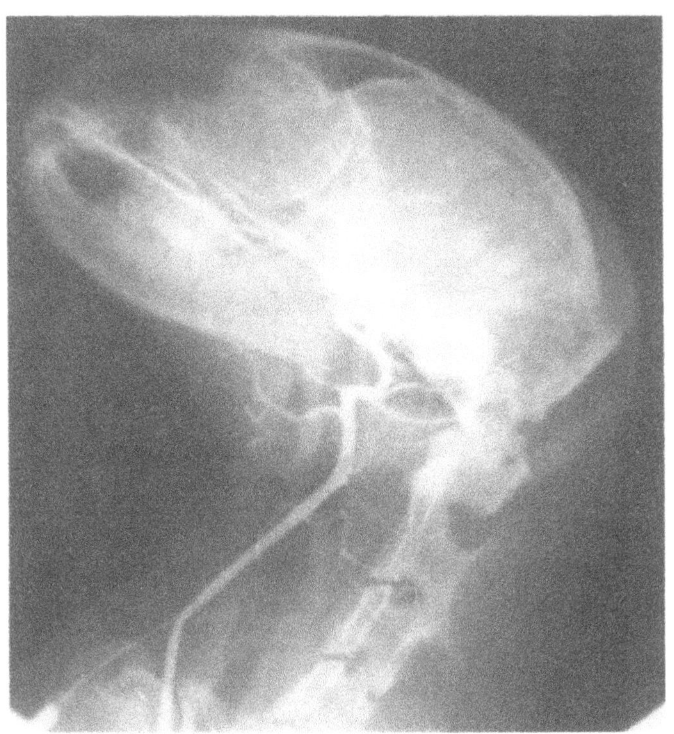
Fig. 3

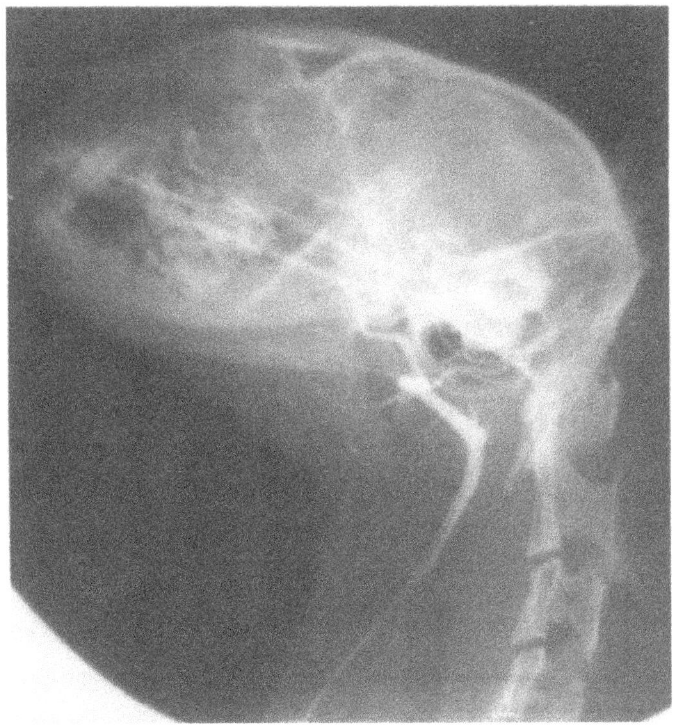
Fig. 4

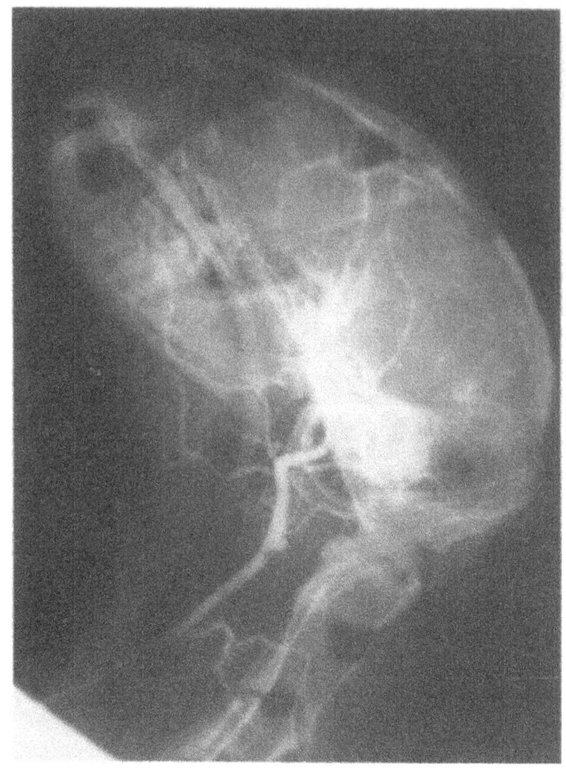

Fig. 5

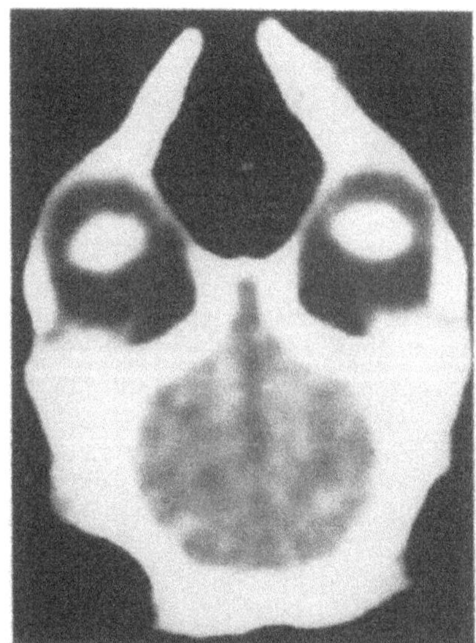

Fig. 6

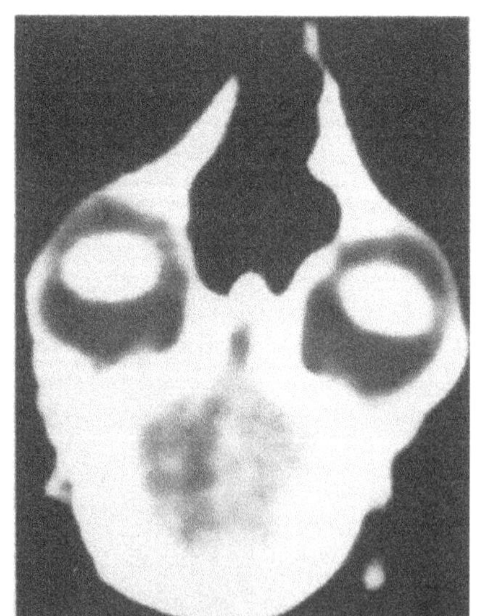

Fig. 7

Early Operations on Ruptured Aneurysms

J. Gilsbach and H. R. Eggert
Neurochirurgische Klinik, Albert Ludwig-Universität Freiburg, Hugstetter Stasse 55, D-7800 Freiburg

Introduction

Good operative results on aneurysms one to three weeks after subarachnoid hemorrhage with a mortality of under 5% (PIA 1969) have to be paid for with high mortality rates due to recurrent bleeding and vasospasm during the waiting period. Mortality with this type of management (ADAMS 1981, HUNT 1980, WEIR 1981) can be as high as 42% (KASSELL 1981). Results with early operations within two to three days published up to now are more favorable, although the operative mortality is about 10% (HUGENHOLTZ 1982, HORI and SUZUKI 1979, HUNT 1980, KASSELL 1981, KOBAYASHI 1979, LJUNGGREN 1981, SANO 1978, TAKAHASHI 1979, WEIR 1981). This is explained by earlier suppression of the risk of rebleeding as well as reduction of complications related to vasospasm, due to removal of clots from the subarachnoid space.

Material and Methods

Prompted by the publications of SANO and SUZUKI (1977, 1978, 1979) we have operated since 1979, on 38 patients in grades I to III, within 48 to 72 hours at the latest (Table 1).

Four vessel angiography was performed under neuroleptic analgesia before the patients were taken to the operating theater. Operations were performed with microsurgical techniques after osmotic diuresis and without lumbar CSF drainage.

Table 1. Case material (1979 - 1982)

No. of cases:	38	M: 16	F: 22
Age: mean	44		
range	19 - 74		
Op. within:	24 h	13	
	25 - 48 h	16	
	49 - 72 h	9	
Grade[a] at operation:		II	10
		III	28
Follow-up:	mean	22 months	
	range	3 - 30 months	

[a] HUNT and HESS

Advances in Neurosurgery, Vol. 11
Edited by H.-P. Jensen, M. Brock, and M. Klinger

During operation CSF was drained as a rule from the basal cisterns, and only exceptionally was the lateral ventricle tapped or the lamina terminalis opened. Brain tissue resections were performed only rarely at the gyrus rectus in anterior communicating artery aneurysms. Nitroprussate hypotension was used only after rupture or in very difficult dissections.

Postoperative CSF drainage was not performed routinely apart from daily lumbar punctures.

Prophylaxis and treatment of symptomatic vasospasm was the main concern of the postoperative management. Apart from steroids we insisted on a positive fluid balance and a relatively high blood pressure.

Results

Although the brain was tense and slightly bulging in three quarters of the patients, osmotic diuresis and drainage of CSF from the basal cisterns always enabled us to obtain enough space to dissect and clip the aneurysm without significant brain compression due to retraction. Rupture occurred in one third of the patients, mostly at the end of the dissection at the site of the previous spontaneous rupture of the aneurysm at the fundus. Rupture did not adversely affect clipping of the aneurysm nor the final result.

Part of the postoperative psycho-organic syndromes in patients with ACoA aneurysms were probably due to lesions of small perforating arteries, which are difficult to identify in the blood-filled subarachnoid space. Thorough cleaning of the accessible cisterns was attempted but could not be achieved in one fifth of the patients due to solid clots which could not be removed without risk of damage to perforating vessels.

Three patients died as a consequence of vasospasm. One of them was operated upon on a retrospectively incorrect indication with already established severe vasospasm due to a first, four days old bleeding (Table 2).

Another patient with an aneurysm of the tip of the basilar artery died after postoperative brain stem infarction and further deterioration due to vasospasm.

Complications due to operation were mainly psycho-organic syndromes in patients with ACoA aneurysms. All but three of them, however, improved. Secondary neurological and neuropsychological deficits related to vasospasm or to excessive brain retraction regressed completely apart from one hemiparesis.

Table 2. Site of aneurysm and results

Outcome	ACoA	ICA	MCA	V.B.	Total No.	%
Good	13	6	2	-	21	55
Fair	6	1	2	2	11	29
Poor	2	-	-	-	2	5
Dead	1	2	-	1	4	11
Total cases	22	9	4	3	38	

Table 3. Surgical results of follow-up review

Preoperative grade	Good	Fair	Poor	Dead	Total
II	7 (70%)	3 (30%)	∅	∅	10
III	15 (54%)	7 (25%)	2 (7%)	4 (14%)	28

Good: fully active; fair: limited activity but independent; poor: dependent

After an average follow-up period of 22 months out of 38 patients, 22 were fully active, ten showed limited activity but were independent, two remained dependent (Table 3).

Discussion

Early operations on patients with ruptured aneurysms are justified only within a time interval of 48 to maximally 72 hours after the hemorrhage (HUNT, HORI). After this delay the brain becomes more vulnerable, vessels tend to react with spasms, and removal of clots from the subarachnoid space in order to prevent vasospasm is ineffective (HORI and SUZUKI).

Although ICP is usually increased, the standard methods of ICP reduction recommended by YASARGIL are sufficient to allow dissection and clipping of the aneurysm. Ventricular drainage as suggested by SANO and progressive temporary clipping of the afferent vessels employed by SUZUKI did not appear necessary to us. This view is also taken by LJUNGGREN and KASSELL.

Most postoperative problems are determined by the subarachnoid hemorrhage and by the reaction of vessels to decomposition products of blood. Although early operation allows partial removal of blood from the subarachnoid space and progress is on average less stormy than with conservative i.e. expectant management (KASSELL), the postoperative result is adversely affected by the remaining cisternal clots. Thus, as long as no new technical or pharmacological means are discovered, the course and the final result still depend to a great extent on the severity of the initial bleeding.

On the other hand the conditions for symptomatic treatment of vasospasm by means of hypertonia and hypervolemia are by far more favorable if the aneurysm has already been clipped.

Conclusion

Three years of experience with operations on aneurysms in the acute phase after SAH gave us, in accordance with other publications, the following impressions:

Early operations do not give rise to particular organisational or technical problems after a short period of familiarization. Standard microsurgical and anesthetic techniques are adequate. Additional measures such as ventricular or lumbar drainage, progressive temporary clipping of afferent vessels and induced hypotension do not seem to be necessary. In general postoperative problems and final results depend to a great extent on the severity and the extension of the initial hemorrhage. Early operations reduce the consequences of the bleeding and improve the conditions for symptomatic treatment.

References

1. Adams, Jr., H.P., Kassell, N.F., Torner, J.C., Nibbelink, D.W., Sahs, A.L.: Early management of aneurysmal subarachnoid hemorrhage. J. Neurosurg. 54, 141-145 (1981)
2. Hori, S., Suzuki, J.: Early intracranial operations for ruptured aneurysms. Acta Neurochirurgica 46, 93-104 (1979)
3. Hugenholtz, H., Elgie, R.G.: Considerations in early surgery on good-risk patients with ruptured intracranial aneurysms. J. Neurosurg. 56, 180-185 (1982)
4. Hunt, W.E., Hess, R.M.: Surgical risk as related to time of intervention in the repair of intracranial aneurysms. J. Neurosurg. 28, 14-20 (1968)
5. Hunt, W.E., Miller, C.A.: Results of early aneurysmorrhaphy in good risk patients. Adv. Neurosurg. 7, 76-80 (1980)
6. Kassell, N.F., Boarini, D.J., Adams, H.P., Sahs, A.L., Graf, C.J., Torner, J.C., Gerk, M.K.: Overall management of ruptured aneurysm: comparison of early and late operation. Neurosurgery 9, 120-128 (1981)
7. Kobayashi, K., Okada, K., Tanimura, K., Nakai, O.: A follow-up study of the patients with ruptured aneurysms submitted to microsurgery within one week. No Shinkei Geka 7, 661-668 (1979)
8. Ljunggren, B., Brandt, L., Kagström, E., Sundbärg, G.: Results of early operations for ruptured aneurysms. J. Neurosurg. 54, 473-479 (1981)
9. Pia, H.W.: Prognosis of operative treatment in cerebral aneurysms. Pia, H.W., Langmaid, C., Zierski, J. (eds.), pp. 512-413. Berlin, Heidelberg, New York: Springer 1979
10. Sano, K., Saito, I.: Timing and indication of surgery for ruptured intracranial aneurysms with regard to cerebral vasospasm. Acta Neurochirurgica 41, 49-60 (1978)
11. Daito, I., Ueda, Y., Sano, K.: Significance of vasospasm in the treatment of ruptured intracranial aneurysms. J. Neurosurg. 47, 412-429 (1977)
12. Tokahashi, Sh., Sonobe, M., Nagamine, y.: Early operations for ruptured intracranial aneurysms. Comparative study with computed tomography. Acta Neurochirurgica 57, 23-31 (1981)
13. Weir, B., Aronyk, K.: Management mortality and the timing of surgery for supratentorial aneurysms. J. Neurosurg. 54, 146-150 (1981)
14. Yasargil, M.G., Fox, J.L., Ray, M.W.: The operative approach to aneurysms of the anterior communicating artery. In: Advances and technical standards in neurosurgery. Krayenbühl, H. (ed.), Vol. 2, pp. 113-170. Wien, New York: Springer 1975
15. Yoshimoto, T., Uchida, K., Kaneko, U., Kayama, T., Suziki, J.: An analysis of follow-up results of 1000 intracranial saccular aneurysms with definitive surgical treatment. J. Neurosurg. 50, 152-157 (1979)

Correlation of Clinical Course, Angiography and CT in Patients After Subarachnoid Hemorrhage

H. M. Mehdorn, F. Brandt, G. Spira, G. Schulte-Mattler, B. Tenfelde, and W. Grote

Neurochirurgische Klinik und Poliklinik, Universitätsklinikum der Gesamthochschule Essen, Hufelandstrasse 55, D-4300 Essen 1

Introduction

The place of cerebral angiography in the investigation of patients suffering from a subarachnoid hemorrhage (SAH) is well defined. Since its invention, cranial computerized tomography (CT) has gained increasing influence upon the management of those patients and is usually the first diagnostic step, instead of or before a lumbar puncture (1-3, 5, 6, 8, 10-21). In this study we try to correlate the clinical course throughout hospitalization with the findings of the two main diagnostic procedures, in order to obtain more information about the pre-operative status in patients with SAH with regard to the post-operative results.

Material and Methods

The study is based upon 110 patients who underwent intracranial operation for a ruptured aneurysm, from 1.1.79 until 31.12.81. There were 41 aneurysms of the internal carotid artery, 24 of the anterior communicating artery (ACoA), 19 of the anterior cerebral artery (ACA) and 18 of the middle cerebral artery (MCA). The remainder were located on the posterior circulation. Their clinical grades were classified according to the Cooperative Aneurysm Study Scale. The angiograms, which had been performed mostly within the first two weeks after SAH were reassessed with specific reference to vasospasm, whether local or multisegmental or diffuse. Pre-operative CT scans were performed mostly within the first week after SAH; they were reviewed independently for intracerebral hemorrhage (ICH), blood clots in the basal cisterns or the subarachnoid spaces, hypodense lesions as a result of cerebral ischemia and form and size of the ventricles. The size of the ventricles was evaluated by measuring the width of the lateral ventricles at the level of the caudate nucleus and was compared to the intra-osseous diameter of the brain at the same level to create the ventriculo-cranial ratio (VCR).

Results

The CT showed blood in the basal cisterns in 70% of all patients examined within three days of SAH; in patients examined on days 4 - 7, blood clot was detected in only 30% of the cases.

Advances in Neurosurgery, Vol. 11
Edited by H.-P. Jensen, M. Brock, and M. Klinger

Both local and diffuse vasospasm occur most frequently in the second week after SAH (Fig. 1). There was a close correlation between the presence of blood clots in the basal cisterns and the occurrence of vasospasm: 73% of the patients that showed blood in the basal cisterns had evidence of vasospasm on the angiograms. On the other hand, 93% of the patients who did not present with cisternal blood clots in the initial CT scan, did not have any vasospasm. Fig. 2 demonstrates the clinical course of the patients throughout their hospital stay, separated according to the occurrence of vasospasm. Initially, both groups present nearly identical grades; however, already on admission to our clinic, patients with angiographically proven vasospasm fare worse than patients without it. This trend continues throughout the clinical evaluation, and at the time of discharge in the second week after operation, the mortality rate is 23.6% in the vasospasm group compared to 9.5% in the non-vasospasm group. Because of the fact that 63% of our patients are admitted to our clinic in the second week or later after a SAH, the negative effects of vasospasm upon clinical grading are already noticeable at the time of admission, although there is generally a tendency towards improvement of the grade, during the interval from admission in the primary care hospital to our clinic.

The presence of a neurological deficit correlates well with the CT findings of a hypodense lesion. While patients in grade I had only a 23% incidence of hypodense lesions on preop CT scans, this figure rose steadily with worsening clinical grades and reached 60% in patients grade IV at time of operation. To define the approximate size of the hypodense lesions, we selected the number of CT slices on which such a lesion was visible. There was a close correlation between the incidence and size of the lesions and the number of attacks of SAH (Fig. 3): patients with only one SAH had infarcted areas in only 30%, while patients with two SAH had such a lesion in 77%.

Patients who had had two SAH showed evidence of ICH on the initial CT scan in 60%; this has to be compared to the 40% incidence of ICH in patients presenting after a single SAH (Fig. 4). Most frequently, an ICH is present among patients harboring aneurysms of the MCA and the proximal ACA/ACoA. This is correlated closely to the clinical course, since patients with MCA aneurysms have the highest mortality rate (38.9%) at the time of discharge, followed by those with aneurysms of the vertebral artery and the ACoA (12.5%) (Fig. 4b). Pre-operative examination of the ventricle width shows a normal bicaudate diameter in 48% of all patients. Initially, 9.3% of the patients presented a moderate increase of the VCR. Postoperatively, in the CT 10% showed a moderate increase of the VDR, and another 10% showed a marked ventricular dilatation. Subsequent to ventricular scintigraphy which demonstrated a pathological absorption pattern, in all of the latter patients a ventriculo-atrial drainage (VAD) was inserted; mainly patients of the initial grade IV required this second operation. We found a very close correlation of the CT finding of marked hydrocephalus and pathological absorption patterns. Therefore ventricular scintigraphy can be ommitted when one has to decide on a VAD for a patient with adequate clinical picture and CT findings. A marked ventricular dilatation occurs more frequently in patients whose angiograms show vasospasm than in those with normal angiograms.

Discussion

The value of CT in recognizing pathological intracranial conditions is unquestionable. There are reports that CT detection of blood clots in the basal cisterns is closely correlated to the consequence of arterial spasm, particularly when CT is performed within the first five days after SAH, and angiograms between day 7 and 17 (5). Our study seems to support this finding, although we did not specifically look for the size of the blood clots nor the time intervals. On the basis of their findings, FISHER et al. suggested (5) that early aneurysm operations might be beneficial by removing the blood clots from the basal cisterns before the induction of vasospasm. This tendency towards early operation, if possible within the first three days after SAH, had been advocated by others (14, 21) based on clinical experience.

An interim analysis of our efforts towards early operation shows that, in the years 1976 - 1978, only 18% of the patients came under our care within the first week after SAH (12); this figure has doubled in the last three years (37% in 1979 - 1981). This may be due to the intention of referring neurologists to obtain an early CT scan in order to assure the diagnosis. It may lead to a slightly "sicker" patient population at time of admission to our clinic (35% grade II patients compared to 27% previously; the percentage of the grades III and IV patients remains unchanged, 38%). Furthermore and more important: previously, the interval between SAH and operation was three weeks or longer for 50% of all patients. This percentage has dropped to less than 25%. While only 4.4% of all patients undergo operation within the first three days after SAH, 47.7% of all patients are operated upon between days 4 and 14. This means a high percentage of patients being operated upon in a phase of developing vasospasm (which, at the time of decision to operate, cannot be detected clinically for any individual patient). The mortality and morbidity may have been somewhat influenced by this development, with a slight increase of mortality from 8% for the period 1976 - 1978 to 11% for the period 1979 - 1981. While previously 42% of all patients had been discharged in grade I, this figure has dropped to 26%, with an increase of the patients discharged in grades II and III. On the other hand, because of the reduced pre-operative time of conservative treatment, it should be possible to save some patients from a recurrent, maybe fatal hemorrhage, which would make up for the somewhat less satisfying surgical results.

As far as the detection of ICH is concerned, the value of CT is known, particularly when one has to decide about the mass effect of a hematoma. Similarly, the detection of a hydrocephalus after SAH, either pre- or postoperative, is facilitated by repeat CT examinations. Our figures are the same as YASARGIL's (20) with a 10% rate of patients requiring a VAD. Because of the good correlation of CT findings, clinical course and ventricular scintigraphy we think that the latter examination can be omitted when the indication for a VAD is discussed.

Conclusions

CT facilitates the management of patients after SAH; however, it is impossible, on the basis of the data presented in this study, to prove any beneficial effect of CT on the outcome of patients after SAH. Such a benefit might be expected if every patient presenting with the clinically based diagnosis of SAH undergoes immediate CT evaluation, to discover patients who risk vasospasm. This examination should be followed by early angiography and, if possible, by early operation. It remains to be seen from the Cooperative Aneurysm Study and the experience of individual neurosurgeons to what extent this aggressive management will reduce the mortality and morbidity figures of aneurysm patients by removing early from the subarachnoid space the agent causing vasospasm, i.e. blood, and by the definitive treatment of the bleeding source.

References

1. Agnoli, A.L., Schoch, P., Bayindir, S., Villagrasa, J.: Computertomographische Befunde bei Gefäßmißbildungen der Hirnarterien. Röntgen-Bl. 34, 55-50 (1981)
2. Davis, K.R., New, P.F.J., Ojemann, R.G., Crowell, R.M., Morawetz, R.B., Roberson, G.: Computed tomographic evaluation of hemorrhage secondary to intracranial aneurysm. Am. J. Roentgenol. 127, 143-153 (1976)
3. Davis, J.M., Davis, K.R., Crowell, R.M.: Subarachnoid hemorrhage secondary to ruptured intracranial aneurysm: Prognostic significance of cranial CT. AJR. 134, 711-715 (1980)
4. Fisher, C.M., Roberson, G.H., Ojemann, R.G.: Cerebral vasospasm with ruptured saccular aneurysm - The clinical manifestations. Neurosurgery 1, 245-248 (1977)
5. Fisher, C.M., Kistler, J.P., Davis, J.M.: Relation of cerebral vasospasm to subarachnoid hemorrhage visualized by computerized tomographic scanning. Neurosurgery 6, 1-9 (1980)
6. Ghoshhajra, K., Scotti, L., Marasco, J., Baghai-Naiini, P.: CT detection of intracranial aneurysms in subarachnoid hemorrhage. AJR 132, 613-616 (1979)
7. Hamer, J.: Häufigkeit und klinische Bedeutung des cerebralen Vasospasmus nach aneurysmatischer Subarachnoidalblutung. Nervenarzt 52, 108-112 (1981)
8. Kendall, B.E., Lee, B.C.P., Claveria, E.: Computerized tomography and angiography in subarachnoid hemorrhage. Br. J. Radiol. 49, 483-501 (1976)
9. Kwak, R., Niizuma, H., Ohi, T., Suziki, J.: Angiographic study of cerebral vasospasm following rupture of intracranial aneurysms: Part I. Time of the appearance. Surg. Neurol. 11, 257-262 (1979)
10. Liliequist, B., Lindquist, M., Valdimarsson, E.: Computed tomography and subarachnoid hemorrhage. Neuroradiology 14, 21-26 (1977)
11. Liliequist, B., Lindquist, M.: Computer tomography in the evaluation of subarachnoid hemorrhage. Acta Radiol. Diagn. 21, 327-331 (1980)
12. Mehdorn, H.M., Grote, W., Tenfelde, B.: Timing of aneurysm surgery after spontaneous subarachnoid hemorrhage. Adv. Neurosurg. 9, 243-249 (1981)
13. Saito, I., Shigeno, T., Aritake, K., Tanishima, T., Sano, K.: Vasospasm assessed by angiography and computerized tomography. J. Neurosurg. 51, 466-475 (1979)
14. Scotti, G., Ethier, R., Melancon, D., Terbrugge, K., Tchang, St.: Computed tomography in the evaluation of intracranial aneurysms and subarachnoid hemorrhage. Radiology 123, 85-90 (1977)
15. Takahashi, Sh., Sonobe, M., Nagamine, Y.: Early operations for ruptured intracranial aneurysms. Comparative study with computed tomography. Acta Neurochirurg. 57, 23-31 (1981)
16. Turnbull, I.W.: Computed tomographic pointers to the prognosis of subarachnoid hemorrhage. Br. J. Radiol. 53, 416-420 (1980)
17. Vassilouthis, J., Richardson, A.E.: Ventricular dilatation and communicating hydrocephalus following spontaneous subarachnoid hemorrhage. J. Neurosurg. 51, 341-351 (1979)

18. Weisberg, L.A.: Computed tomography in aneurysmal subarachnoid hemorrhage. Neurology 29, 802-808 (1979)

19. Wenig, Ch., Huber, G., Emde, H.: Hydrocephalus after subarachnoid bleeding. Eur. Neurol. 18, 1-7 (1979)

20. Yasargil, M.G., Yonekawa, Y., Zumstein, B., Stahl, H.-J.: Hydrocephalus following spontaneous subarachnoid hemorrhage. J. Neurosurg. 39, 473-479 (1973)

21. Yoshimoto, T., Uchida, K., Kaneko, U., Kayama, T., Suzuki, J.: An analysis of follow-up results of 1000 intracranial saccular aneurysms with definitive surgical treatment. J. Neurosurg. 50, 152-157 (1979)

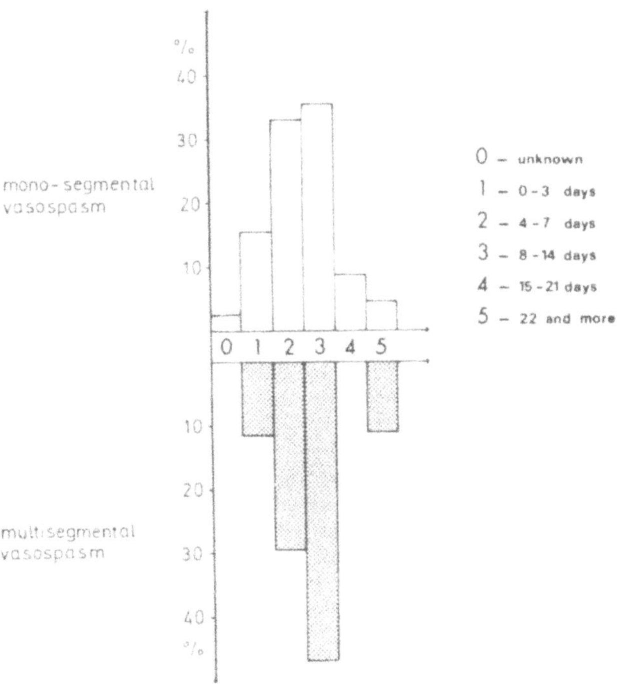

Fig. 1. Development of vasospasm with reference to interval from SAH to angiography

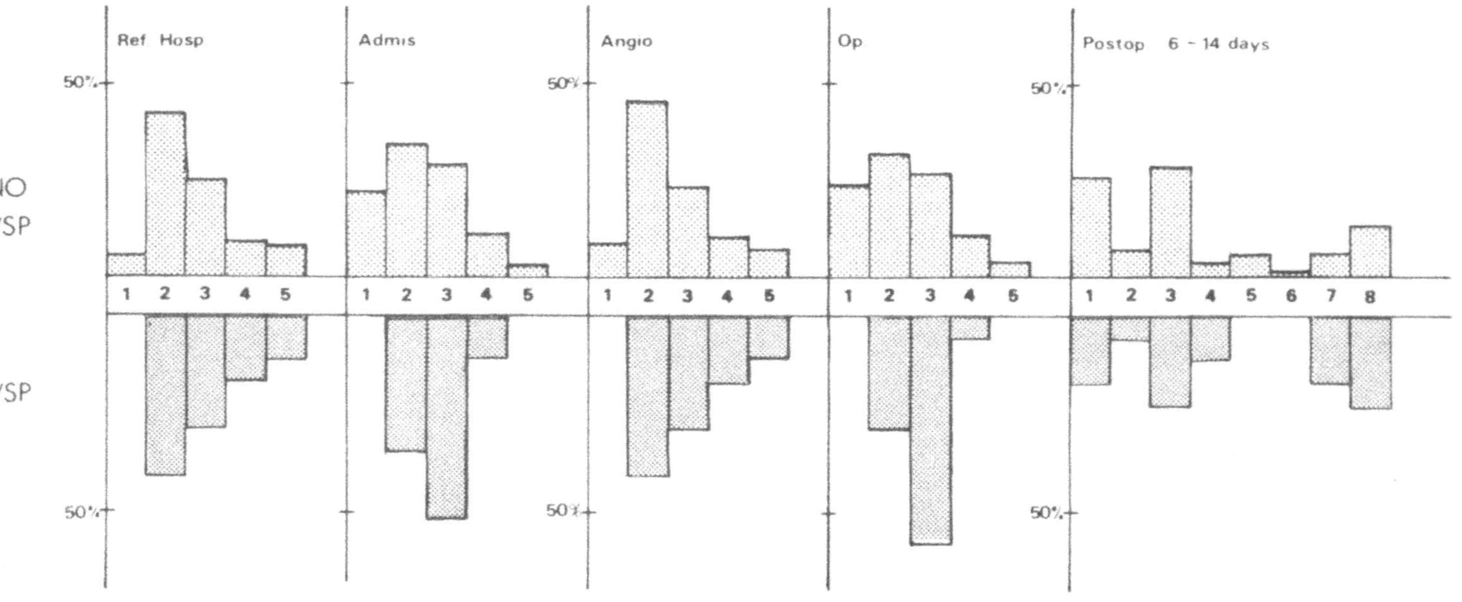

1-6: GRADE I-VI CAS SCALE; 7: DEAD; 8: FRONTAL LOBE DISORDERS

Fig. 2. Clinical course following SAH with reference to vasospasm (VSP)

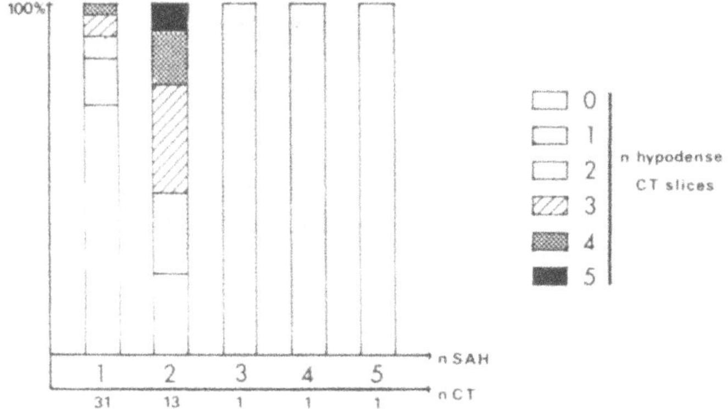

Fig. 3. Size of hypodense lesion depending on incidence of SAH

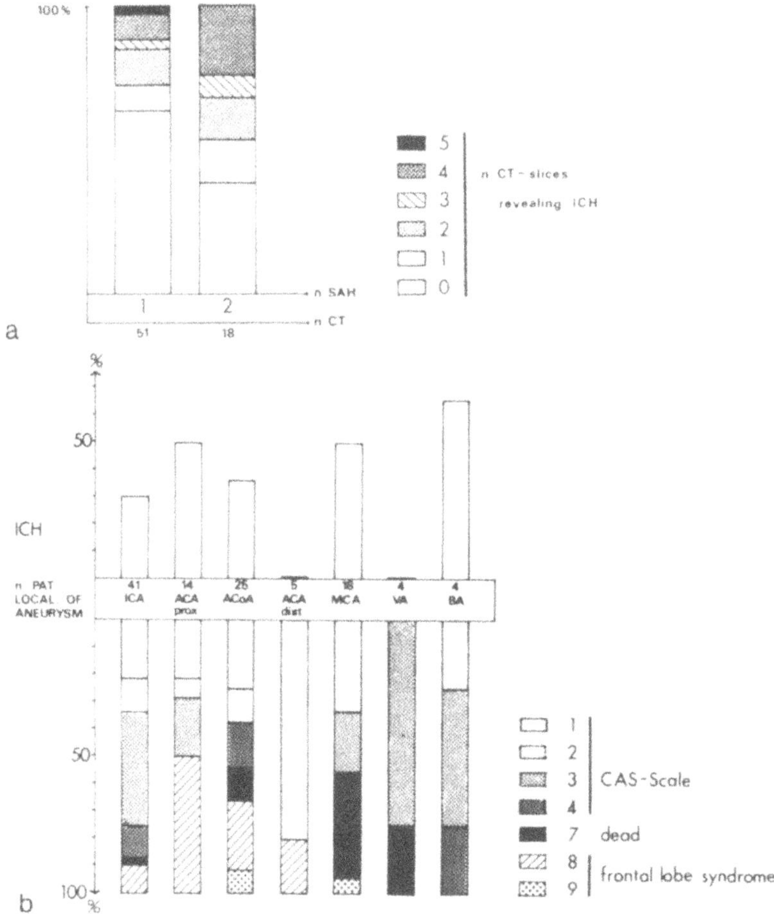

Fig. 4. a Size of haematoma depending on incidence of SAH, b influence of ICH on clinical grade at discharge

Spontaneous Intracranial Hemorrhage, Intracranial Pressure and Indications for Operation: A Clinical Follow-Up Study*

A. Brawanski[1], M.R. Gaab[2], U. Fuhrmeister[1], and I. Haubitz[3]

[1] Neurochirurgische Klinik der Universität, Josef-Schneider-Strasse 11, D-8700 Würzburg
[2] Neurochirurgische Universitäts-Klinik, Alserstr. 4, A-1040 Wien
[3] Rechenzentrum der Universität Würzburg, Hubland, D-8700 Würzburg

Introduction

The indications for operation in spontaneous intracranial hemorrhage (ICH) are often questionable, especially in typical white-matter or basal ganglia hemorrhages in stroke patients where spontaneous resorption may yield a better functional result than an acute operation (9). However, any delay in the operation until clinical signs of decompensation appear, adversely affects the prognosis by additional brain compression.

We therefore investigated patients with ICH (with the exception of hematomas caused by ruptured aneurysms), which did not demand an immediate surgical intervention. We studied the course of the ICP and the EEG and tested the efficiency of these factors in the planning of treatment.

Patients and Methods

Patients

We investigated *24 patients* aged between 28 and 74 years, with a clear history of a spontaneous ICH. The clinical diagnosis was confirmed by a CT scan. Patients with hemorrhage from an aneurysm were excluded. In addition, only those patients were accepted in the study, who had the attack at least one day before admission and who did not need an immediate operation. In comparison with this group, we studies seven patients with ischemic stroke, caused by the occlusion of branches of the middle cerebral artery.

After improvement in the clinical condition angiography was performed on each patient to clarify the cause of the hemorrhage.

Method

After admission and initial investigations an *epidural pressure transducer* was implanted frontally via a burrhole (5). The course of ICP was recorded continuously. Beside the simple ICP monitoring we made an analysis of the ICP-modulations and calculated the regression slopes of the

*Supported by grants from DFG (Ga 273/1), BMFT (MMT 19)

Edited by H.-P. Jensen, M. Brock, and M. Klinger

width of the pulse-amplitude (Amp) and the systolic ICP (Amp/ICP_S) in most of the patients. In addition, the Fourier power spectrum was done. The EEG was monitored continuously and the power spectrum was calculated.

At least twice a day the clinical condition of the patients was checked according to a modified Glasgow coma-scale. The course of the ICP, EEG, the Amp/ICP_S regression and the clinical condition were compared with each other, the most important factor being the clinical status of the patient.

Results

Course of ICP

In nearly all patients (90%) with *hypertensive ICH* we observed critical rises of the ICP (40 mm Hg) (Fig. 1a). The most frequent occurrence of these ICP elevations was recorded between the second and the eighth day after the attack, with a maximum at the fourth day.

In most cases (64%), the plateau waves reacted initially only to conservative treatment, and an operation was necessary in the further development (Fig. 1 a-c). Only rarely (2%) did see a spontaneous recovery after critical ICP rises in these patients, but in them, the ICP levels mostly remained below 40 mm Hg, and any acute conservative treatment by dehydration was not necessary.

In all cases we found a good correlation between the clinical condition and the ICP course (Fig. 1 a-c). A rise of ICP up to critical levels was always accompanied by a clinical deterioration, whereas a clinical improvement was sometimes delayed despite low ICP levels.

In those patients suffering from an *ICH caused by an angioma* the ICP behaved quite differently. In these cases we also saw critical rises of ICP (Fig. 2a, b), but they could well be treated conservatively. As Fig. 2 a-c demonstrate, here also there is a good correlation between the ICP and the clinical status. The occurrence of several plateau waves (29.4.) (Fig. 2a) is accompanied by a dramatic clinical deterioration (Fig. 2c).

As Fig. 3a shows, the slopes of the Amp/ICP_S regression lines increase with rising ICP. Similarly the relation of the cardiac and respiration induced ICP modulations change in a characteristic manner: With rising ICP the respiration associated amplitude (RA) decreases, whereas the cardiac-induced amplitude (PA) increases.

Table 1. The ICP-course, the incidence of operation and the mortality of the investigated patients according to diagnosis

Diagnosis	n	ICP > 40 mm Hg	ICP < 40 mm Hg	OP	no OP	+
ICH	19	17	2	11	6	5
Angioma	5	3	2	-	5	-
Strokes	7	0	7	-	7	-
	31	20	11	11	18	5

EEG

The course of the EEG showed no differences between patients with hypertensive ICH and those with ICH caused by an angioma. On the contrary, the EEG power-spectrum failed to give any evidence of incipient clinical deterioration, as Fig. 3b and 3c demonstrate. Despite several plateau waves (Fig. 3c), the corresponding EEG power-spectrum shows no clear shift to slower frequencies (Fig. 3b). We only observed a distinct deterioration in the EEG after a more permanent rise in ICP.

Statistical Evaluation

As Table 1 shows, five of the 24 patients with spontaneous ICH had an angioma as cause for the hemorrhage. None of these five patients had to be operated on because of a decompensation of the ICP. All rises in ICP could be treated conservatively.

In clear contrast to this group, of the 19 patients with hypertensive ICH nearly 90% had critical ICP rises. Eleven of these patients (n=17) required operation, because of clinical decompensation accompanied by untreatable plateau waves. However, five of the operated patients died later on.

In comparison those patients with ischemia restricted to the area of the middle cerebral artery did not show any critical rises of ICP.

Discussion

The management of spontaneous ICH is still controversial. Some authors plead for early operative intervention in spite of the additional risk of destroying intact brain structures (3, 8). Others prefer a primarily conservative treatment (9). We tested the ICP and the EEG as possible factors for giving additional information for the timing of the operation.

In all cases investigated, the ICP course correlates with the clinical condition (Fig. 1 a-c, 2 a-c). This corresponds to the findings of other authors (1, 7). The patients with *hypertensive ICH* had a maximum of critical rises of ICP between the second and eighth day after the attack, with a maximum on the fourth day. These data demonstrate, that not only the blood clot itself, but also the edematous, necrotic and vasoparalytic reaction of the surrounding brain tissue plays an important role in the further clinical development, as SUZUKI was able to show experimentally (11). If critical ICP levels are reached, conservative treatment usually fails to show more permanent therapeutic effects (4). Despite intensive conservative treatment, 80% of the patients with hypertensive ICH had to be operated on later (Table 1). About 50% (Table 1) of the operated patients died, these results being in agreement with the statistical analyses of other authors (10).

On the contrary, in patients with spontaneous ICH caused by *an angioma*, most critical rises of ICP could be treated conservatively (Fig. 2a-b). An acute operation was not necessary (Table 1). This observation can obviously be explained by the fact, that we never saw any edema in the surrounding brain tissue, e.g. caused by vasospasm. Here the ICP rise is only due to the blood clot itself, and a state of compensation can be reached without progressive vasoparalysis. Using the *ICP* as additional observation, early operative intervention could be *avoided* in these patients. A delayed microsurgical operation brought much better results.

By means of the analysis of ICP modulations, we gained additional knowledge of the ICP course (Fig. 3a). Imminent rises of the ICP can be easily detected by the increasing steepness of the Amp/ICP_S regression slope, although the mean ICP is still low (2). Similarly, the relation of the expiratory and the cardiac amplitude in the Fourier power-spectrum support these results.

On the contrary, the EEG monitoring gave *no additional* information (Fig. 3 b, c). An increase of ICP and the concomitant clinical deterioration are not shown by this. Sometimes the power-spectrum shifts to faster frequencies in the beginning, simulating a clinical improvement (6). Only after a more permanent deterioration of the clinical status does the EEG power-spectrum show a corresponding decrease of the frequencies.

In patients with ischemic stroke caused by the occlusion of branches of the middle cerebral artery the ICP course was nearly always normal, so that in these cases an intensive therapeutic reduction of the ICP did not have to be attempted.

Conclusions

The ICP course gives a clear evidence of decompensation in a spontaneous ICH and helps to decide the date for any operative intervention, whereas the EEG fails to give such information. With compensated ICP a further traumatizing acute operation can be avoided.

In agreement with other authors (1, 7) we see the ICP when considered in association with the clinical condition as a very important deciding factor for the treatment of patients with spontaneous ICH.

Summary

We tested the value of continuous ICP and EEG monitoring in patients with spontaneous ICH. The ICP course showed a clear correlation with the clinical condition, whereas the EEG failed to give any hints of an imminent clinical deterioration. Also, no clear correlation of the rising ICP and EEG power-spectra could be found.

The ICP data of patients with *hypertensive ICH* differed from those of patients with ICH caused by an *angioma*. If ICP elevations occurred in the group of patients with hypertensive ICH, they could often be treated conservatively only initially. In 90% of the patients examined, a decompensation of ICP was observed and an operative intervention was necessary.

Patients with *ICH caused by angiomas*, however, often had only mild rises of ICP, which could well be treated conservatively. An operative intervention could be delayed until a better clinical state was achieved.

Beside the clinical condition, we therefore consider the ICP as a very important factor in deciding the treatment of patients with spontaneous ICH.

References

1. Beks, J.W.: Intracranial pressure in ICH. In: Spontaneous Intracerebral haematomas, Pia, H.W., Langmaid, C., Zierski, J. (eds.), pp. 116-119. Berlin, Heidelberg, New York: Springer 1980
2. Brawanski, A., Gaab, M.R.: Continuous analysis of vasoregulation and intracranial compliance by ICP amplitude analysis. In: Neurological Surgery, Abstracts of the 7th International Congress of Neurological Surgery, Dietz, H., Metzel, E., Langmaid, C. (eds.), pp. 176. Stuttgart, New York: Thieme 1981
3. Cuatico, W., Adip, S., Gaston, Ph.: Spontaneous intracerebral hematomas. J. Neurosurg. 22, 569-575 (1965)
4. Gaab, M.R.: Habil. Schrift 1981
5. Gaab, M.R., Knoblich, O.E., Dietrich, K.: Miniaturisierte Methoden zur Überwachung des intrakraniellen Druckes. Techniken und klinische Ergebnisse. Langenbecks Arch. Chir. 350, 13-31 (1979)
6. Ingvar, D.H.: Cerebral metabolism, cerebral blood flow and EEG. Electroenceph. clin. Neurophysiol. Suppl. 25, 102-106 (1967)
7. Janny, P., Colnet, G., Georget, A.-M., Chazal, J.: Intracranial pressure with intracerebral hemorrhage. Surg. Neurol. 10, 371-375 (1978)
8. Lazorthes, G.: Surgery of cerebral hemorrhage. Report on the results of 52 surgically treated cases. J. Neurosurg. 16, 355-364 (1959)
9. McKissock, W., Richardson, A., Taylor, J.: Primary intracerebral haemorrhage, a controlled trial of surgical and conservative treatment in 180 unselected cases. Lancet 2, 221-226 (1961)
10. Pertuiset, B., Yacoubi, J., Sichez, J.P., Cardeur, D.: Prognostic factors in the first 48 hours in spontaneous ICH and spreading haemorrhages. In: Spontaneous Intracerebral Haematomas. Pia, H.W., Langmaid, C., Zierski, J. (eds.), pp. 294-301. Berlin, Heidelberg, New York: Springer 1980
11. Suzuki, J., Ebina, T.: Sequential changes in tissue surrounding ICH. In: Spontaneous intracerebral haematomas. Pia, H.W., Langmaid, C., Zierski, J. (eds.), pp. 121-128. Berlin, Heidelberg, New York: Springer 1980

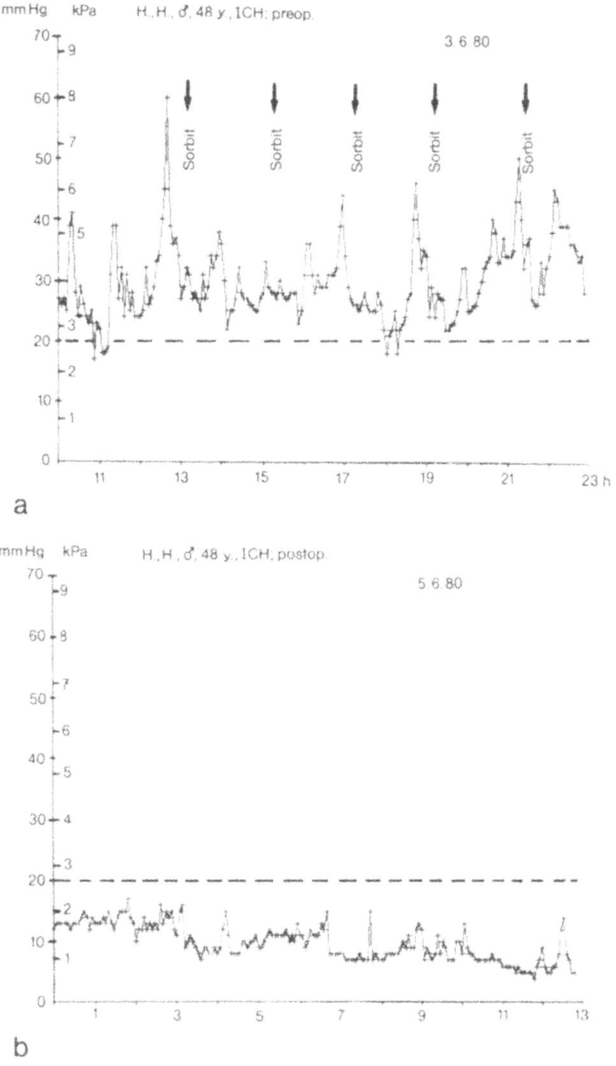

Fig. 1a, b. Legend see page 96

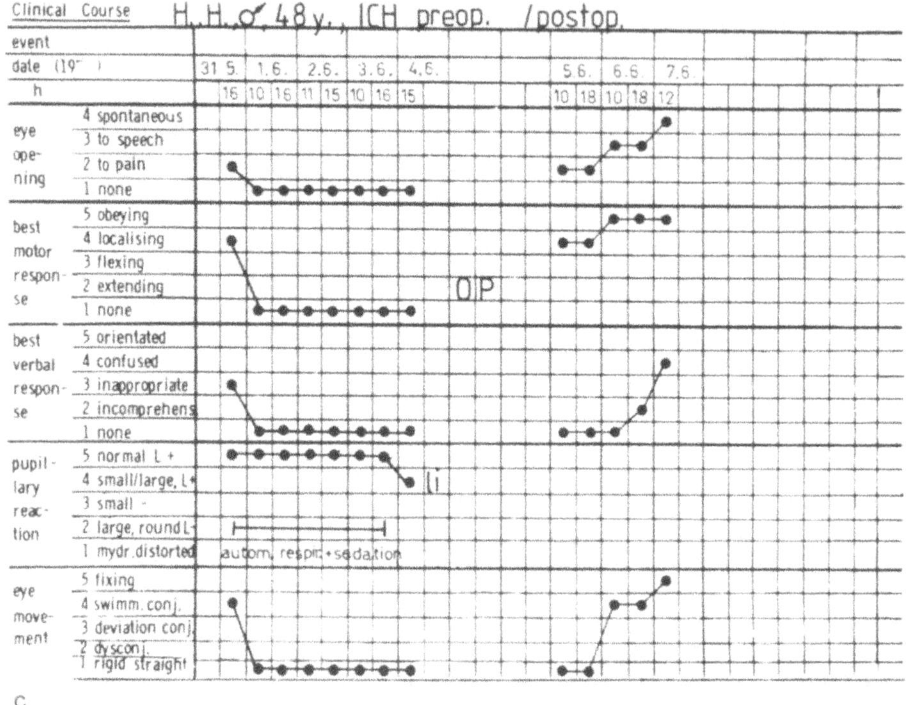

Fig. 1 a-c. Spontaneous ICH in the left basal ganglia; increasing plateau waves cannot be controlled conservatively (a). After operation on the haematoma the ICP pattern returned to normal (b). The clinical condition corresponds to the ICP, as the coma scale (c) demonstrates

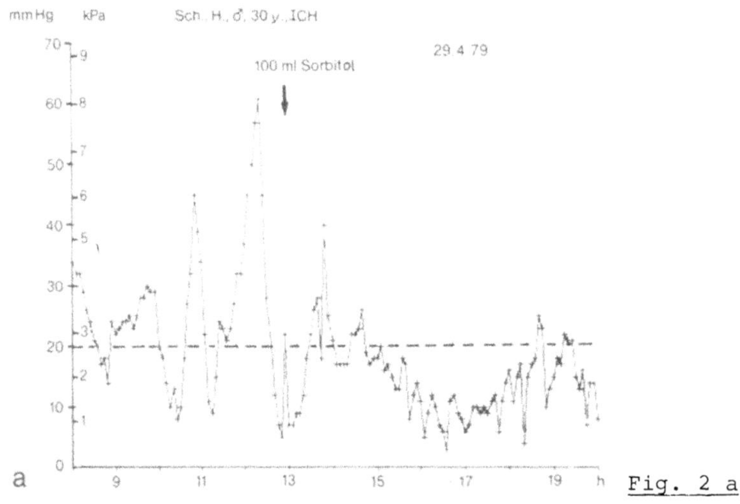

Fig. 2 a

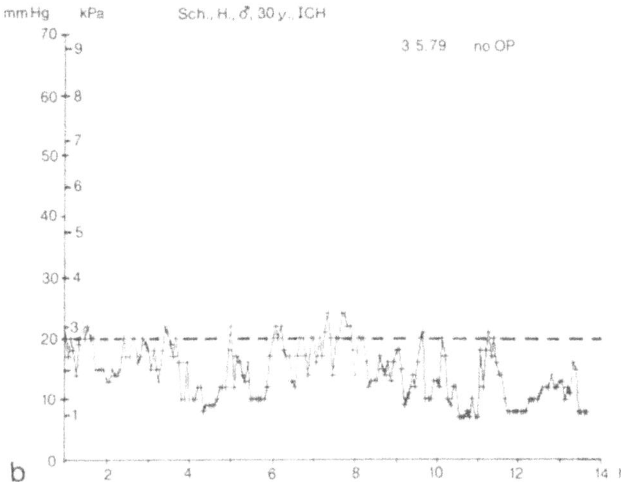

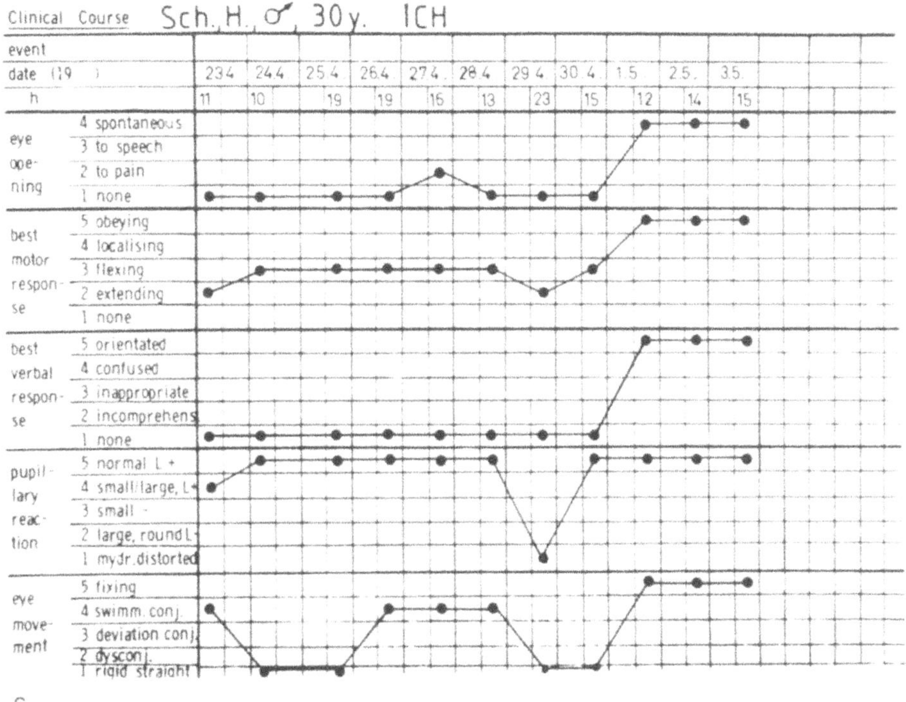

Fig. 2 a-c. Patient with a spontaneous ICH in the left hemisphere, caused by an angioma. After a critical rise in ICP (a), which could be treated conservatively, the ICP values remain low (b). The patient recovered without operation (c)

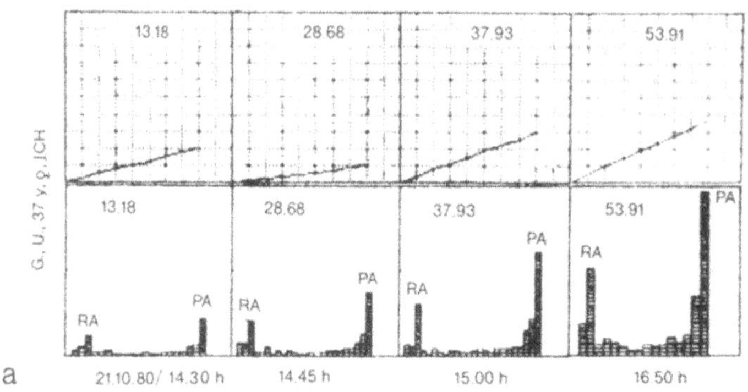

a

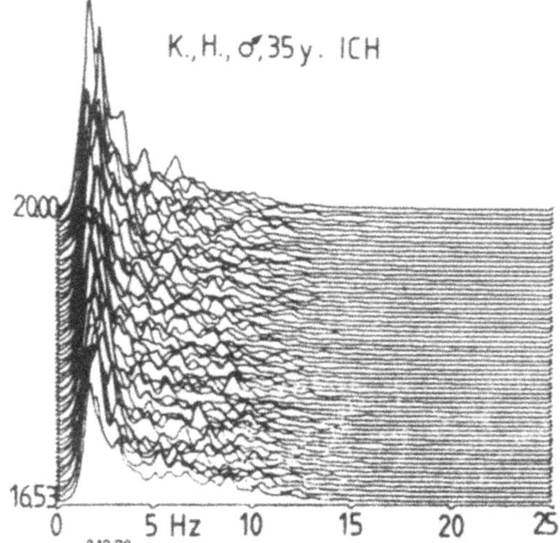

b

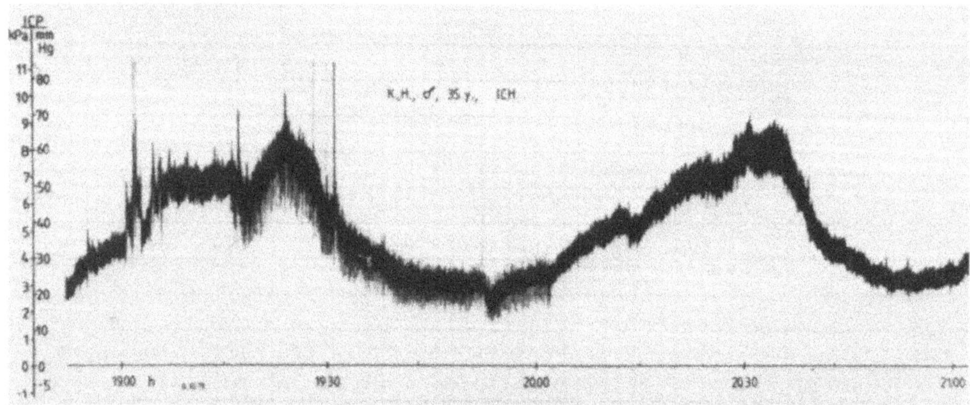

c

Fig. 3 a-c. The Amp/ICP_S-regression shows an increasing steepness of the slopes with rising pressure (a). In the Fourier spectrum of the same ICP data the pulse amplitude (PA) increases, whereas the respiratory component (RA) decreases
The EEG power-spectrum (b) does not correspond to an acute rise of ICP (c). Only after a more permanent rise in ICP, a shift of the power-spectrum to lower frequencies can be observed

The Significance of the Size of an Intracerebral Hematoma for Treatment and Prognosis

W. I. Steudel[1], G. Hopp[2], and H. Hacker[1]

[1] Abteilung für Allgemeine Neurochirurgie, Zentrum der Neurologie und Neurochirurgie der Johann Wolfgang Goethe-Universität, Schleusenweg 2–16, D-6000 Frankfurt 71

[2] Abteilung für Neurologie, Zentrum der Neurologie und Neurochirurgie der Johann Wolfgang Goethe-Universität, Schleusenweg 2–16, D-6000 Frankfurt 71

Introduction

In many diseases of the brain such as traumatic lesions and tumors, it is important for the prognosis and the therapeutic procedure to take into account the size of the space-occupying lesion.

Before the introduction of computer-tomography (CT), the prognostic significance of the size of the intracerebral hematoma (ICH) was disputed. McKISSOCK et al. and CUATICO et al. considered that the level of consciousness and the site but not the size of the ICH are decisive for the prognosis. STEINER et al. (1975) observed a relation between prognosis and the volume of the hematomas in cubic centimeters as calculated by CT.

We have calculated the hematoma volume in 352 patients with intracerebral hematomas of various etiology (202 spontaneous and 150 traumatic hematomas). The relationship between the size of the ICH and the level of consciousness, the location, age, etiology, treatment, and midline shift are investigated and the results discussed.

Methods

The progress of all patients with an intracerebral hematoma examined by CT in the period from July 1974 to June 1980 was recorded. There were 352 patients with hematomas of varying etiology (150 brain injuries, 92 hypertensive ICH, 41 aneurysms, 12 angiomas, 8 microangiomas, 9 anticoagulants, 9 miscellaneous, 32 unknown). The following conditions were not investigated: gunshot injuries, small hemorrhages (less than 3 ml), tumor bleeding, hemorrhages due to an operation and children under 14 years old.

Follow-up examinations of the 158 surviving patients were carried out six months to three years after the hemorrhage in the period from 1978 to December 1980.

The size of the hematoma is determined by addition of the values calculated planimetrically in the individual slices (thickness 1 cm). The intracerebral and intraventricular portion of the ICH is calculated separately. The most important factors (etiology, level of consciousness, age, treatment, size, site, midline shift) are investigated by means of an analysis of variance with regard to prognosis and degree

Advances in Neurosurgery, Vol. 11
Edited by H.-P. Jensen, M. Brock, and M. Klinger

of disability. The level of consciousness was classified into ten categories; the degree of disability into five categories.

Results

The total size of the ICH comprised the size of the actual intracerebral hematoma and of the intraventricular portion. The total size of the hematomas ranged from 3 to 245 ml.

Of the 352 hematomas, 243 measured from 3 to 40 ml, 70 between 41 and 70 ml and 39 had a volume in excess of 71 ml.

With regard to survival, four groups were distinguished (Fig. 1). The first group comprised 243 cases with a hematoma size of 3 to 41 ml; of 117 operated patients, 67 survived whereas 63 out of 126 nonoperated patients survived. In this first group, a non-significant difference was shown with regard to survival between the operated and the nonoperated patients. In the second group with a total size of the ICH of 41 to 70 ml, ten out of 42 operated patients survived and only three out of 28 nonoperated patients. In the third group)from 71 to 120 ml), six out of ten operated patients survived, the nonoperated patients died. Out of 13 patients with a hematoma volume of 121 to 250 ml (fourth group), none survived.

Figure 2 shows the size of the hematomas in the cerebral hemispheres. Of the traumatic ICH, those operated patients with an ICH size up to 60 ml and of the nonoperated patients up to 30 ml survived. In the spontaneous ICH, operated patients with a volume up to 120 ml survived, and with one exception only up to 40 ml in the nonoperated patients. Of the 16 cerebellar ICH, seven survived (five of these were operated on); hematomas in excess of 20 ml did not survive (Fig. 3).

The size of the 84 hematomas originating from the basal ganglion region (9 traumatic, 75 spontaneous) ranged from 10 to 200 ml. Of the nine traumatic ICH, three survived (with only one hematoma under 10 ml). Of the spontaneous ICH (19 operated, 56 nonoperated), 26 out of 43 cases survived with hematomas up to 60 ml. Larger hematomas did not survive without operation. Of the 13 operated patients with an ICH size up to 60 ml, five survived. Of four operated patients with an ICH between 60 to 90 ml, three survived. Only "lateral types" were operated on (KWAK et al.).

The relationship between the level of consciousness and the size of the hematoma was examined statistically (FISHER test). With regard to the traumatic ICH, no significant relation was shown at a probability level of $p \leq 0.01$. In the spontaneous ICH, there is a high significance ($p > 0.01$). With increasing size of the hematoma, the depth of the disturbance of consciousness increases. The results of the analysis of variance with regard to the prognosis are contained in Table 1. The level of consciousness (coma / no coma) clearly had a very great significance. The other factors (size, operation, age) had less significance. Etiology, location and midline shift had no demonstrable influence on the prognosis.

The ranking sequence of the effects with regard to the degree of disability is shown in Table 2. The location of the ICH was decisive for the degree of disability. The size, age, and level of consciousness then followed. Compared to the results in terms of the prognosis, it is noticeable that the etiology and operation had no demonstrable influence on the degree of disability.

Table 1. Results of the variance analysis: ranking sequence of the effects with regard to prognosis (n = 352)

Effects (factors/interactions)	Classes	N (Mortality in %)	P
Level of consciousness	n_1 Non comatose	145 (24%)	0.000 01
	n_2 Comatose	207 (78%)	
Operation / level of consciousness			0.019
Age	n_1 $\leq$ 40 years	144 (47%)	0.024
	n_2 > 40 years	208 (61%)	
Size	n_1 $\leq$ 30 cm^3	203 (43%)	0.026
	n_2 > 30 cm^3	149 (72%)	
Operation	n_1 yes	174 (47%)	0.043
	n_2 no	178 (63%)	

Table 2. Results of the variance analysis: ranking sequence of the effects with regard to the degree of disability (grade 1 to 5) in the survivors (n = 158)

Effects	Classes	N	Means value	P
Site	n_1 Lobar	105	2.23	0.000 001
	n_2 Central			
Size	n_1 $\leq$ 30 cm^3	116	2.32	0.000 003
	n_2 > 30 cm^3	42	3.00	
Age	n_1 $\leq$ 40 years	78	2.25	0.000 003
	n_2 > 40 years	80	2.72	
Level of consciousness	n_1 Non comatose	111	2.36	0.000 003
	n_2 Comatose	47	2.80	

Discussion

We have calculated the volume of the ICH (which ranged from 3 to 245 ml) on the basis of the method described by STEINER. According to STEINER, the measurement error is $\pm$ 3 to 9% with a slice thickness of 13 mm (in our measurements 10 mm); others report the error in volumetry as 7% (PENTLOW et al.).

As regards the prognosis the analysis of variance shows that the size of the ICH follows in the ranking of various factors after the level of consciousness. In the surviving patients the size of the hematoma affects the extent of the disability.

In the hemispheres, the size of the ICH is of prognostic significance when it is over 30 to 40 ml; in hematomas in excess of 70 ml, only operated patients survived. The size of the ICH in those who do not survive is about one tenth of the normal brain volume. Acute space occupying lesions can evidently be compensated by reserved spaces of the CSF (about 120 ml) up to one tenth of the brain volume (FURUSE et al., 1976).

There is a statistically demonstrable difference between the size of the hematoma and the disturbance of consciousness in spontaneous ICH, but not in traumatic hematomas. Extensive edema may be present in traumatic lesions. The perifocal and generalized brain edema which, together with the hemorrhage, determine the size of the space-occupying lesion and the intracerebral pressure were not measured quantitatively.

Conclusion

It is important to take into account the ICH size when deciding the indications for operation and in prognostic evaluation.

References

1. Cuatico, W., Adib, S., Gaston, P.: Spontaneous intracerebral hematomas. J. Neurosurg. 22, 596-675 (1965)
2. Furuse, M., Hasuo, M., Brock, M., Dietz, H.: The role of CSF resorption on the intracranial pressure, volume relationship. In: Behr, J.W.F., Bosch, D.A., Brock, M. (eds.). Intracranial Pressure III, pp. 20-24. Berlin, Heidelberg, New York: Springer 1976
3. McKissock, W., Richardson, A., Taylor, J.: Primary intracerebral hemorrhage. A controlled trial of surgical and conservative treatment in 180 unselected cases. Lancet 2, 221, 226 (1961)
4. Kwak, S., Inou, S., Nagashima, T., Sano, K.: Classification of hypertensive intracerebral hemorrhage by means of computed tomography. Neuroradiology 16, 159-161 (1978)
5. Orthner, H., Sendler, W.: Planimetrische Volumetrie an menschlichen Gehirnen. Fortschr. Neurol. Psychiat. 43, 191-209 (1975)
6. Pentlow, K.S., Rottenberg, D.A., Deck, M.D.F.: Partial Volume Summation: A simple approach to ventricular volume determination from CT. Neuroradiology 16, 130-132 (1978)
7. Pia, H.W.: Discussion. In: Spontaneous Intracerebral Haematomas, Pia, H.W., Langmaid, C., Zierski, J. (eds.), pp. 279-281. Berlin, Heidelberg, New York: Springer 1980
8. Steiner, L., Bergvall, U., Zwetnow, N.: Quantitative estimation of intracerebral and intraventricular hematoma by computer tomography. Acta Radiol. (Suppl.) (Stockh.) 36, 143-154 (1975)

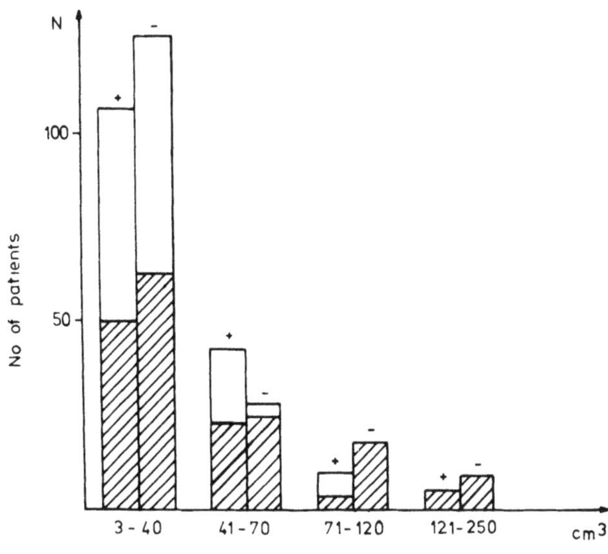

Fig. 1. Size of ICH with regard to treatment and survival

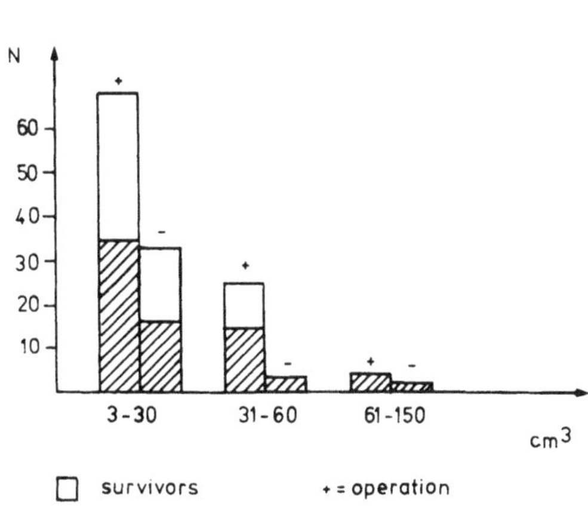

Fig. 2. Size of traumatic ICH in the cerebral hemispheres with regard to treatment and survival

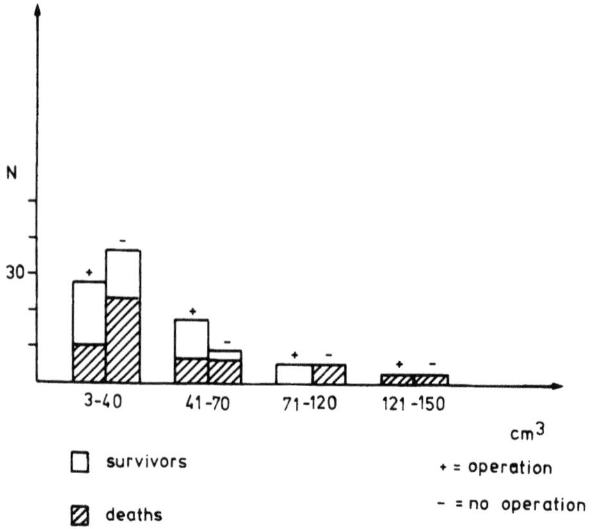

Fig. 3. Size of spontaneous ICH in the cerebral hemispheres with regard to treatment and survival

Clinical Course of 160 Operated and 45 Conservatively Treated Patients with Spontaneous Intracerebral Hematoma

A. Karimi-Nejad[1] and A. Möller[2]

[1] Neurochirurgische Universitätsklinik Köln, Joseph-Stelzmann-Strasse 9, D-5000 Köln 41
[2] Universitäts-Nervenklinik Köln, Joseph-Stelzmann-Strasse 9, D-5000 Köln 41

An analysis of results of surgical treatment in spontaneous intracerebral hematomas (ICH) before the era of computer tomography (CT) shows that the indication for operation should be restricted especially in cases of deep location and hypertensive etiology (KARIMI-NEJAD, HAMEL 1980). Also those ICH of different depth and extent which could not be seen on angiography can be verified by CT. This improvement in diagnostic possibilities may mistakenly lead to surgical treatment even of those ICH which may be expected to show a better outcome under conservative treatment. By comparing the clinical course of 160 operated and 45 conservatively treated cases of ICH with particular consideration of the CT findings, guiding principles either for conservative or surgical treatment should be worked out.

Table 1 shows the total number of ICH, their etiology and age distribution. ICH of unknown etiology usually occur in young people, those of hypertensive and coagulopathic etiology in older people.

Table 2 shows the location of ICH in relation to their etiology. A preference for the temporal region is evident in cases of hypertensive etiology.

Table 1. Spontaneous intracerebral hematomas - etiology and age distribution

Age	Hypertensive		Coagulopathic		Unknown etiology	
	No.	%	No.	%	No.	%
0 - 10						
11 - 20					3	5
21 - 30	6	5			35	55
31 - 40	13	12			4	6
41 - 50	31	28	2	7	1	1
51 - 60	42	37	7	24		
61 - 70	15	13	11	38		
71 -	6	5	9	31		
	n = 113		n = 29		n = 64	

Advances in Neurosurgery, Vol. 11
Edited by H.-P. Jensen, M. Brock, and M. Klinger

Table 2. Spontaneous intracerebral hematomas - etiology and location

	Hypertensive		Coagulopathic		Unknown etiology		Total	
	No.	%	No.	%	No.	%	No.	%
Frontal	12	11	9	33	18	28	39	19
Temporal	81	72	15	56	27	42	123	60
Parietal	17	15	2	7	17	27	36	18
Occipital	3	2	1	4	2	3	6	3

Factors Which Influence the Clinical Course

Prognosis was influenced considerably by the state of alertness either pre-operative or at time of admission. Surgically as well as conservatively treated patients show a high mortality of 82% and 91% if they are unconsciousnes at this time. It must be emphasized that conservatively treated patients who are unconscious represent a group of negative selection because of their usual preterminal general condition. In contrast to these unfavourable results alert patients seem to have good prognosis with a low mortality rate of 17% and 8% and often, in addition a complete recovery in their later clinical course (see Fig. 1). So, in cases of ICH without any clouding of consciousness no indication for operative intervention can be seen.

Location

Table 3 shows the clinical course in relation to the location of hemorrhage. The unsatisfactory results in cases of temporal lesions which has already been pointed out by other authors (PIOTROWSKI 1980) must be confirmed.

Table 3 a, b. Outcome in relation to location of hemorrhage

a Surgically treated

	Frontal		Temporal		Parietal		Occipital	
	No.	%	No.	%	No.	%	No.	%
Died	21	58	57	66	8	25	3	50
Need care	3	8	12	14	10	31		
Disabled	6	17	5	6	8	25	2	33
Fit	6	17	12	14	6	19	1	17
	n = 36		n = 86		n = 32		n = 6	

b Conservatively treated

	Frontal		Temporal		Parietal		Occipital	
	No.	%	No.	%	No.	%	No.	%
Died	1	33	24	67	1	33	1	50
Need care	1	33	5	14	2	67		
Disabled	1	33	4	11				
Fit			3	8			1	50
	n = 3		n = 36		n = 3		n = 2	

Extent of Hemorrhage

Figure 2 demonstrates the relationship between mortality rate and extent of intracerebral hemorrhage determined as its greatest diameter in the CT picture. Especially with an ICH of more than 5 cm in greatest diameter, the mortality also increases to a high level of 66% and 82% under surgical and conservative treatment.

Depth of Hemorrhage

Figure 3 demonstrates the even closer relationship between mortality rate and depth of hemorrhage as it can be verified in CT. In ICH of unknown or coagulopathic etiology and at a depth of three cm or less, no case of a fatal outcome could be seen. In deep ICH the mortality rate increases to 75% and is more independent of its etiology and treatment.

Pareses

We studied the clinical course of pareses as verified on admission or pre-operatively and graded them as follows:

0: without pareses
1: mild hemiparesis
2: moderately disabled because of hemiparesis
3: severe hemiparesis, needs help
4: hemiplegia.

Figure 4 shows that in surgically as well as in conservatively treated patients an improvement in the severity of hemiparesis could be seen in the later clinical course.

In 6 of 31 (" 1+%) surgically treated patients and 2 of 16 (= 13%) conservatively treated patients no change in the degree of paresis could be found. An improvement of one degree as explained above had been observed in 18 (= 58%) surgically treated and 14 (= 87%) conservatively treated patients. In four cases of operated ICH a mild hemiparesis developed as a result of the surgical intervention. Patients with a fatal outcome have been left out of consideration here.

Midline Shift in CT

In cases of a midline shift between 6 and 10 mm verified by CT we found a fatal outcome in 54% of our surgically treated patients (n = 13) and in 62% of our conservatively treated patients (n = 8). If there is a greater midline shift (of more than 10 mm), mortality rate increases to 62% in surgically treated (n = 19) and 83% in conservatively treated patients (n = 6). So, comparing these results, there is a trend to a better outcome for an operated ICH with distinct signs of midline shift in space-occupying intracerebral haematoma.

Tendency of Clinical Course

Comparing the clinical course of operated and conservatively treated patients we judged its tendency by the following symptoms:

1. hemiparesis
2. anisocoria
3. clouding of consciousness.

If there is a tendency to deterioration in the clinical course with progressive clouding of consciousness and progress in hemiparesis, a more favourable outcome may be expected from surgical treatment (see Fig. 5).

On the other hand if patients with a tendency to deterioration of hemiparesis or state of alertness were operated on, their mortality rate was determined above the general average of mortality in surgically treated patients. Those patients treated conservatively show a mortality rate corresponding to the general average of mortality in non-operated patients. It should be pointed out again that this group of patients represents a negative selection because of their initial bad general condition which is a contraindication to operation.

Summary

The clinical course of 160 operated and 45 conservatively treated patients has been analysed. Hypertensive etiology of ICH, temporal location, great extent and especially depth as could be verified by CT, and clouding of consciousness are common signs of unfavourable prognosis. The results of our study demonstrate that in case of space-occupying hemorrhage and edema in the hemispheres with distinct signs of midline-shift, operative intervention shows the more favorable outcome. Focal neurological signs such as hemiparesis cannot provide a reason for operative intervention except in case of progressive deterioration when, in general, operation shows the better outcome. In case of an improvement in the patients condition no further indication for operation can be seen.

Conclusion

Comparing the clinical outcome of 160 operated and 45 conservatively treated patients the following guiding principles for operation can be worked out:

hemorrhage in the hemispheres in patients with clouded consciousness
hemorrhage and oedema with definite signs of midline-shift
deterioration in the later clinical course under conservative treatment.

Focal neurological signs such as hemiparesis cannot by themselves be a reason for operation because their recovery can also be expected under conservative treatment. Also in alert patients the results of conservative treatment are better. In case of improvement in the patients condition under conservative treatment as can sometimes be seen even in deep comatose patients (grade III or IV) operative intervention shows the more unfavorable outcome.

References

1. Karimi-Nejad, A., Hamel, A.: Comparative study of the results in spontaneous and traumatic ICH. In: Spontaneous intracerebral haematomas. Pia, H.W., Langmaid, C., Zierski, J. (eds.), pp. 310-316. Berlin, Heidelberg, New York: Springer 1980

2. Piotrowski, W.: Spontaneous ICH - Prognosis and Discussion. In: Spontaneous intracerebral haematomas. Pia, H.W., Langmaid, C., Zierski, J. (eds.), pp. 262-263. Berlin, Heidelberg, New York: Springer 1980

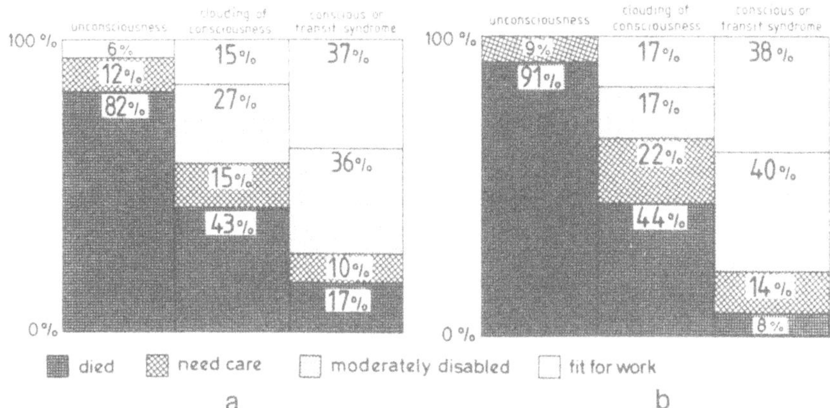

Fig. 1. Results in relation to state of consciousness; a surgically treated (n = 160); b conservatively treated (n = 45)

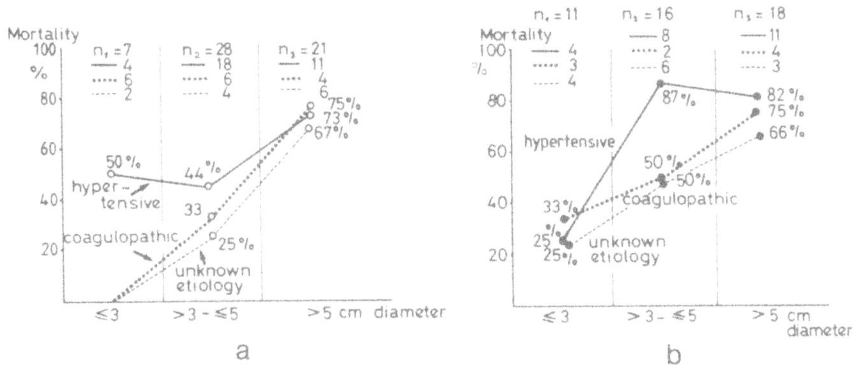

Fig. 2. Results in relation to diameter of spontaneous intracerebral hematomas; a surgically treated (n = 56); b conservatively treated (n = 45)

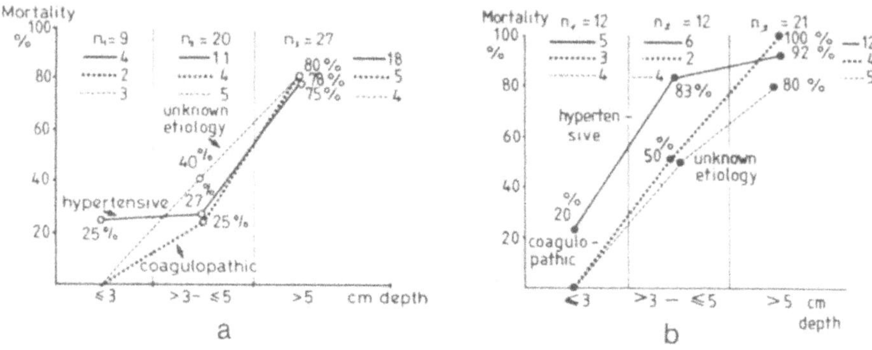

Fig. 3. Results in relation to depth of spontaneous intracerebral hematomas; a surgically treated (n = 56); b conservatively treated (n = 45)

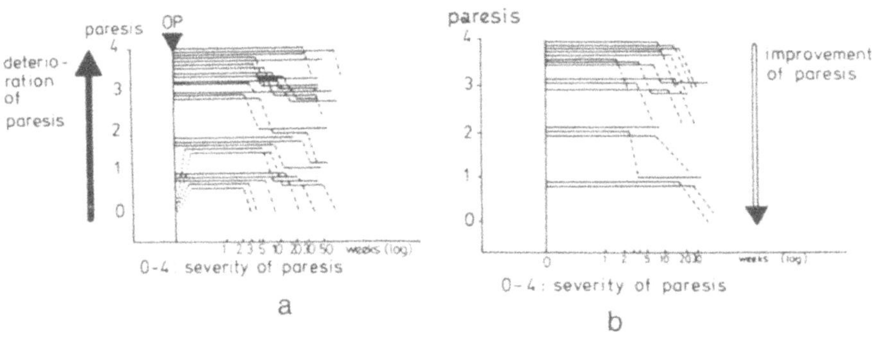

Fig. 4. Clinical course of pareses in patients with ICH; a surgically treated (n = 31); b conservatively treated (n = 16)

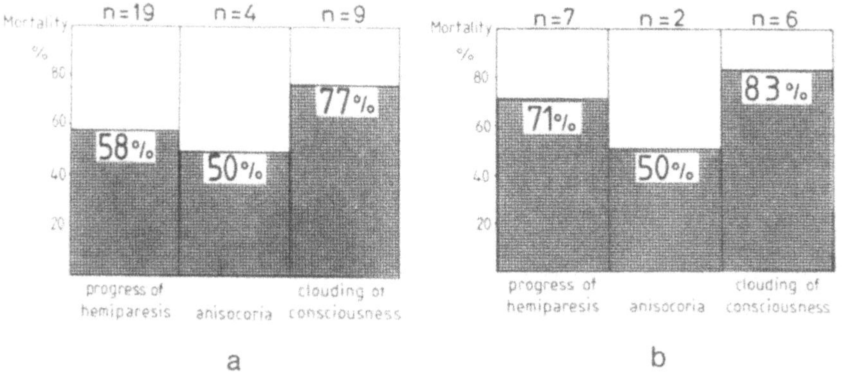

Fig. 5. Results in deterioration of clinical course; a surgically treated (n = 32); b conservatively treated (n = 15)

Value of CT Scan in the Prognosis of Spontaneous Intracerebral Hemorrhage into the Ventricular System

J. Piek and Ch. Lumenta
Abteilung für Neurochirurgie, Medizinische Einrichtungen der Universität Düsseldorf, Moorenstrasse 5, D-4000 Düsseldorf 1

Introduction

A direct demonstration of spontaneous intracerebral hemorrhage into the ventricular system is not possible by conventional neuroradiologic methods. By using CT scan, the site, size, and volume of such bleeding can be measured and determined exactly, as well as the type of distribution within the ventricular system. However, the clinical state and natural course produced by such hemorrhages vary and are strongly influenced by individual factors.

The purpose of this study was to establish an exact prognosis of spontaneous intraventricular hemorrhage based on the CT scan findings on admission.

Material and Methods

The diagnosis of spontaneous intraventricular hemorrhage (IVH) was made by means of the CT scan in 72 patients admitted to our clinic from 1977 - 1981. In 58 patients the primary CT scan was performed within 8 hours of the spontaneous IVH. Twenty-four patients beyond this eight hour limit were excluded from this study. The clinical state was determined using the Glasgow coma score. CT findings were analysed as shown in Fig. 1. The results obtained by this CT classification were compared with the rates of survival.

Neuropathological examinations were carried out in all fatal cases.

Results

Epidemiology

Age and sex distribution showed a sex ratio of nearly 1 : 1 (30 males to 28 females in our patients) with a preference for the sixth and seventh decade.

Etiology

Underlying diseases included hypertension (30 patients; 51.5%), vascular malformations (five angiomas and six aneurysms; 19.5%), and disorders of coagulation - all due to Cumarine therapy (five patients; 9.5%). In 11 patients the etiology of IVH remained unknown.

Advances in Neurosurgery, Vol. 11
Edited by H.-P. Jensen, M. Brock, and M. Klinger

Clinical State

Using the GCS it varied from 3 - 14 points with a mean score of 5.65 points. In cases of IVH only the lowest mean score found was 4.7 points. Patients with supratentorial hemorrhage had a mean score of 5.8 points whereas the best mean score of 6.2 points was found in cases of infratentorial bleedings.

CT Findings

Location and Volume. Out of our 58 patients there were 42 cases (72.5%) of supratentorial hemorrhages, 34 of which were located in the temporal (most frequently deep gangliobasal) region. Three occipital, three parietal, and two frontal intracerebral hemorrhages had also ruptured into the ventricular system. Ten hemorrhages (17.25%) showed a purely intraventricular distribution and six (10.25%) were in the posterior fossa.

The volume of intracerebral blood was measured mathematically and showed a mean volume of 51.3 ml (min. volume 6.5 ml; max. volume 130.8 ml) in supratentorial and of 6.7 ml (min. volume 1.2 ml; max. volume 14.5 ml) in infratentorial hemorrhages.

Intraventricular Distribution. The type of intraventricular distribution is shown in Fig. 2. A quite homogeneous type is found in cases of IVH only. Supratentorial hemorrhages show a diminishing frequency from the ipsilateral lateral ventricle down to the fourth. The same type of distribution but in reverse order is found in cases of infratentorial hemorrhages.

Mass Effects. Indirect signs of space-occupying lesions in our 42 cases of supratentorial ICH with IVH include local edema (100%) and compression of the lateral ventricle (96%) in the ipsilateral hemisphere as well as a shift of the midline (69%), contralateral edema (69%) and compression of the contralateral lateral ventricle (32%). In an axial direction mass effects are observed ipsilateral (82%) and contralateral (70%) demonstrated by compression of the parapontine cisterns.

Survival Rates

Of all our patients 32% died within the first 24 hours after admission. Another 34% did not survive the first seven days after the IVH had occurred. Out of 34% of our patients surviving longer than seven days another quarter died during the next four weeks, most frequently from secondary complications such as severe pulmonary infections, gastrointestinal bleeding, and thromboembolism. Thus the total survival rate was 25.5% after four weeks.

This survival rate includes those patients treated by operation (extermal CSF drainage in three patients, operation for removal of the hematoma in six cases) and also those managed conservatively. Operative treatment was performed only on patients in a good or excellent condition (GCS 8 points or higher).

Correlation Between CT Findings and Survival Rates

Correlation between CT findings on admission and survival rates is shown in Figs. 3 and 4. Almost two-thirds of our patients in each group of supratentorial, infratentorial and intraventricular hemorrhage died.

The mean volume of the hematoma in our 42 patients with supratentorial hemorrhage not surviving the first day was 74.81 ml. Patients dying during the first week after IVH showed a mean volume of intracerebral blood of 37.89 ml. Patients surviving the first week had a mean volume of 18.89 ml. Only one patient with a volume higher than 74.81 ml survived longer (three days) than 24 hours; another one with a blood volume of more than 37.89 ml survived the first week after IVH (death occurred after nine days).

Correlation made between the type of intraventricular distribution and survival rate showed that patients in those cases in which only one lateral ventricle was involved survived longer than the first week in 58% of our cases. Involvement of two lateral ventricles and/or the third ventricle reduced survival rates to 40%. With the fourth ventricle involved in IVH only 28% of our patients survived the first week. Side-to-side mass effects of the space-occupying supratentorial hemorrhage gave a mortality rate of 16.4% only (2 of 13 patients) if these signs could be seen only on the side of the hematoma. Mortality-rate was much higher in patients when mass-effects were also seen in the contralateral hemisphere (85% of 29 patients). Unilateral compression of the upper brain stem does not seem to influence mortality very much, as unilateral compression of the parapontine cisterns of almost the same frequency was found in both groups.

Discussion

Using the present CT classification analysis of primary CT scan findings can in our opinion help to establish an exact prognosis in patients with spontaneous IVH on admission.

The site of the intracerebral hemorrhage does not seem to have a great influence on the prognosis in such patients. The volume of intracerebral blood, distribution of intraventricular blood, and side-to-side mass-effects are of much more importance.

Operation does not seem to be helpful in cases which are expected to die within the first 24 hours, according to their CT scan findings on admission. On the other hand operation may be helpful in cases expected to survive this 24 hour limit. To obtain exact indications for operation in such cases, analysis of many more cases seems to be necessary as regards CT findings, epidemiological data, etiology, and clinical course.

Hemorrhage

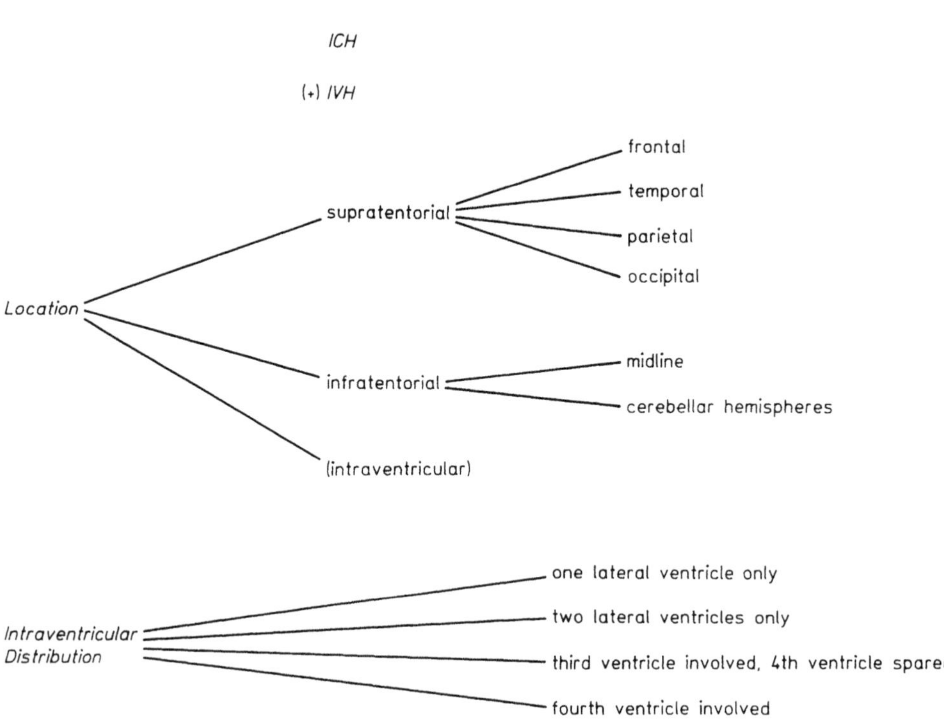

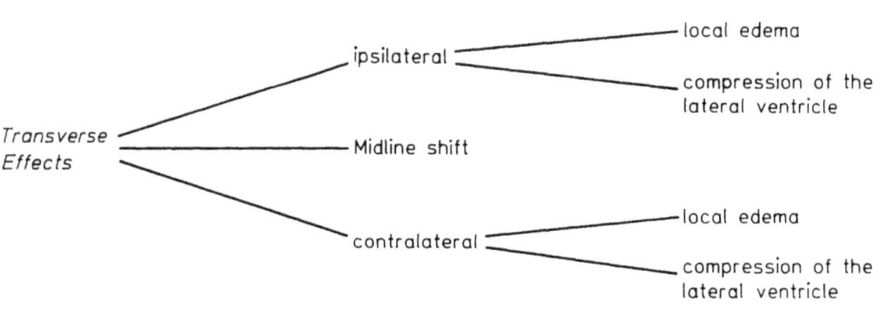

Axial Effects —— Compression of the parapontine cisterns (ipsilateral-contralateral)

Fig. 1. CT classification used in patients with spontaneous intraventricular hemorrhage (IVH)

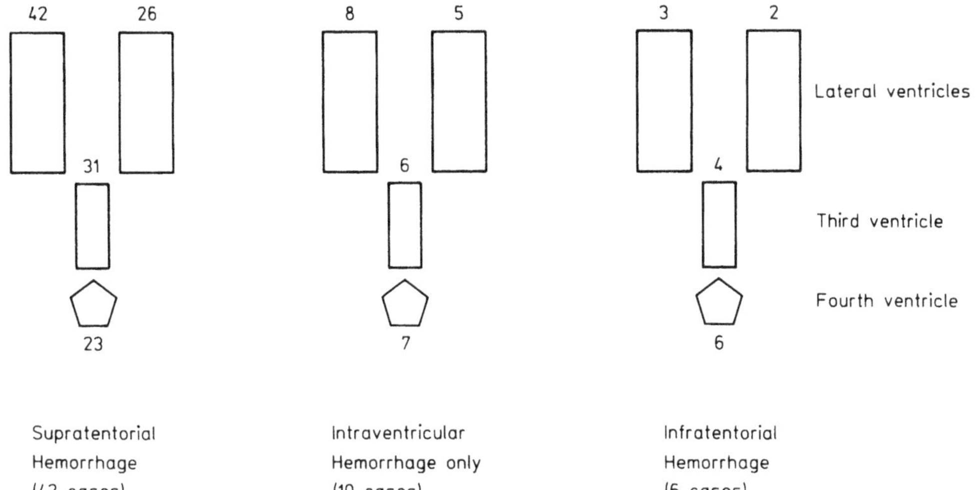

Fig. 2. Intraventricular distribution of IVH in 58 patients (Neurosurg. Clinic of Düsseldorf, 1977-1981)

		+ 1st day	+ 1st week	+ =	survival
Total survival		n = 18 (32%)	n = 20 (34%)	n = 38 (66%)	n = 20 (34%)
Location	supratentorial			n = 28 (66%)	n = 14 (33%)
	infratentorial			n = 4 (66%)	n = 2 (33%)
	intraventricular			n = 6 (60%)	n = 4 (40%)
Intraventricular distribution	1 lat. ventr. only			n = 3	n = 4
	2 lat. ventr. only			n = 3	n = 2
	3rd ventr. involved			n = 6	n = 4
	4th ventr. involved			n = 26	n = 10
Mean blood volume in hematoma		74.81 ml	37.89 ml		18.89 ml

Fig. 3. CT findings, mortality and survival rates in patients with spontaneous IVH (Neurosurg. Clinic of Düsseldorf, 1977-1981, n = 58)

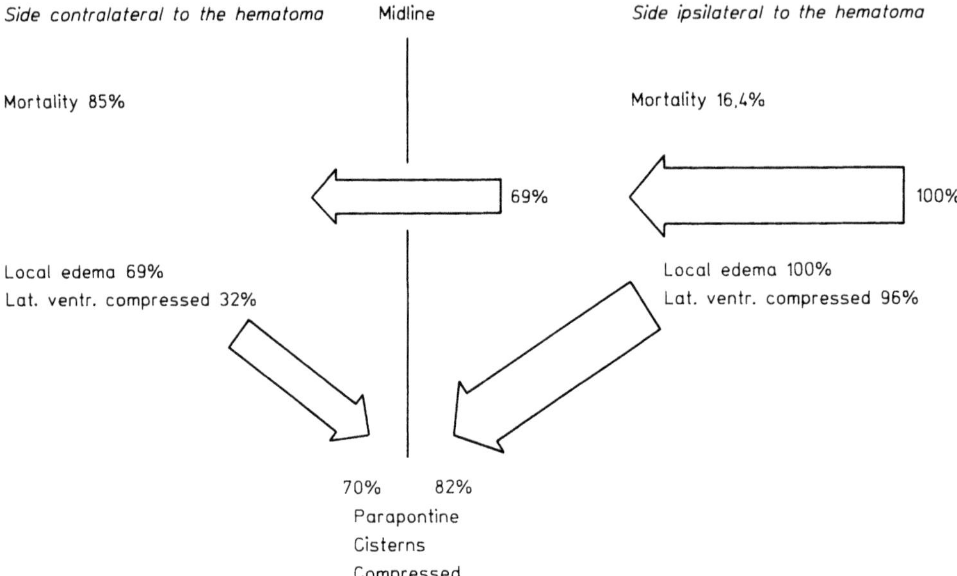

Fig. 4. Mass-effects and mortality rates in patients with supratentorial ICH and IVH (Neurosurg. Clinic of Düsseldorf, 1977-1981, n = 42)

Spontaneous Infratentorial Hematomas: CT and Angiographic Findings-Prognosis

G. Scharek and H. Vogelsang

Abteilung für Neuroradiologie, Zentrum Radiologie der Medizinischen Hochschule Hannover, Karl-Wiechert-Allee 9, D-3000 Hannover 61

Today CT is able to demonstrate intracranial hemorrhage quickly and with certainty. This also applies to the posterior cranial fossa. However, angiography is required to clarify the etiology (13). Prognostic inferences from CT are only possible to a limited extent.

Investigation Material

Among 324 spontaneous intracranial hemorrhages, 28 (8.6%) were localized infratentorially. The age of the patients ranged from neonates to 81 years old (average 43.7 years). Thirteen patients were female, and 15 male. In about 22 patients who had been admitted as in-patients to the University Medical School in Hanover (MHH), the patient records gave information on the further course. The period of observation, including possible out-patient follow-up care, averaged six months, but was in some cases up to 4 years. In six patients who had returned to the referring hospital after CT, we were not informed of their further course.

Results

The site of the hemorrhage in the posterior cranial fossa was in the cerebellum itself in 11 cases, in the brain stem in nine cases, with a combination in five cases. The subarachnoid space alone was filled with blood in three cases. Sixteen of the 28 patients had a rupture into the ventricle in addition, and 14 patients had developed an internal hydrocephalus (Table 1).

Table 1. Location of hemorrhage in the posterior fossa

	N	Extension into ventricles	Hydrocephalus
Cerebellum	11	7	9
Brain stem	9	3	3
Combination	5	4	2
Subarachnoid space	3	2	
Total	28	16	14

Advances in Neurosurgery, Vol. 11
Edited by H.-P. Jensen, M. Brock, and M. Klinger

The maximum diameter of the hematoma was between 0.5 and 7.0 cm. Those in the region of the cerebellar vermis tended to invade the ventricular system. The same applied with increasing size of the hematoma.

Angiographic display of the vertebal-basilar arterial circulation was regarded as clinically indicated in only 11 patients. A cerebrovascular malformation (an aneurysm in one case and an arteriovenous angioma in another case) could be demonstrated as the cause of the hemorrhage. There was a marked vascular lesion in one case. There were signs of infratentorial space-occupation in five cases and normal findings in three cases.

Clinically, hypertension (eight cases) and disturbance of coagulation (three cases) were the most frequent causes amongst the 22 cases admitted as in-patients to the MHH. However, the etiology of the hemorrhage remained unclear in eight patients.

For treatment, a surgical intervention with clearance of the hematoma was performed three times, and a ventricular drainage system applied five times because of developing hydrocephalus. Otherwise, treatment was conservative.

Prognostically the isolated brain stem hemorrhages and combined brain stem/cerebellar hemorrhages were the most unfavorable (Table 2). In all, almost 50% of the patients followed up died, the survival time amounting to a few hours up to three weeks.

Discussion

Fresh intracranial hematoma can be demonstrated and localized with certainty by computerized tomography (3, 17, 20). This also applies to hemorrhages in the posterior cranial fossa (16), whereas in this region the evaluation of other lesions is limited (12). Nevertheless, the delimitation of a hemorrhage in the deep basal tomography can be impeded by artefact formation when registration is not optimal. The density uptake of a hematoma focus up to hypodensity does not correspond in every case to complete resorption of the coagulated blood with a cystic residue, but may also be due to a residual hematoma (15). This could also be confirmed in one of our own cases (Fig. 3). Otherwise, the infratentorial resorption processes proceed in time and in a qualitatively similar way to that of the supratentorial hematomas (19) as can be seen from Fig. 1 and 2. In contrast to the supratentorial hemorrhages, there is frequently a secondary rise in pressure resulting from increasing CSF obstruction and the development of hydrocephalus (10) in the special topographical conditions in infratentorial hemorrhages. This was the case in 63% of the patients who could be checked in our own material. Here, the computer tomogram is a decisive

Table 2. Hemorrhage in the posterior fossa; location - treatment - prognosis

Location	N	Treatment			Neurological deficit		Death
		Shunt	Operative	Conservative	Minor	Major	
Cerebellum	10	5	3	3	4	3	3
Brain stem	5	1		4	1	1	3
Combination	4			4		1	3
Total	19	6	3	11	5	5	9

aid in progress control, especially when considering measures for CSF drainage.

Even if it is indispensable to demonstrate the cause of a hemorrhage, the results of vertebral/basilar angiography when performed was disappointing. A cerebrovascular malformation could be diagnosed in only two cases. As known for the intracerebral hemorrhages (8), the most frequent cause of spontaneous infratentorial hematomas is likewise hypertension, followed by disorders of coagulation. The extent to which microangiomas play a role, especially in childhood (14) must be left open, since the patient material investigated was too small and the angiographies were also frequently carried out in emergency situations and without later controls.

In the parenchymatous hemorrhages, as expected the mortality rises with increasing size of hematoma. The site of the hemorrhage plays an important role in the prognosis. Isolated brain stem hemorrhage (60%) and combined brain stem-cerebellar hemorrhages (75%) generally had a poor prognosis, but they could be survived. A 15-year-old boy survived a brain stem hematoma 2.5 cm in size, with rupture into the ventricle and CSF obstruction without appreciable residues after a pressure-relieving operation. Favorable progress courses with and without operation has been reported in many cases (1, 2, 4, 22), so that even brain stem hemorrhages are not generally to be rated as hopeless from the start. The same also applies to extensive cerebellar hemorrhages (18, 21) (Fig. 1), in which even cases treated with purely conservative measures have a chance of survival (6, 9). In neonates and children with and without operation, especially favorable outcomes have been reported (5, 7, 11, 14). There is the case of a neonate with primary coma and subsequent respiratory arrest which initially appeared to be hopeless (Fig. 2). This patient survived with an astonishing remission under conservative intensive medical treatment alone.

To summarize, it can be stated that around 9% of spontaneous intracranial mass hematomas are located infratentorially. Angiography clarifies the etiology only in a very small percentage of cases and the mortality is almost 50%. However, even massive hemorrhages into the cerebellum, hemorrhages into the brain stem with and without rupture into the ventricle are survived and can heal up with relatively slight neurological deficits.

References

1. Brismar, J., Hindfelt, B., Nilsson, O.: Benign brain stem hematoma. Acta. Neurol. Scand. 60, 178-182 (1979)
2. Cioffi, F.A., Tomasello, F., D'Avanzo, R.: Pontine hematomas. Surg. Neurol. 16, 13-16 (1981)
3. Davis, D.O., Pressman, B.D.: Computerized tomography of the brain. Radiol. Clin. N. Amer. 12, 297-313 (1974)
4. Dhopesh, V.P., Greenberg, J.O., Cohen, M.M.: Computed tomography in brain stem hemorrhage. J. Comput. Assist. Tomogr. 4, 603-607 (1980)
5. Erenberg, G.: Cerebellar hemorrhage in children (letter). Neurology (Minneap) 28, 308 (1978)
6. Feijo de Freixo, M., Garcia, M.J., Alcelay, L.G.: Cerebellar hemorrhage: nonsurgical forms. Ann. Neurol. 6, 84 (1979)

7. Fishman, M.A., Percy, A.K., Cheek, W.R., Speer, M.E.: Successful conservative management of cerebellar hematomas in term neonates. J. Pediatr. 98, 466-468 (1981)

8. Gänshirt, H., Keuler, R.: Intrazerebrale Blutungen. Ursachen, Klinik, Verlauf, Spätprognose. Nervenarzt 51, 201-206 (1980)

9. Heimann, T.D., Satya-Murti, S.: Benign cerebellar hemorrhages. Ann. Neurol. 3, 366-368 (1978)

10. Janny, P., Colnet, G., Georget, A.-M., Chazal, J.: Intracranial pressure with intracerebral hemorrhages. Surg. Neurol. 10, 371-375 (1978)

11. Kazimiroff, P.B., Weichsel, M.E., Grinnell, V., Young, R.F.: Acute cerebellar hemorrhage in childhood: etiology, diagnosis, and treatment. Neurosurg. 6, 524-528 (1980)

12. Kazner, E., Grumme, Th., Lanksch, W., Wilske, J.: Limitations of computerized tomography in the detection of posterior fossa lesions. In: Advances in Neurosurgery, Vol. 5. Frowein, R.A., Wilcke, O., Karimi-Nejad, A., Brock, M., Klinger, M. (eds.), pp. 160-170. Berlin, Heidelberg, New York: Springer 1978

13. Kretzschmar, K., Grumme, Th., Kazner, E.: Möglichkeiten und Grenzen der Computer-Tomographie und Angiographie in der Diagnostik zerebraler Gefäßmißbildungen. Radiologe 20, 109-112 (1980)

14. Mauersberger, W., Fuchs, E.C., Ebhardt, G.: Spontane intrazerebelläre Hämatome im Kindesalter. Acta. Neurochir. 36, 255-264 (1977)

15. Messina, A.V., Chernik, N.L.: Computed tomography: the resolving intracerebral hemorrhage. Radiology 118, 609-613 (1975)

16. Müller, H.R., Wüthrich, R., Wiggli, U., Hünig, R., Elke, M.: The contribution of computerized axial tomography to the diagnosis of cerebellar and pontine hematomas. Stroke 6, 467-475 (1975)

17. Nadjmi, M., Piepgras, U., Vogelsang, H.: In: Kranielle Computertomographie. Neuroradiologische Atlanten. Nadjmi, M., Piepgras, U., Vogelsang, H. (eds.), 66 pp. Stuttgart, New York: Thieme 1981

18. Scharfetter, F.: Das intrazerebrale Hämatom - Die Indikation zur Operation. Ther. Umsch. 38, 762-766 (1981)

19. Schumacher, M., Rossberg, Ch., Stoeter, P.: CT-Differentialdiagnose und Verlaufsuntersuchungen bei intrazerebralen Blutungen. Arch. Psychiatr. Nervenkr. 231, 171-185 (1982)

20. Scott, W.R., New, P.F.J., Davis, K.R., Schnur, J.A.: Computerized axial tomography of intracerebral and intraventricular hemorrhage. Radiology 112, 73-80 (1974)

21. Yoshida, S., Sasaki, M., Oka, H., Gotoh, O., Yamada, R., Sano, K.: Acute hypertensive cerebellar hemorrhage with signs of lower brain stem compression. Surg. Neurol. 10, 79-83 (1978)

22. Zuccarello, M., Iavicoli, R., Pardatscher, K., Scanarini, M., Fiore, D., Andrioli, G.C.: Primary brain stem hematomas. Diagnosis and treatment. Acta. Neurochir. (Wien) 54, 45-52 (1980)

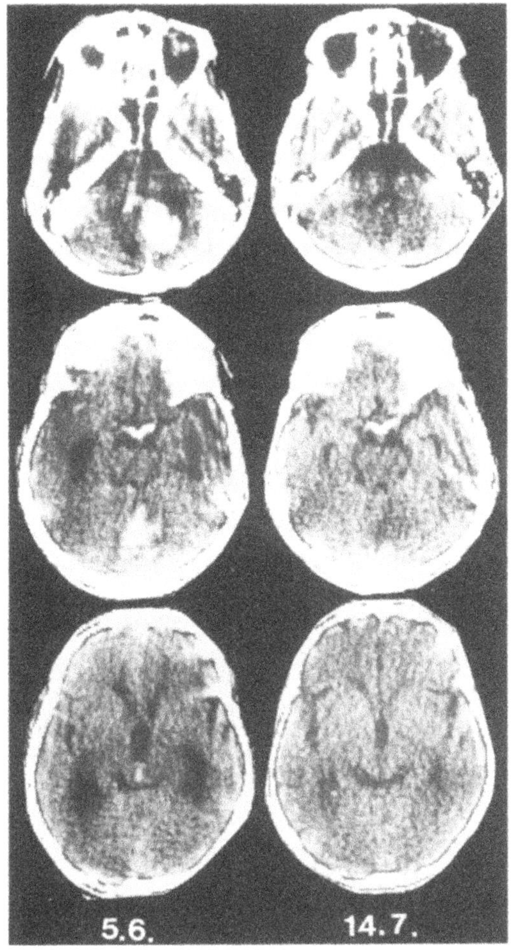

Fig. 1. Male, 62 years. Spontaneous cerebellar hemorrhage, right side with penetration of the fourth ventricle and aqueduct, hydrocephalus. Treated conservatively. Control CT six weeks later: normal findings. No neurological deficit

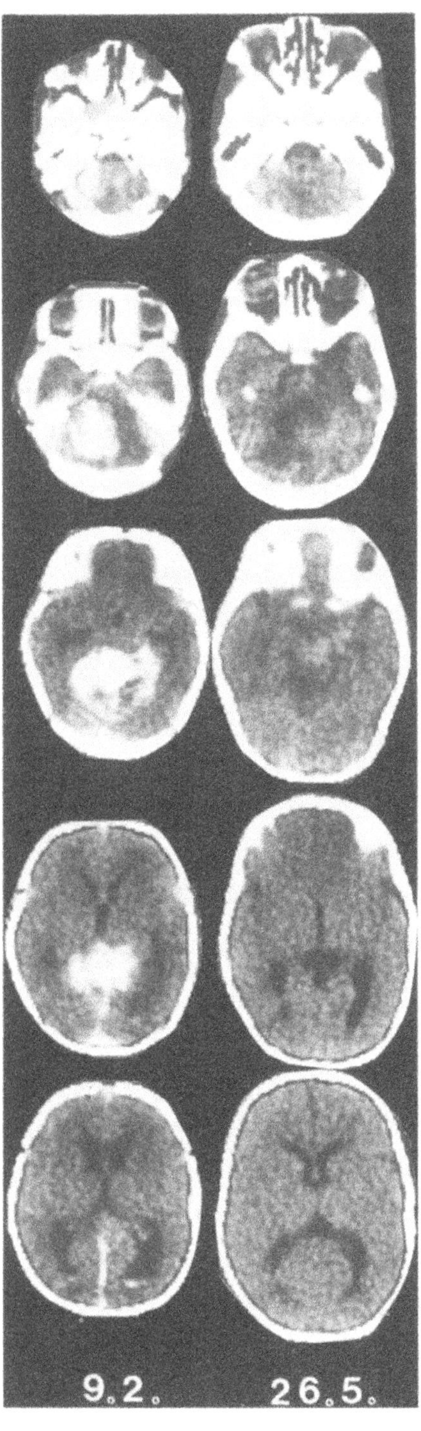

Fig. 2. Newborn, female. Massive hemorrhage in the posterior fossa, coma and respiratory arrest. After conservative treatment 3.5 months later in the control CT only a minor defect in the region of hemorrhage. No neurological deficit, but retardation of development

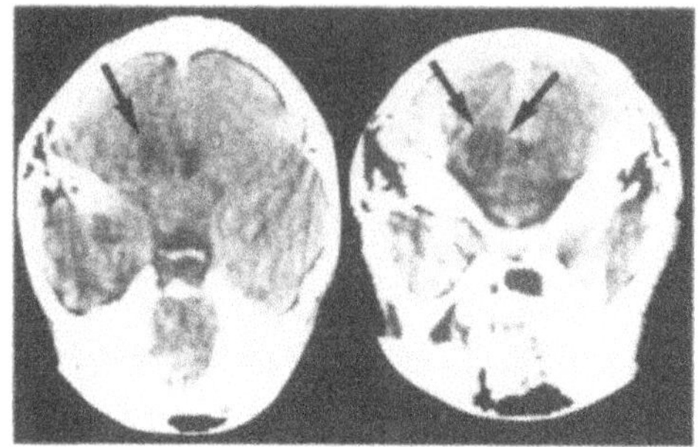

Fig. 3. Female, 42 years. Two months ago acute disease with cerebellar signs. CT shows a cystic lesion at the right site of cerebellum, near fourth ventricle. An old hemorrhage was seen at operation. Very good rehabilitation in the last 5 years

Acute Intracranial Hemorrhage Following Thrombosis of Cerebral Veins and Dural Sinuses

S. Tiyaworabun[1], H. Grau[1], W. I. Steudel[1], and H. Gräfin Vitzthum[2]

[1] Abteilung für Allgemeine Neurochirurgie, Zentrum der Neurologie und Neurochirurgie, Klinikum der Johann Wolfgang Goethe-Universität, Schleusenweg 2–16, D-6000 Frankfurt 71

[2] Neurologisches Institut (Edinger-Institut) der Johann Wolfgang Goethe-Universität, Schleusenweg 2–16, D-6000 Frankfurt 71

Knowledge of the sequelae of thrombosis of dural sinuses was first reported by RIBES in 1825. Since then, despite many investigations on the entity, it still remains a challenging diagnostic and therapeutic problem with a high mortality rate, 40 - 81%, especially in cases associated with intracranial hemorrhage (5, 6, 8, 11, 13, 15, 19). Since the introduction of cerebral computerized tomography (CT), many of these cerebral venous sinus thromboses (CVST) and their sequelae can be promptly suggested or diagnosed during life, when they are taken in conjunction with the clinical picture and supplemented with cerebral angiography (1, 3, 5, 10, 14). In spite of the advances in the medical and surgical treatment of vascular thrombosis, their effect on the CVST remains controversial, limited and questionable so that there is a need for further investigations (2, 6-9, 11-13, 15, 19, 22, 23).

During the past six years, five cases of CVST associated with acute intracranial hemorrhages (ICH), have been seen and treated in the departments of neuroradiology and of neurosurgery, at the University hospital Frankfurt. The purpose of this communication is to analyse the clinical course of our cases, comparing them with those in the literature, and discussing the recent problems and pitfalls in the diagnosis, and treatment, as well as the autopsy findings. Two example cases are briefly described below.

Case 1: A 46-year-old male was admitted after ten hours of severe headache and vomiting. On admission he was in coma, had a dilated pupil on the right, loss of corneal reflexes, left hemiparesis and decerebrate rigidity. B.P. 130+180 mm Hg, pulse 80/min. The findings of the coagulation studies were unremarkable. CT revealed multiple hemorrhages, right parieto-temporo-occipital with compression of the right lateral ventricle and a mass effect. Cerebral angiography showed an avascular right temporo-parietal mass with stasis of the venous outflow. Craniotomy was performed and bloot clot removed. In addition, a decompressive right temporal lobectomy was carried out. He was then treated with dexamethasone, frusemide and anticoagulants. His clinical course deteriorated and he died ten days later. Autopsy confirmed thrombosis of the right parietal cortical veins with massive multiple confluent hemorrhagic ischemic changes right parieto-occipital, as well as hemorrhagic infarction in the left occipital region.

Case 2: This 43-year-old male had suffered from deep vein thrombosis in the left leg for ten months. He had a herniotomy performed ten days ago and was discharged one week later in good general condition. He was

Advances in Neurosurgery, Vol. 11
Edited by H.-P. Jensen, M. Brock, and M. Klinger

admitted to us after he had developed disorientation, retrograde amnesia, angularis syndrome, dilated left pupil and hemiparesis. The coagulation studies were unremarkable. CT revealed a right parietal intracerebral hemorrhage. Angiography showed an avascular mass with stasis of the venous drainage and probably thrombosis of the cortical veins. The hematoma was surgically evacuated. Thrombosed cortical veins were observed at operation. Histology confirmed congested venous thrombosis in the hematoma. He was then discharged with improvement in his neurological status. He received Colfarit 3 x 1 gm daily. Three weeks after his discharge, he complained for one day of a sudden right chest pain which subsided spontaneously. Two days later, deep venous thrombosis was observed in the right leg. He was treated with streptokinase 100,000 U/hourly by continuous i.v. perfusion for two days and then with heparin sodium 1000 U/hourly by i.v. perfusion for one more day. This regimen had to be reduced and discontinued as the patient became gradually somnolent with meningismus. Cerebral angiography revealed thrombosis of the superior sagittal sinus and cortical veins. CT also confirmed the findings of thrombosis and left parietal ICH with intraventricular hemorrhage. He was then treated with dexamethasone, spironolactone, calcium heparin and anticonvulsants. CT examination ten days later revealed hemorrhagic infarction in the right parietal region. He was discharged one month later with improvement in his general condition and in the CT findings. He died two months later in a provincial hospital after a further ICH. Autopsy was refused (Fig. 1).

Discussion

The true incidence of CVST is difficult to determine for most of the descriptions are of sporadic cases. REDDY (1968) and TOWBIN (1973) found in 1.6% and 9% of 1396 and 189 autopsies respectively (24). Mc CORMICK (1973) and JELLINGER (1977) found it in 2% and 3% of autopsies with ICH. We do believe that the incidence is more common than the published reports, since it may closely simulate other diseases in both its clinical and radiological features; spontaneous recovery after recanalization or establishment of sufficient collateral anastomosis; relative low incidence of associated massive ICH; or it may be overlooked at autopsy unless specifically looked for.

The known etiologies, which may be primary, secondary or associated with CVST, are: septic thrombosis; aseptic thrombosis after head trauma, hemodynamic alterations and hemodycrasia; and idiopathic.

Despite the many etiologies known, the cause sometimes remains obscure. Three cases in our series had a fulminant onset passing rapidly into coma and death, so that a full evaluation was impossible and the cause remained unknown. One case was due to post partum coagulopathy. The second example was after a herniotomy and was associated with thrombosis of other part of the body.

The clinical manifestations of this very grave disorder, that affects all ages, are variable. It is often very difficult to recognize on purely clinical grounds and can be misinterpreted as due to some other cause, especially when it is associated with other serious diseases. It usually runs a progressive, intermittent relapsing or gradual course, but can also be acute and fulminant in cases with excessive raised ICP or massive hemorrhage, as were observed in our patients (6, 8, 9, 12-14, 17, 19). The factors, which affect the broad spectrum of clinical presentations, are the rapidity, extent and location of occlusion, the subsequent sequelae as well as the nature of the underlying disease process.

Before the introduction of cerebral angiography, this disorder was only detected at autopsy; ante mortem diagnosis was impossible owing to the lack of clear cut clinical symptoms and signs. However, it is sometimes very difficult to make a differential diagnosis from other pathological processes by cerebral angiography, as hemodynamic changes, tumor, edema, or anomaly, can even be ignored or misinterpreted if one is not alert for the entity, or due to inadequacy of the angiographic technique. Much of the angiographic diagnosis is retrospective after CT or autopsy (10, 14, 17, 21). CT in the diagnosis of CVST was first reported by WENDLING in 1978 (26). Although it provides a rapid, accurate and non-invasive technique for the detection of the disorder, as do those of the static and dynamic radionuclide scanning (1, 3), it should be stressed that, in the early stages or milder form of CVST, it may remain normal or almost normal. Without doubt, in such instances angiography may be the key to early and precise diagnosis (3, 10, 14, 17, 20, 26).

In spite of advances in the management of venous diseases (4), the treatment of CVST remains controversial. This may be due to its low incidence, delay or misinterpretation and the difficulties of comparing the observation of cases reported in the literature, which may differ in the underlying clinical condition as well as the methods of medical and surgical treatment (2, 6, 7, 9, 11-13, 15, 22, 23). The proposed treatment schedule of CVST is shown in Table 1. Owing to the conflicting points of view the clinical use of anticoagulants, fibrinolytic activators and acetylsalicylic acid has not yet been universally accepted. However, we do believe that the correlation between time factors, the stage of the lesion and the use of the appropriated medication, with monitoring by CT and coagulation studies, may play a role in the success of treatment (6, 8, 9, 13-15, 25). Various methods in the surgical treatment of CVS injuries after head trauma have been tried and reported (12, 16, 22, 23). However, the actual role of surgical treatment of CVST is difficult to assess (2, 7, 11, 13, 16, 22, 23). It is generally opposed by the surgeons in cases without ICH, hydrocephalus, abscess or empyema. In the acute, preorganization phase with insufficient collateral circulation, it seems of benefit to attempt thrombectomy, autogenous venous grafting or venous sinus anastomosis.

KRÜCKE (1971) described two forms of ICH after CVST; the typical form of multiple small diapedesis bleedings in the grey and white matter; the atypical form of confluent massive hemorrhages in the white more than grey matter simulating those seen in massive spontaneous intracerebral rhexis hemorrhages (18). Furthermore, he divided the pathological changes of CVST associated with ICH into three stages: stage 1 as hemorrhages and necrosis, stage 2 as resorption and stage 3 as scar or cyst formation. This observation is extraordinarily important, especially in cases with a fulminant course and death within five days of the onset, as in two patients of our series; the microscopic findings of the fresh thrombus cannot help in deciding its nature or the relation between the hemorrhages and the thrombus, whether the latter is the primary or secondary, but the clinico-pathological picture of the brain lesions may suggest the primary cause.

The prognosis of CVST is very variable. It depends on the basic clinical lesion, the location and extent of thrombosis, the age of the patients, the rapidity of diagnosis and institution of adequate treatment, the establishment of sufficient collateral circulation or rate of recanalization, the subsequent sequelae and the damage to the brain tissues and lastly the prevention of a recurrence of the thrombosis.

Table 1. The proposed treatment schedule of CVST

Clinical symptoms and signs Computerized tomography Cerebral angiography ↓ Cerebral Venous sinus Thrombosis	
Basic diseases	Eradication if possible
Sequelae	Brain edema and increased ICP infarction or hemorrhagic infarction infection epileptic seizure dehydration cerebral and systemic circulatory disturbances respiratory disturbance hydrocephalus
Thrombosis	(CT and coagulation study)
Acute (pre-organization) sufficient collateral circulation, without hemorrhagic infarct	fibrinolytic activators anticoagulants
insufficient collateral circulation	thrombectomy autogenous venous grafting veno-sinus anastomosis irradiation
Subacute or chronic (organization and recanalization)	anticoagulants acetylsalicylic acid (Aspirin) surgical treatment

In conclusion, it could be said that, owing to the low incidence of the condition, no comparative studies in the efficacy of various types of treatment exist. The striking incidence of spontaneous recovery make it additionally difficult to evaluate. Further experimental and clinical investigations are required.

References

1. Barnes, B.D., Winestock, D.P.: Dynamic radionuclide scanning in the diagnosis of thrombosis of the superior sagittal sinus. Neurology. 27, 656-661 (1977)
2. Bonnal, J., Buduba, C.: Surgery of the central third of the superior sagittal sinus. Experimental study. Acta. Neurochirurg. 30, 207-215 (1974)
3. Buonanno, F.S., Moody, D.M., Ball, M.R., Laster, D.W.: Computed cranial tomography findings in cerebral sinovenous occlusion. J. Comput. Assist. Tomogr. 2, 281-290 (1978)
4. Breddin, K.: Symposium on "Thrombolytische Therapie venöser Thrombosen". 5.5.1982. Frankfurt am Main

5. D'Avella, D., Greenberg, R.P., Mingrino, S., Scanarini, M., Pardatscher, K.: Alterations in ventricular size and intracranial pressure caused by sagittal sinus pathology in man. J. Neurosurg. 53, 656-661 (1980)

6. Di Rocoo, C., Iannelli, A., Leone, G., Moschini, M., Valori, V.M.: Heparin-Urokinase treatment in aseptic dural sinus thrombosis. Arch. Neurol. 38, 431-435 (1981)

7. Donaghy, R.M.P., Wallman, L.J., Flanagan, M.J., Numoto, N.: Sagittal sinus repair. Technical note. J. Neurosurg. 38, 244-248 (1973)

8. Gettelfinger, D.M., Kokmen, E.: Superior sagittal sinus thrombosis. Arch. Neurol. 34, 2-6 (1977)

9. Hacker, H., May, B.: Zur Behandlung der Hirnvenenthrombose. Der Nervenarzt. 40, 440-443 (1969)

10. Hacker, H.: Das Röntgenbild der Großhirnvenen. Encyclopedia of medical radiology. Vol. XIV. Part. 1B. Diethelm, L., Wende, S. (eds.). Berlin, Heidelberg, New York: Springer 1981

11. Heinz, E.R., Geeter, D., Gabrielsen, T.O.: Cortical vein thrombosis in the dog, with a review of aseptic intracranial venous thrombosis in man. Acta. Radiol. 13, 105-114 (1972)

12. Hensell, V.: Traumatische Hirnsinus- und Venenthrombosen. Acta. Neurochir. Supplement 7: 363-366 (1961)

13. Huhn, A.: Die Thrombosen der intracranial Venen und Sinus. Klinische und pathologisch-anatomische Untersuchungen. Stuttgart: Schattauer 1965

14. Ingunza, I., Hacker, H.: Die cerebral Venen- und Sinusthrombose im neurochirurgischen Krankengut. Der Radiologe. 12, 48-52 (1972)

15. Kalbag, R.M., Woolf, A.L.: Cerebral venous thrombosis. London, Oxford: University Press 1967

16. Kapp, J.P., Gielchinsky, I., Petty, C., McClure, C.: An internal shunt for use in the reconstruction of dural venous sinuses. Technical note. J. Neurosurg. 35, 351-354 (1971)

17. Krayenbühl, H., Yasargil, M.B.: Cerebral angiography. Philadelphia: Lippinott & Co. 1968

18. Krücke, W.: Pathologie der cerebral Venen- und Sinusthrombosen. Der Radiologe. 11, 370-377 (1971)

19. Noetzel, H., Jerusalem, F.: Die Hirnvenen und Sinusthrombosen. Monographie aus dem Gesamtgebiet der Neurologie und Psychiatrie, H. 106. Berlin, Heidelberg, New York: Springer 1965

20. Norman, D., Price, D., Boyd, D., Fishman, R., Newton, T.H.: Quantitative aspects of computed tomography of the blood and cerebrospinal fluid. Radiology. 123, 335-338 (1977)

21. Patronas, N.J., Duda, E.E., Mirfakhraee, M., Wollmann, R.L.: Superior sagittal sinus thrombosis diagnosed by computed tomography. Surg. Neurol. 15, 11-14 (1981)

22. Rish, B.L.: The repair of dural venous sinus wounds by autogenous venorrhaphy. J. Neurosurg. 35, 392-395 (1971)

23. Sindau, M., Mazoyer, J.F., Fischer, G., Pialat, J., Fourcade, C.: Experimental bypass for sagittal sinus repair. Preliminary report. J. Neurosurg. 44, 325-330 (1976)

24. Towbin, A.: The syndrome of latent cerebral venous thrombosis. Its frequency and relation to age and congestive heart failure. Stroke. 4, 419-430 (1973)

25. Vines, F.S., Davis, D.O.: Clinical radiological correlation in cerebral venous occlusive disease. Radiology. 98, 9-21 (1971)

26. Wendling, L.R.: Intracranial venous sinus thrombosis: Diagnosis suggested by computed tomography. Am. J. Roentgenol. 130, 978-980 (1978)

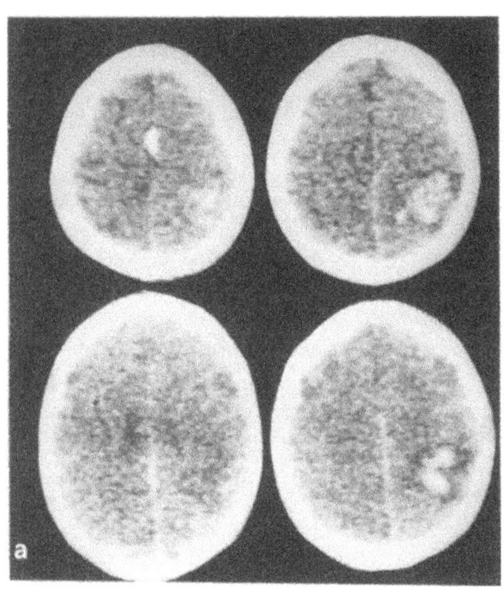

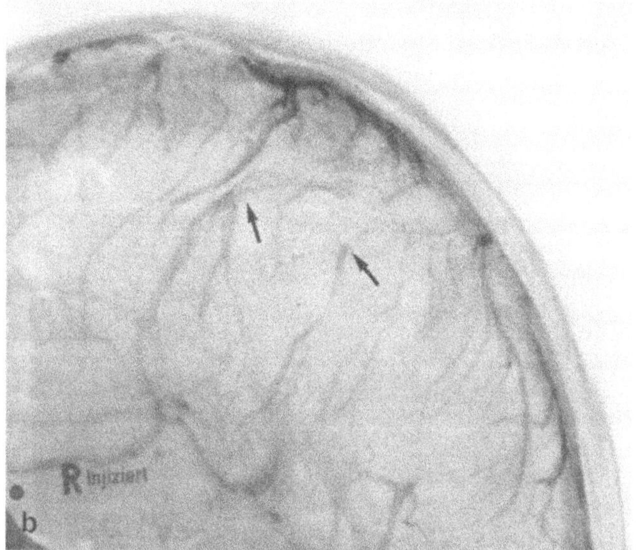

Legend see p. 130

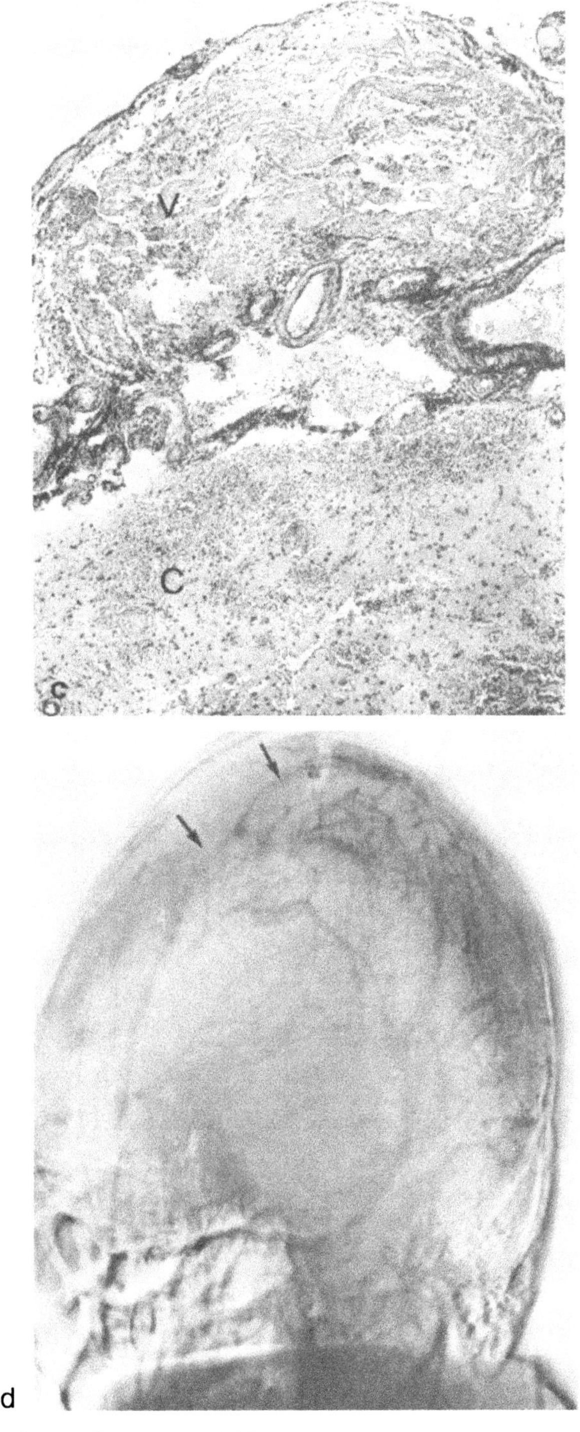

Legend see p. 130

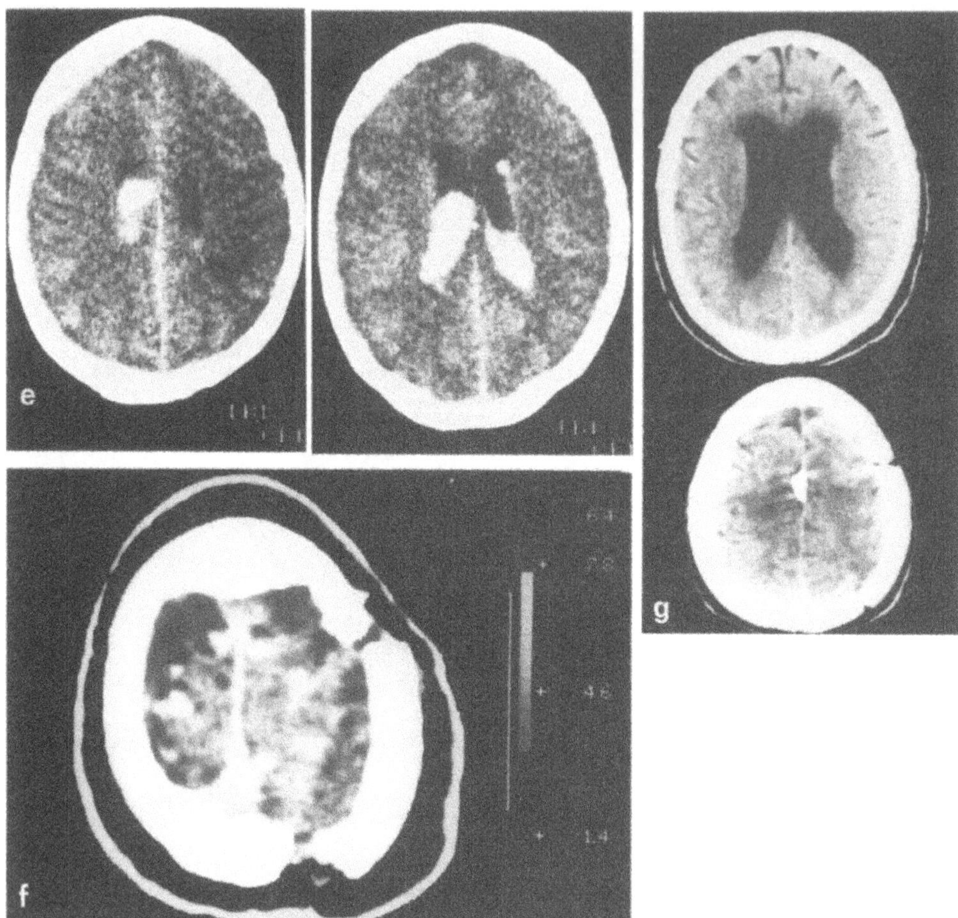

Fig. 1 a-g. Case 2. a CT (two days before operation): Confluent intracerebral hemorrhage in the right parietal lobe. b Angiography: Avascular mass with stasis of the venous drainage and occlusion of the cortical veins. c Histopathology of biopsy: dilated leptomeningeal vein with fibroblasts, which are organizing the lamellated fibrin clot of the occluding thrombus (V). Underlying hemorrhagic cortex with confluent small erythrocytic diapedeses (C). d Angiography at the second admission (six weeks after operation (p.o.)): thrombosis of superior sagittal sinus and cortical veins. e CCT (six weeks p.o.): besides thrombosis, additional parietal ICH on the left with ventricular bleeding. f CT (eight weeks p.o.): hemorrhagic infarction in the right parietal lobe with VST. g CT (12 weeks p.o.): Slight dilatation of ventricular system and biparietal hypodense area

Antifibrinolytic Therapy of Intracranial Hemorrhage with Tranexamic Acid

V. Hossmann, H. Auel, H. Bewermeyer, and H.-D. Heiss

Abteilung Innere Medizin II der Universität Köln und Max-Planck-Institut für Hirnforschung, Ostmerheimer-Strasse 200, D-5000 Köln 91

Introduction

Controlled double-blind studies have shown, that the antifibrinolytic agent tranexamic acid significantly reduces both the mortality and the rate of rebleeding from a fresh ruptured subarachnoid aneurysm (1-3). KASTE and RAMSAY (4) did not observe this favourable result in a Finnish study. Rebleeding of a subarachnoid aneurysm occurs most often during the initial two weeks, i.e. at a time, where mortality is highest. Therefore, antifibrinolytic therapy is often carried out as a pre-operative procedure in order to overcome this critical period. While fibrinolytic activation is rarely observed in the blood after subarachnoid hemorrhage, it is generally seen in the cerebrospinal fluid. Here, raised concentration of fibrin degradation products indicate enhanced fibrinolytic activity, which may be responsible for a dissolution of the peri-aneurysmal fibrin plug and subsequent rebleeding. Tranexamic acid crosses the blood-brain barrier and inhibits the activation of plasminogen to plasmin.

Continuous administration of tranexamic acid subsequently reduces the concentration of fibrin degradation products in the cerebrospinal fluid (5, 6). The safety and efficacy of this treatment clearly depends on careful controls, in order to obtain effective concentrations of tranexamic acid in the blood and the CSF. Thereby, ineffectiveness or complications of a critically reduced fibrinolysis with systemic thrombosis, disseminated intravascular coagulation or hydrocephalus may be prevented.

The aim of this study, therefore, was to examine whether simple laboratory methods allow an effective survey of the treatment with an antifibrinolytic agent in patients with acute intracranial hemorrhage.

Patients and Methods

Twenty patients (aged between 20 and 81 years, mean age 48 years) with an acute intracranial hemorrhage, ascertained by a detailed general and neurological examination, lumbar puncture and cerebral angiography, were included in the study. In 11 patients subarachnoid hemorrhage was diagnosed (eight hemorrhages from saccular aneurysms of the large cerebral arteries at the cerebral basis, seven hemorrhages from sources that were not demonstrable by cerebral angiogram), in one patient a typical intracranial mass bleeding in the territory of the basal ganglia, and in another patient a cerebellar hemorrhage were demon-

Advances in Neurosurgery, Vol. 11
Edited by H.-P. Jensen, M. Brock, and M. Klinger

strable. According to the classification of HUNT et al. (7), seven patients were in stage I - II, eight patients in stage III, and three patients in stage IV - V. Tranexamic acid was continuously infused by a peristaltic pump via an intravenous line, for up to 15 days (mean 6.6 ± 4.7 days) at a dose of 5 g daily.

Before, at the end, and daily during therapy a battery of coagulation and hematological tests were performed: Recalcification time, PPT, TT, fibrinogen, factor V, X, and XII were measured with a coagulometer of SCHNITGER and GROSS (Amelung, Lemgo-Brake) using commercially available test kits (Boehringer, Mannheim, Behringwerke, Marburg); antithrombin III, α_2 antiplasmin and factor XIIf by chromogenic substrate assays (Boehringer, Mannheim, Kabi, Stockholm); fibrin degradation products (FDP) immunologically with the TRCHIA (Wellcome, Beckenham, England); erythrocyte, leucocyte and thrombocyte count with the Cellcounter (Analys Instrument, Stockholm); hematocrit with a microcentrifuge (Heräus Christ, Osterode) and hemoglobin photometrically with a standard kit. The available plasmin activity was measured photometrically by use of the chromogenic substrate S-2251. In addition a modified bioassay according to MARKWARDT (8) for measurement of antifibrinolytic activity of tranexamic acid was developed. 100 µl of 0.2% bovine fibrinogen, 50 µl streptokinase (500 U/ml) and 50 µl thrombin (50 U/ml) were added to 50 µl of citrated plasma. The duration until complete lysis of the formed fibrin plug was then measured thrombelastographically. A standard curve was obtained from the citrated plasma before therapy, where tranexamic acid was added in increasing concentrations. Lysis time was measured in the same way as described above after an incubation time of 30 min at 37° C. This curve gives a straight line in a double logarithmic scale. From each individual standard curve the biological activity of tranexamic acid in plasma may be derived and expressed as antifibrinolytic equivalent in µg/ml of tranexamic acid.

Results

Nine patients could be discharged from hospital after treatment without any complaints, three patients with mild persistent neurological symptoms, one with severe neurological deficit. Seven patients died, one as a consequence of rebleeding, one following a postoperative stroke, two because of general complications. The treatment with continuous intravenous infusion of tranexamic acid 5 g daily did not reveal any complication in patients below 65 years of age. The available plasminogen activity fell from 99.2 ± 3.1% to values of about 50-60%, whereby this level was rather consistent during treatment and increased to 83.2 ± 4.2% 24 hours after the termination of the infusion. The antifibrinolytic equivalent determined by the described bioassay corresponded to 140 - 160 µg tranexamic acid/ml plasma during treatment and showed only minor day to day variation. The optimal therapeutic concentration was, however, only obtained 48 hours after commencing the infusion. The other factors of blood coagulation, fibrinolysis and hematology did not reveal significant differences before, during, and after infusion of tranexamic acid. Two patients above 65 years of age who were treated with the same dose regimen of 5 g tranexamic acid daily revealed clear evidence of activation of the coagulation system on day 6 and day 10 with reduction of the concentration of factors XII, X, V, of the thrombocyte count and a sustained fall of the available plasminogen activity, which made the termination of treatment with tranexamic acid mandatory.

Discussion

The determination of the available plasminogen activity with the chromogenic substrate S-2251 as well as the thrombelastic measurement of the fibrinolytic equivalent of tranexamic acid are simple and fast methods, allowing an accurate control of antifibrinolytic treatment. Earlier described methods such as the determination of the "streptokinase clot lysis time" as well as a fluorimetric method (9) need several hours and in the latter case are more expensive. The results of this study indicate, that a loading dose of about 2 g of tranexamic acid over 30 min may be desirable in order to obtain therapeutic levels as early as possible. This may be especially important in patients where the acute bleeding has already happened 48 hours before treatment, since the danger of rebleeding due to lysis of the fibrin plug is highest from the third day on. In patients over 65 years of age in whom a reduction of the fibrinolytic and increase in the procoagulant activity is often observed, antifibrinolytic treatment should not be performed, or only under very careful supervision, whereby measurement of FDP in cerebrospinal fluid would further help to optimize the dosage of tranexamic acid. Whether a dose regimen of 5 g tranexamic acid daily is optimal in younger patients is questionable and has to be examined in further controlled studies.

References

1. Chandra, B.: Treatment of subarachnoid hemorrhage from ruptured intracranial aneurysm with tranexamic acid: a double-blind clinical trial. Ann. Neurol. 3, 502-504 (1978)
2. Maurice-Williams, R.S.: Prolonged antifibrinolysis: an effective non-surgical treatment for ruptured intracranial aneurysms? Brit. Med. J. 2, 945-947 (1978)
3. Fodstad, H., Liliequist, B., Schannong, M., Thulin, C.-A.: Tranexamic acid in the preoperative management of ruptured intracranial aneurysms. Surg. Neurol. 10, 9-15 (1978)
4. Kaste, M., Ramsay, M.: Tranexamic acid in subarachnoid hemorrhage. A double-blind study. Stroke 10, 519-522 (1979)
5. Tovi, D., Nilsson, I.M., Thulin, C.A.: Fibrinolytic activity of the cerebrospinal fluid after subarachnoid hemorrhage. Acta Neurol. Scand. 49, 1-9 (1973)
6. Fodstad, H., Pilbrant, A., Schannong, M., Strömberg, S.: Determination of tranexamic acid (AMCA) and fibrin/fibrinogen degradation products in cerebrospinal fluid after aneurysmal subarachnoid hemorrhage. Acta neurochir. 58, 1-13 (1981)
7. Hunt, W.E., Hess, R.M.: Surgical risk as related to time of intervention in the repair of intracranial aneurysms. J. Neurosurg. 28, 14-20 (1968)
8. Markwardt, F.: Synthetic inhibitors of fibrinolysis. In: Fibrinolytics and antifibrinolytics. Markwardt, F. (ed.), pp. 511-577. Berlin, Heidelberg, New York: Springer 1978
9. Burchiel, K.J., Schmer, G.: A method of monitoring antifibrinolytic therapy in patients with ruptured intracranial aneurysms. J. Neurosurg. 54, 12-15 (1981)

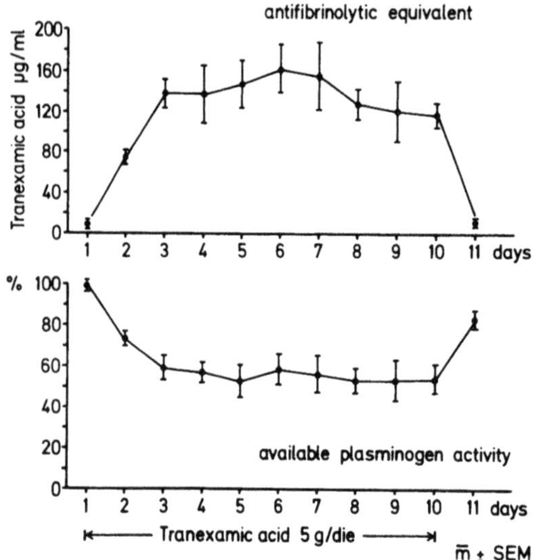

Fig. 1. Mean changes of the available plasminogen activity (APA) and antifibrinolytic equivalent in patients with acute intracranial hemorrhage before, during, and after treatment with tranexamic acid 5 g daily

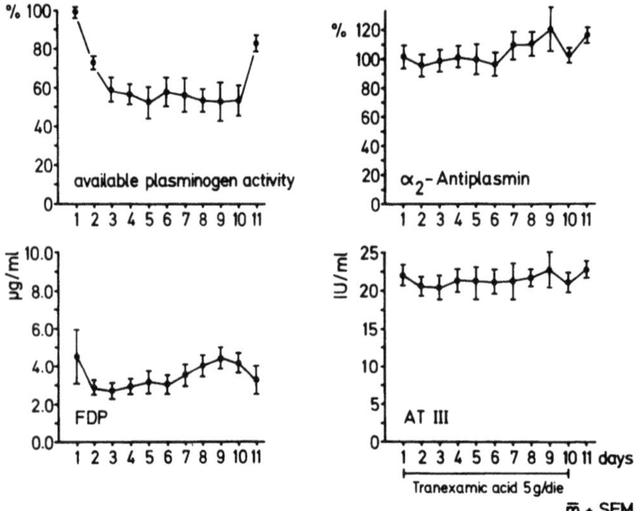

Fig. 2. Mean changes of APA, α_2-antiplasmin, fibrin degradation products (FDP) and antithrombin III (AT III) in patients with acute intracranial hemorrhage before, during, and after treatment with tranexamic acid 5 g daily

The Long-Term Prognosis of Patients with Subarachnoid Hemorrhage

R. Henkel and S. Poser

Abteilung für Neurologie, Medizinische Einrichtungen der Universität Göttingen, Robert-Koch-Strasse 40, D-3400 Göttingen

The present study reports the long-term prognosis of patients who experienced a non-traumatic subarachnoid hemorrhage (SAH). In the period from January 1970 until February 1979 139 patients with SAH were admitted to the Neurological Department of Göttingen University. Hemorrhages due to angiomas and hypertension were excluded. In addition to clinical methods a cranial computed tomography (CT) was included in the follow-up examination to assess the frequency of a permanent hydrocephalus.

Fifty of the 139 patients (Fig. 1) died in the acute phase in the hospital, 52 of the remaining 89 patients agreed to a follow-up examination. In 18 further cases it is known that the patients are still alive, in 12 of these the family physicians could give details on the present condition. Thirteen patients died during the observation period, and no information was available in six cases.

The length of the follow-up period for three groups of patients is given in Fig. 2:

1. patients who agreed to a follow-up examination,
2. patients who died in the meantime (except for three patients with unknown date of death),
3. patients who are still alive, but could not be examined.

There was no difference between the average duration of observation period of 5.4 years for the primary survivors and the patients we could re-examine.

Table 1 summarizes the long-term results of the study: 32 of the 64 patients were free of any symptoms, 28 further patients showed minimal symptoms (slight hemiparesis, disturbance of cranial nerves or of memory, depressive mood).

Three patients were partially disabled but working; in one case an extracranial ligation of the internal carotid artery was performed instead of a clipping of the aneurysm. A permanent complete hemiparesis makes the patient dependent on assistence for her household duties, though she is independent in respect of her personal care.

Table 1 shows in addition a classification of the patients as regards whether an aneurysm could be found or not. The group with unknown cause of hemorrhage includes 12 patients in whom an angiography could

Advances in Neurosurgery, Vol. 11
Edited by H.-P. Jensen, M. Brock, and M. Klinger

Table 1. Degree of recovery

	Total	SAH due to aneurysm			Unknown cause of hemorrhage
		Surgically treated	Conservatively treated	Total	
Symptom-free	32	3	4	7	25
Minimal symptoms	28	9	1	10	18
Partially disabled but working	3	2	-	2	1
Unable to work but caring for self	1	1	-	1	-
Requires nursing care	-	-	-	-	-
	64	15	5	20	44

(64 = 52 patients at follow-up examination and 12 patients known from referring physicians)

not be performed because they did not agree to it or the age or the general condition did not allow an operation.

For this reason an aneurysm was proven in only 20 patients. In 15 of these it was surgically treated, two of the remaining patients refused an operation, the other three could not be operated on for other reasons.

One of the patients who were operated on suffered a second SAH six years after the primary event. A second aneurysm was found and operated on successfully. After a total observation period of nine years he is still alive and shows minimal symptoms.

In 13 cases we got the information that in the meantime the patient had died. Five of these patients had been operated on. In seven cases the cause of death was known: late rebleeding in four cases, other causes in three cases.

Table 2 summarizes the results of the CT-examinations: minimal ventricular dilatation was seen in 11 out of 50 cases, hypodense areas in 17 cases and a definite internal hydrocephalus in four further cases.

Table 3 compares the follow-up tomograms with the scans after the acute event. In three cases the development of an internal hydrocephalus was apparent:

In one case the acute hydrocephalus had disappeared, the only pathological finding being a slight dilatation of the fourth ventricle after a period of 2 1/2 years; in one further case the acute hydrocephalus persisted for six months; in the last case a hydrocephalus developed later in the course of two years.

Table 2. Results of follow-up computed tomography

No pathological findings	17 cases
Minimal ventricular dilatation	11 cases
Internal hydrocephalus	4 cases
Hypodense areas	17 cases
Hyperdense area	1 case
	50 cases

Table 3. Comparison of the follow-up computed tomograms with those of the acute event

Stable without pathological findings	5 cases
Improvement	4 cases
Deterioration	2 cases
Permanent hydrocephalus	1 case
	12 cases

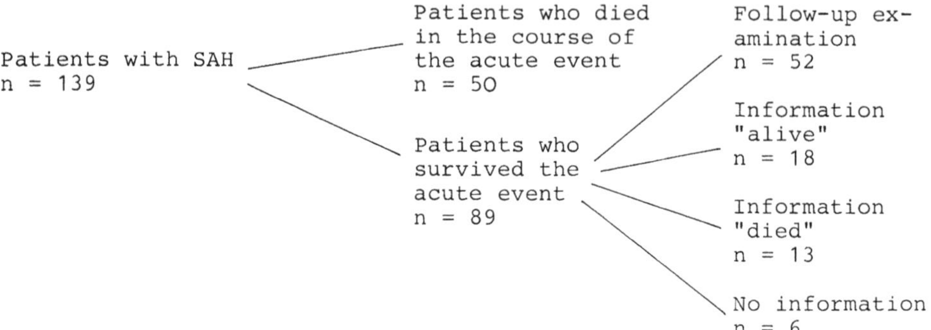

Fig. 1. Follow-up information on patients with subarachnoid hemorrhage (SAH)

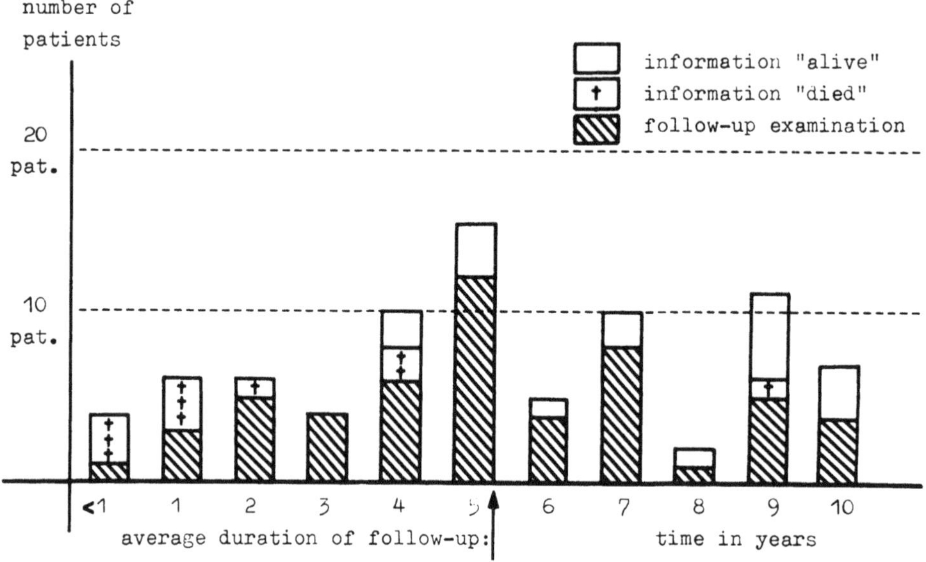

Fig. 2. Duration of follow-up for different groups of patients

Intracranial Bleedings Associated with Anticoagulant Therapy

H. Reinhardt and E. Huber

Abteilung für Neurochirurgie, Neurochirurgische Universitäts-Klinik, Kantonsspital Basel, CH-4031 Basel

Bleeding in anticoagulated patients is not a rare occurrence in Basle. In 1966 four neurosurgical cases were reported by KLINGER (5). In 1971 LEVI and STULA (6) found as many as 27 cases of cerebral and spinal bleeding. SCHAERER and co-workers (8) described complications following cerebral bleeding in 24 patients in 1976. In recent years we have observed a further increase in these potentially dangerous intracranial hematomas (1-4) which have a 66% mortality rate.

All acute neurosurgical cases in the relatively self-contained North-West region of Switzerland, which has a population of ca. 400,000 are referred to the Kantonspital Basle for CT diagnosis and, if necessary, surgical treatment. It consequently seemed appropriate to review these cases, to ascertain the cause of these too-frequent complications and - if possible - to suggest appropriate preventive measures.

Our study covers exactly 50 patients, seen during the calendar year 1980, in whom intracranial bleeding has been diagnosed either computer-tomographically (39 patients) or during autopsy (11 patients). We reviewed the history of the bleeding retrospectively and made a statistical analysis. The average age was 69.1 years: The youngest patient - a 19-year-old - was being anticoagulated on account of Cimino shunt, the oldest patient was 91. There were 29 male patients, and 21 female; 92% were over 60; 54% over 70.

We found 31 intracerebral hematomas, ventricular breakthrough occurring in 21 of these. In 14 cases bleeding was subdural, in four cases subarachnoid (two ruptured aneurysms) and in one case it was epidural. Bleeding was spontaneous in 33 patients; in 12 cases a more or less adequate trauma was present, in five patients only a slight trauma was found.

On entering the clinic, one quarter of the patients were fully conscious, half were suffering from impaired consciousness or were disoriented and a quarter were comatose. Twenty-six patients were treated in a conventional manner (mannitol, corticoids, antiepileptic prophylaxis, treatment of concomitant disorders). Nine patients underwent operation immediately. No treatment was given in 15 cases due to the unfavourable nature of the prognosis.

Only 20% of the patients were completely cured; 16% still showed symptoms of greater or lesser severity; 64% died. The recovery rate was identical in the surgical patients. Although mortality was lower (one third of the total), almost half those surviving did not make a complete recovery.

Advances in Neurosurgery, Vol. 11
Edited by H.-P. Jensen, M. Brock, and M. Klinger

Table 1

Indications for anticoagulation therapy	50 patients
Periph. arterial occlusive disease	21
Pulmonary embolism	8
Myocardial infarction	7
Deep thrombophlebitis	5
Obstr. cerebral vascular disease	2
Others	7

The reasons for employing anticoagulant therapy are set out in Table 1. There is a noticeable preponderance of peripheral arterial occlusive disease, a fact that is partially explained by the large-scale "Basle Study" involving preventive treatment of degenerative vascular disease in a group of 6400 apparently healthy professional people. Over 90% were being treated with coumarins, the remainder being in the critical transition phase between heparin and coumarins, or were only taking heparin. In two patients bleeding occurred during streptokinase treatment. Although a special watch was kept for this, we saw no case of intracranial bleeding with thrombocyte aggregation inhibitors. Duration of anticoagulant therapy varied greatly between a minimum of five days and a maximum of 20 years, the average being 3 1/2 years. In patients presenting no compelling indications, the mean value rose to 6.8 years. In those patients receiving coumarins brought in to us immediately after the bleeding incident, the prothrombin time (after Quick) lay within therapeutic values in over half the patients, but was lower in 20% and higher in 15%.

It is difficult to assess indications for anticoagulation retrospectively (7). It is definitely contraindicated in ulcer patients, in severe hypertension, in pre-existing coagulopathy and in unreliable patients. Advanced and very old age is, in itself, no contraindication, but is an important factor due to the general increase in concomitant illnesses (hypertension, diabetes etc.) and due to diminished responsibility brought about by cerebral sclerosis as well as due to augmented risk of falls. The patients were independently assessed by an angiologist, by a cardiologist and by ourselve as neurosurgeons. Whereas certain differences of opinion were apparent in the case of a few patients, the consensus was that anticoagulation treatment was only justified in about one fifth of the patients seen. In two fifths the need of therapy was doubtful, whilst treatment was not (or no longer) necessary in the remainder, or was even unjustifiable because of clear contraindications. By far the most common risk factor of over 50% was arterial hypertension, which necessitated treatment.

Cautious estimates based on sales figures for coumarin derivatives and on knowledge of their average consumption suggest that about 4% of the population in the Basle region is being anticoagulated. This high figure may be accounted for on the one hand by the rise in the percentage of old people in the population and, on the other hand, by the high doctor:patient ratio (1:400) bringing with it the danger of over-care. Our calculations suggest that one must expect a complication rate of 0.5% per year. This corresponds to one severe cerebral bleeding complication in 200 treatment years and one fatal complication every 312 treatment years. Since we also found that bleeding complications also arise with correct stabilisation (P.T. values between 15 and 25%) it must, in principle, be clear from the outset that this complication must be taken into account (3, 9, 10). Clear-cut faulty dosage schedules tend to be an exception.

Although anticoagulation complications are increasingly to be expected in predominantly older patients, it would be wrong to reject anticoagulation out of hand. After six months treatment however one should, as a matter of principle, consider whether anticoagulation is still indicated. Treatment should only be continued on the basis of strictly selected criteria. The fact that over half the patients were over 70 years of age at the time of the potentially dangerous bleeding does, however, suggest that this age should be taken as a critical limit. Cases should then be considered individually and in the light of the biological age to determine whether or not the increased risk of continuing anticoagulation therapy is justified.

Summary

The authors report on over 50 intracranial bleeding complications in anticoagulated patients in Basle, Switzerland over a 12-month period. Mortality was very high (66%). It was striking that a strict indication only remained in 20% of the patients and that the average age - 69 years - was high. It is suggested that the indication for continuation of treatment should be assessed critically after six months and that treatment should only be continued on the basis of strict criteria, especially in patients above 70 years of age.

References

1. Christiaens, J.L. et al.: Hémorragies intra-craniennes et intrarachidiennes chez des malades sous traitement anticoagulant. Lille Médical 1980 Mai 25(5): 178-82
2. Gaudel, C., Fouchard, J.: Complications du traitement anticoagulant. Sem. Hôp. Paris 1979 Sep. 8-15; 55(27-30): 1340-5
3. Gehrmann, G.: Neurologische Komplikationen bei Antikoagulantienbehandlung. Med. Welt 1971 22/Heft 9: 327-9
4. Kanoff, R.B., Ruberg, R.L.: Bilateral spontaneous intracerebral hematomas following anticoagulation therapy. JAOA 1979 Nov; 79(3): 174-8
5. Klingler, M.: Intrakranielle Blutungen bei Antikoagulantientherapie. Schweiz. Arch. Neurol. Neurochir. Psychiat. 1966 (98): 20
6. Lévy, A., Stula, D.: Neurochirurgische Aspekte bei Antikoagulantienblutungen im Zentralnervensystem. Dtsch. Med. Wschr. 1971, 96(24): 1043-8
7. Petrov, V., Bonnel, J.: Complications neurologiques des anticoagulants. Acta Neurol. Belg. Sept-Oct 75; 75(5): 205-18
8. Schärer, H.U. et al.: Schwere Blutungen unter Antikoagulation. In: Neuhaus, K., Duckert, F. (eds.), Blutgerinnung und Antikoagulation, pp. 19-23. Stuttgart, New York: Schattauer 1976
9. Silverstein, A.: Neurological complications of anticoagulation therapy. Arch. Intern. Med. 1979 Feb.; 139(2): 217-20
10. Snyder, M., Renaudin, J.: Intracranial hemorrhage associated with anticoagulation therapy. Surg. Neurol. 1977, Jan.; 7(1): 31-4

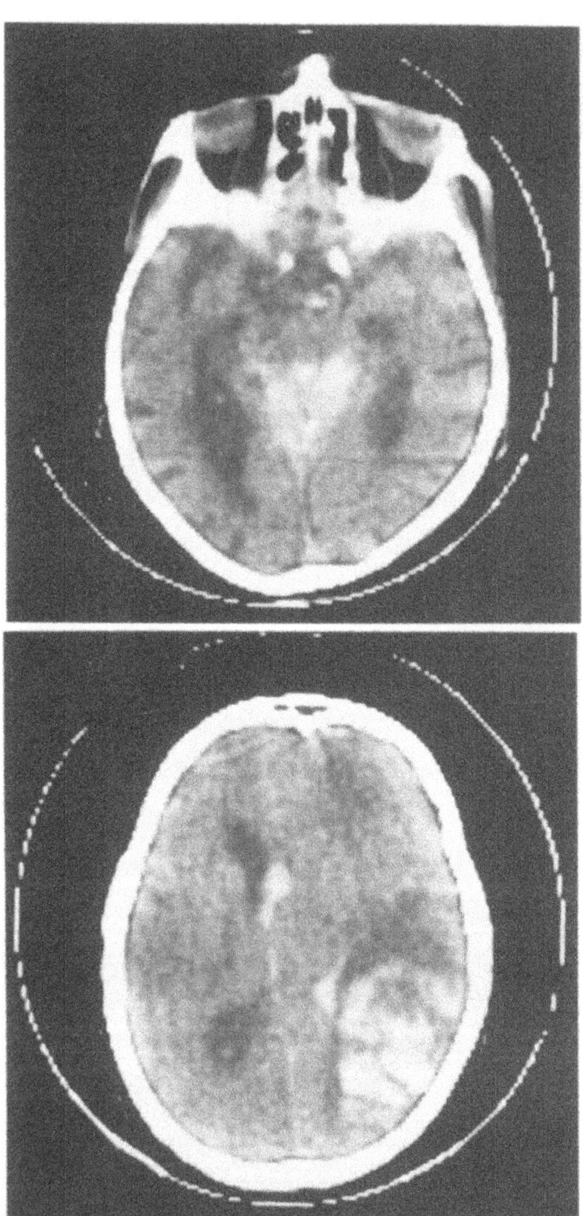

Fig. 1. Two CT-examples of intracranial anticoagulant hemorrhages: *Above*: A lethal hemorrhage in the midbrain/cerebellar region, partially cisternal (slight trauma); *below*: A "typical" intracerebral bleeding (spontaneous) in the right hemisphere with ventricular breakthrough

Long-Term ECG Investigation in Subarachnoid and Intracerebral Hemorrhage

T. Stober, T. Anstätt, S. Sen, L. Burger, G. Rettig, and K. Schimrigk

Abteilung für Neurologie und Innere Medizin III, Medizinische Fakultät, Klinikum im Landeskrankenhaus der Universität des Saarlandes, D-6650 Homburg-Saar

From sporadic communications on conventional ECG recordings it is known that ECG alterations occur in intracranial hemorrhage as exemplified by the occurrence of false positive diagnoses of acute myocardial infarction. Alterations are by no means infarct-specific, but are mainly manifested by arrhythmias, ST segment and T wave changes, as well as prolongation of the QT interval. A systematic prospective study with continuous ECG recording has not been carried out so far. An investigation on the association between the nature, location, dynamics and complications of hemorrhages on the one hand and the various ECG-findings and arrhythmias on the other hand is likewise lacking.

To clarify these questions, a prospective study in subarachnoid hemorrhage and primary intracerebral hemorrhage was commenced in spring 1981 in Homburg. So far, 50 patients have been included in the study. Of these, 11 with subarachnoid hemorrhage and ten with primary intracerebral hemorrhage have been fully evaluated. The results from these patients will be presented in this report. Besides daily neurological investigation and conventional electrocardiography, a continuous Holter ECG recording over an average of 120 hours was performed in addition.

In order to demonstrate the possible interrelationship between the central nervous disturbance caused by the hemorrhage and the ECG findings, the results are treated in the following grouping and order:

- subarachnoid hemorrhage without complication
- subarachnoid hemorrhage with secondary intracerebral bleeding
- subarachnoid hemorrhage with arterial spasm
- primary intracerebral hemorrhage without complications and only local symptoms
- primary intracerebral hemorrhage with signs of brain stem compression and/or rupture into the ventricular system.

The report will focus on the subarachnoid hemorrhages with complicated course and a few representative studies will be highlighted.

Of the patients with a subarachnoid hemorrhage, three with uncomplicated course (Botterell grade I) displayed no appreciable pathological ECG changes. The situation was quite different in five patients with subarachnoid hemorrhage in whom intracerebral bleeding was present in addition (Fig. 1). Besides the ST depression present in all cases, three patients had displaced T wave alterations, four a prolongation of the QT interval and three patients exhibited marked arrhythmias with pauses over 2000 msec and ventricular extrasystoles. Figure 2 illus-

Advances in Neurosurgery, Vol. 11
Edited by H.-P. Jensen, M. Brock, and M. Klinger

trates the time course of the rate corrected QT interval (QT_c) in patient no. 5. We assume an upper limit of the normal range of 450 msec. In the first days after the initial hemorrhage the QT_c is in the upper normal range. After the secondary bleeding on the fifth day, it rises to 491 msec and remains in the pathological range up to the death of the patient. Otherwise, no further alterations of the ST segment can be detected here. Figure 3 shows details of the long-term ECG recording in the same patient on the third and fourth day after the initial hemorrhage; there are appreciable variations of the T wave with excessively high amplitudes on the second and third strip on the left, with even slight inversion on the left. The QT_c duration as well as the heart rate, display considerable fluctuations. These abnormalities were not detected by the daily recording (Fig. 2) of the conventional routine ECG.

In two out of five patients with subarachnoid hemorrhages and intracerebral bleeding, T wave inversion developed intermittently (Fig. 4a and b). Two hours before the second hemorrhage, the precordial ECG was normal. Six minutes after the first secondary bleeding, there was a high T wave and a QT_c prolongation. Two and a half hours later, an inverted T had developed in V5 and V6 (Fig. 4a), and this became more pronounced in the following period and reached a maximum 30 hours after the event (Fig. 4b). The QT_c duration also displayed maximal prolongation at this time. On the 5th day after the secondary bleeding these abnormalities had completely subsided (Fig. 4b).

Figure 5 gives a summary of the results in those patients who developed complicating arterial spasms. QT prolongation and ST segment alterations were the most prominent features, whereas no significant arrhythmias were observed in this setting.

In five cases with primary intracerebral hemorrhage without complications and only local symptoms, no pathological ECG results could be found. On the other hand, five patients with signs of brain stem compression and/or rupture of the hemorrhage into the ventricular system displayed arrhythmias and alterations of the ST-T segment (Fig. 6). However, a T inversion was not observed.

The following conclusions can be drawn:

1. In uncomplicated subarachnoid hemorrhage and primary intracerebral hemorrhage without signs of brain stem compression and/or rupture into the ventricular system no appreciable pathological ECG findings are to be expected.
2. The most consistent and most frequent finding in all hemorrhages with a complicated course is a prolongation of the QT interval.
3. In correlation with the QT prolongation, arrhythmias are frequently present in subarachnoid hemorrhage with additional intracerebral hemorrhages as well as in primary intracerebral hemorrhage with signs of brain stem compression and/or rupture into the ventricular system.
4. On the other hand, in subarachnoid hemorrhage with spasm the alterations of the ST-T segment appear to be the most prominent feature.

Concerning the pathophysiology of the ECG abnormalities described these are commonly attributed to an autonomic dysfunction especially of the sympathetic nervous system. The following clinical consequences of our finding should be pointed out: patients with complicated intracranial hemorrhage should be subjected to continuous cardiac monitoring. Thus developing arrhythmias of potentially prognostic significance can be adequately treated. In particular marked QT prolongation deserves special attention since this syndrome signifies significant electrical instability of the myocardium with the potential for serious life threatening arrhythmias such as ventricular tachycardia and fibrillation. Because of their tendency to encourage prolongation of the QT interval, phenothiazines should be avoided. In addition to these therapeutic considerations useful diagnostic information on the natural course of the cerebral hemorrhage may be obtained, in that ECG abnormalities seem to reflect major complications such as brain stem compression and/or extension of the hemorrhage.

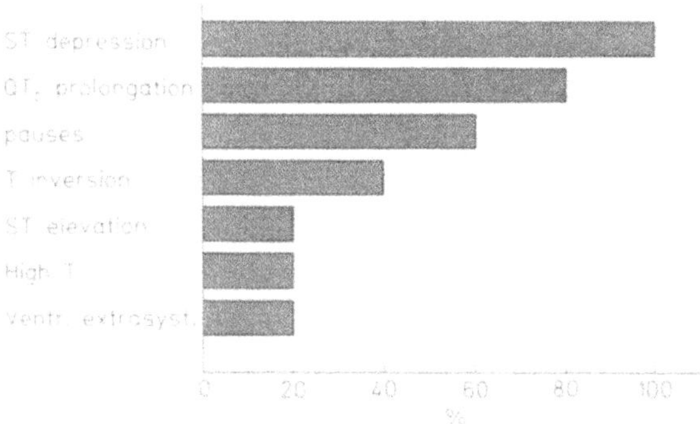

Fig. 1. ECG findings in five patients with SAH and intracerebral hemorrhage

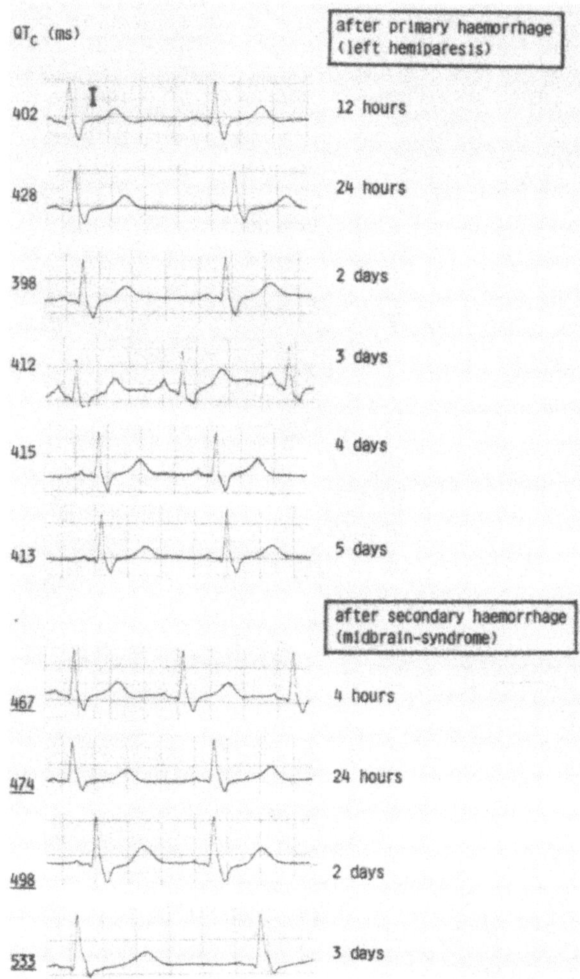

Fig. 2. Observation on the QT_c pattern in a 46-year-old male with subarachnoid hemorrhage and complicating intracerebral bleeding with left side hemiparesis. A second hemorrhage on the fifth day with signs of brain stem compression. Initial QT_c in the upper normal range. Second hemorrhage: consistent pathological QT_c prolongation until death three days later

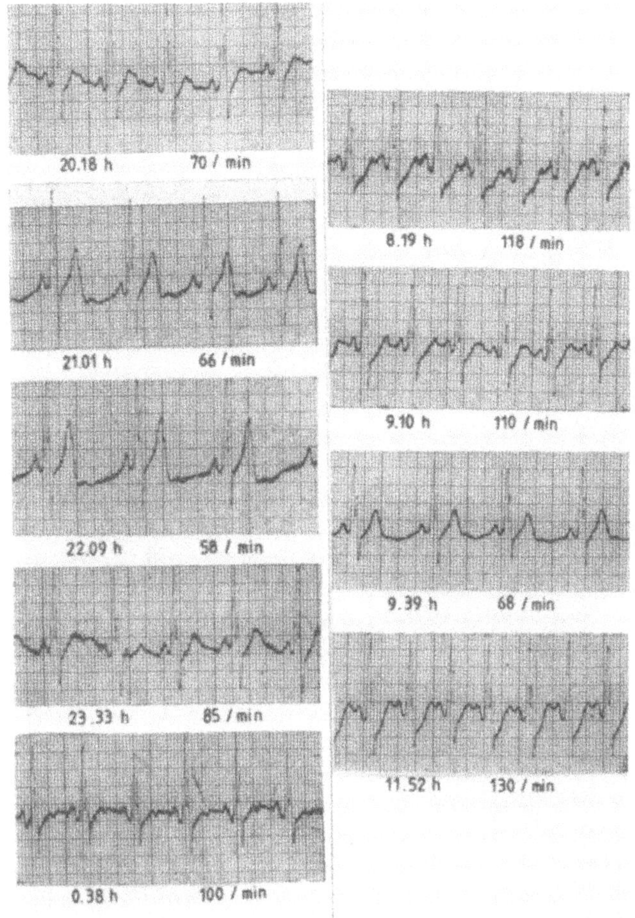

Fig. 3. Details of the continuous Holter ECG recording on the third and fourth days after the primary hemorrhage in the same patient as in Fig. 2. Note the appreciable fluctuations in heart rate, T amplitude and ST segment

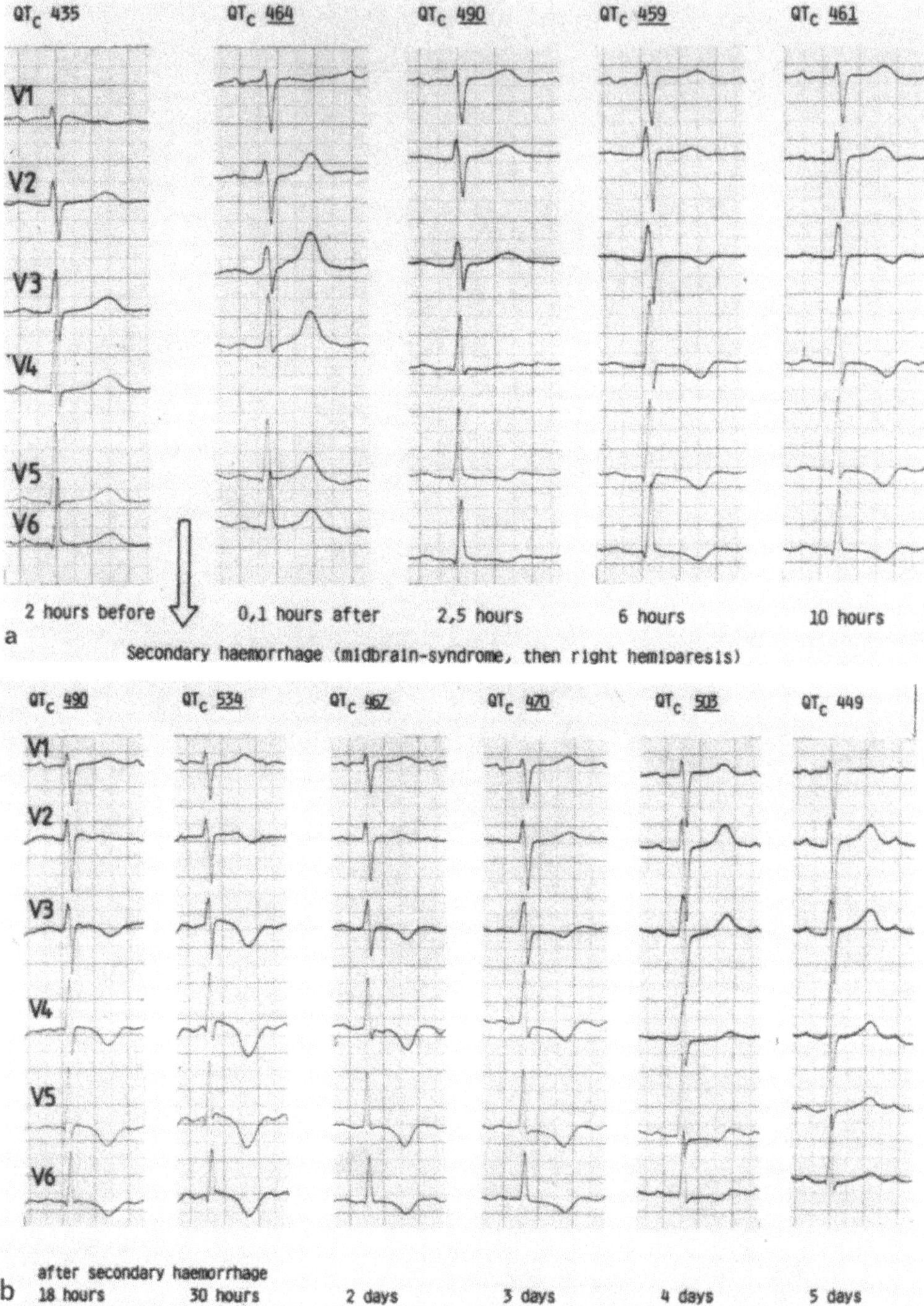
QT$_c$ 435
QT$_c$ 464
QT$_c$ 490
QT$_c$ 459
QT$_c$ 461
V1
V2
V3
V4
V5
V6
2 hours before
0,1 hours after
2,5 hours
6 hours
10 hours
a
Secondary haemorrhage (midbrain-syndrome, then right hemiparesis)
QT$_c$ 490
QT$_c$ 534
QT$_c$ 467
QT$_c$ 470
QT$_c$ 503
QT$_c$ 449
after secondary haemorrhage
b
18 hours
30 hours
2 days
3 days
4 days
5 days

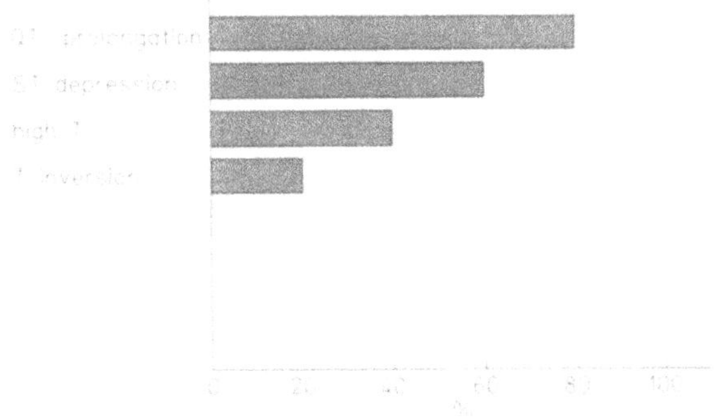

Fig. 5. ECG findings in five patients with SAH and spasm

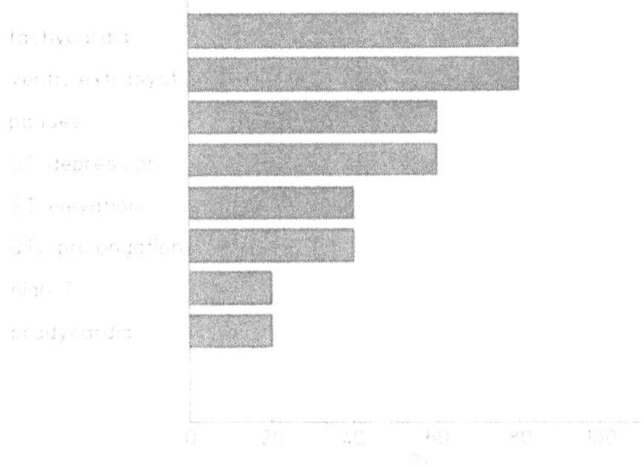

Fig. 6. ECG findings in five patients with complicated intracerebral hemorrhage (midbrain-syndrome and/or ventricular hemorrhage)

◄

Fig. 4. a Serial routine ECG in a 42-year-old female patient. Two hours before a second hemorrhage, there were no neurological deficits and the ECG was normal. Six minutes after the secondary hemorrhage (2nd strip from the right), increasing T amplitude, QT_c duration in the pathological range. Two and a half hours later (middle strip), negative T waves in V5 - V6, extending to V3 and V4 after six and ten hours and increasing in amplitude. b Serial ECG in the same patient as in a, continued 18 hours until five days after the second hemorrhage. Maximum T inversion after 30 hours (2nd strip from the left) comprising V2 - V6. Thereafter regression of the T inversion until return to normal on the fifth day. Concomitant QT_c alterations can be observed

Perinatal Cerebral Hemorrhage: Morphology, Clinical Diagnosis, Treatment and Follow-Up

H. C. Nahser[1], H. E. Nau[1], L. Gerhard[2], V. Reinhardt[2], I. Stude[3], and C. Roosen[1]

[1] Neurochirurgische Klinik und Poliklinik, Universitätsklinikum der Gesamthochschule Essen, Hufelandstrasse 55, D-4300 Essen 1

[2] Neuropathologisches Institut des Universitätsklinikum der Gesamthochschule Essen, Hufelandstrasse 55, D-4300 Essen 1

[3] Kinderklinik des Universitätsklinikum der Gesamthochschule Essen, Hufelandstrasse 55, D-4300 Essen 1

Introduction

It is usual to classify perinatal cerebral hemorrhage according to its etiology. FRIEDE (2) as well as PAPE and WIGGLESWORTH (6), differentiated a traumatic group from a hypoxic one (Table 1). They do however stress that different etiologies can lead to similar morphological changes. The clinical classification can be made nowadays by the newer methods of neuroradiology (1, 3-5, 7, 9). In this contribution we concentrate on hemorrhages into the germinal matrix. These hemorrhages are characteristic of asphyxia in the immature newborn.

Material and Methods

Between 1978 and 1981 we investigated 56 children with germinal matrix hemorrhages. These newborns were examined clinically, and by means of a second generation CT scanner and B-scan ultrasonography. Further neuroradiological investigations were carried out in cases of neurological disturbances. The CT was analysed according to our classification (5).

Autopsy and morphological investigations were carried out in the majority of the non-survivors.

Table 1. Perinatal cranial hemorrhages (FRIEDE, 1975)

Traumatic causes	Hypoxic causes
Extracranial	Subependymal and intraventricular hemorrhage
Caput succedaneum Subaponeurotic hematoma	Hemorrhage of choroid plexus
Intracranial	Subpial hemorrhage
Intradural hemorrhage Rupture of tentorium and falx Intracerebral hemorrhage Subarachnoid hemorrhage	Hemorrhage from falx
	ASH

 Advances in Neurosurgery, Vol. 11
Edited by H.-P. Jensen, M. Brock, and M. Klinger

Results

Out of 56 children only three had an uneventful gestation, four were complicated by toxemia, 49 were preterminated. A pathological APGAR score was seen in 37 newborns. Eleven had a birth weight less than 1000 g, 36 less than 1500. The following disorders and complications were observed: congenital heart malformation in six cases, hyaline membranes in 30, renal failure in four (Table 2).

CT revealed subependymal hemorrhages (stage I) in 12 cases (Fig. 1). In these cases morphology showed hemorrhages of different size in and underneath the ependymal germinal matrix. Late sequelae have been cysts or scars in this location. Although hemorrhages may occur in all sites with germinal matrix cells, the preferred site is the region of the foramina of Monro. Out of 12 patients in this group five died during the first month. The survivors are under clinical control with a favorable development in only two cases. In the CT follow-up no ventricular dilatation was observed.

In stage II there were 21 newborns. CT (Fig. 1) showed subependymal and combined intraventricular hemorrhages (Fig. 2). Histological investigations showed that in some cases the source of hemorrhage may be the choroid plexus, but in most however it was the subependymal matrix. Sequelae are damage and scarring of the ventricular wall as well as aqueduct stenosis and closure of the foramen of Magendie. Seven patients of this group died within 35 days. Late death occurred in two cases after 11 and 17 months. All patients except one developed hydrocephalus. Treatment consisted of shunting in seven cases during the first few weeks, after resorption of the intraventricular blood clot, or in three cases by serial lumbar punctures. Ten patients are still under control and two of these have no neurological deficit.

In 14 cases of stage III, the patients' CT (Fig. 1) revealed subependymal, intraventricular, and parenchymal hemorrhages. Histology showed large hemorrhages in the brain substance with a special preference for the fronto-parietal and occipital white matter. These intracerebral hematomas lead to cystic defects of the white matter (Fig. 3). Nine newborns out of 14 died within the first month, one patient after five months. All patients developed hydrocephalus. Three underwent shunting and one had serial lumbar puncture. Four are still under clinical control. All show severe defects of cerebral function (Table 3).

Table 2. Perinatal cerebral hemorrhages (n = 56)

Clinical data and incidence	
Mother	
Toxemia	4
Normal date of birth	3
Normal delivery	18
Twin birth	8
Newborn	
Normal birth weight	3
Hyaline membrane disease	30
Congenital heart malformation	6
Reduced APGAR score	37
Hb loss, blood-stained CSF	32
Convulsion, pathological EEG	30
Abnormal neurological state	34

Table 3. Perinatal cerebral hemorrhages (n = 56)

Stage	Survivors	Non survivors (%)	Duration of survival: Weeks <1	<2	<3	<4	Months <4 (we)	<2	<3	<5	<12	>12
I	12	5 (41.66%)	1		2	2	2					
II	21	10 (47.61%)	5	1				1			1	1
III	14 (71.42%)	10	2	4	2	2		1		1		
IV	9 (77.77%)	7	4		2				1			

In stage IV (subependymal), intraventricular, intracerebral, and subarachnoid hemorrhage) we found nine cases. Histology revealed all the disturbances mentioned in addition to subarachnoidal hemorrhages. Late consequences are siderosis and fibrosis of the pia mater. Six patients died within 17 days, one after two and a half months. There are two survivors, one underwent shunting, one serial lumbar puncture.

Discussion

Our results demonstrate that the most frequent stage is stage II. There seems to be a connection between the severity of CT findings and the number of survivors, but there are only small differences in the neurological deficits of the survivors in all four groups. This can only be due to additional disturbances. The most astonishing fact was the high incidence of hydrocephalus in all cases of ventricular hemorrhage. It seems that the non-infectious etiology of hydrocephalus in the perinatal period is more frequent than is supposed. It may even be that the non-infectious cause has increased on the one hand by the control of infectious diseases and on the other hand by the increase of immature newborns who can survive only by means of modern intensive medical therapy.

According to our experience so far we believe that ultrasonography may prove its value in follow-ups, especially in cases of ventricular dilatation. CT, however, seems to be necessary for the initial staging, especially because further detailed pathological conditions can be detected.

Long-term follow-ups will be necessary to define definitely how gross the final neurological defects of the survivors really are. The results of these investigations will enable us to assess the value of shunting or puncture procedures (8). This question is underlined by the findings that the ventricular dilatation precedes the increase in the circumference of the head (5).

Summary

We investigated 56 mature and premature newborns with germinal matrix hemorrhages. The clinical data, computer tomographic, ultrasonic, and morphological findings are discussed. Special hints are given about the complications which arise and the different forms of treatment.

References

1. Flodmark, O., Becker, L.E., Harwood-Nash, D.C., Fitzhardinge, P.M., Fitz, C.R., Chuang, S.H.: Correlation between computed tomography and autopsy in premature and full-term neonates that have suffered perinatal asphyxia. Neuroradiology 137, 93-103 (1980)
2. Friede, R.L.: Developmental Neuropathology. Wien, New York: Springer 1975
3. Harwood-Nash, D.C., Flodmark, O.: Diagnostic imaging of the neonatal brain: review and protocol. AJNR 3, 103-115 (1982)
4. Leblanc, R., O'Gorman, A.M.: Neonatal intracranial hemorrhage. A clinical and serial computerized tomography study. J. Neurosurgery 53, 642-651 (1980)
5. Nasher, H.C., Hanssler, L., Nau, H.-E., Löhr, E.: Computertomographische Befunde in der Diagnostik und Verlaufskontrolle von germinalen Matrixblutungen bei Neugeborenen. Radiologie 21, 516-520 (1981)
6. Pape, K.E., Wigglesworth, J.S.: Haemorrhage, ischaemia and the perinatal brain. London: Heinemann 1979
7. Papile, L., Burstein, J., Burstein, R., Koffler, H.: Incidence and evolution of subependymal and intraventricular hemorrhage: A study of infants with birth weights less than 1500 g. J. Pediatr. 92, 529-534 (1978)
8. Papile, L., Burstein, J., Burstein, R., Koffler, H., Johnson, J.D.: Posthemorrhagic hydrocephalus in low-birth-weight infants: Treatment by serial lumbar punctures. J. Pediatr. 97, 273-277 (1980)
9. Rumack, C.M., McDonald, M.M., O'Meara, O.P., Sanders, B.B., Rudikoff, J.: CT detection and course of intracranial hemorrhage in premature infants. Am. J. Roentgenol. 131, 493-497 (1978)

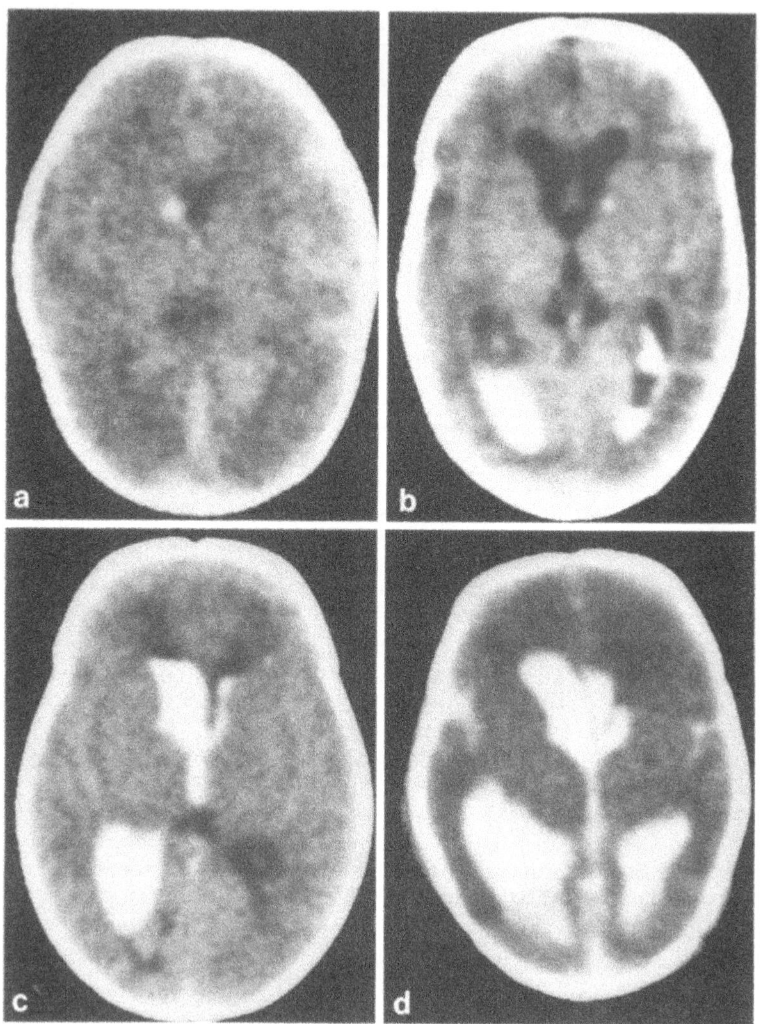

Fig. 1. a Subependymal bleeding (grade I) adjacent to the ventricular wall of the frontal horn; b Subependymal and intraventricular hemorrhage (grade II): Subependymal bleeding at the foramen of Monro (preferred location) and blood clots in the posterior horns. c Subependymal, intraventricular and parenchymal hemorrhage (grade III); d Combination of grade III with subarachnoid hemorrhage (grade IV)

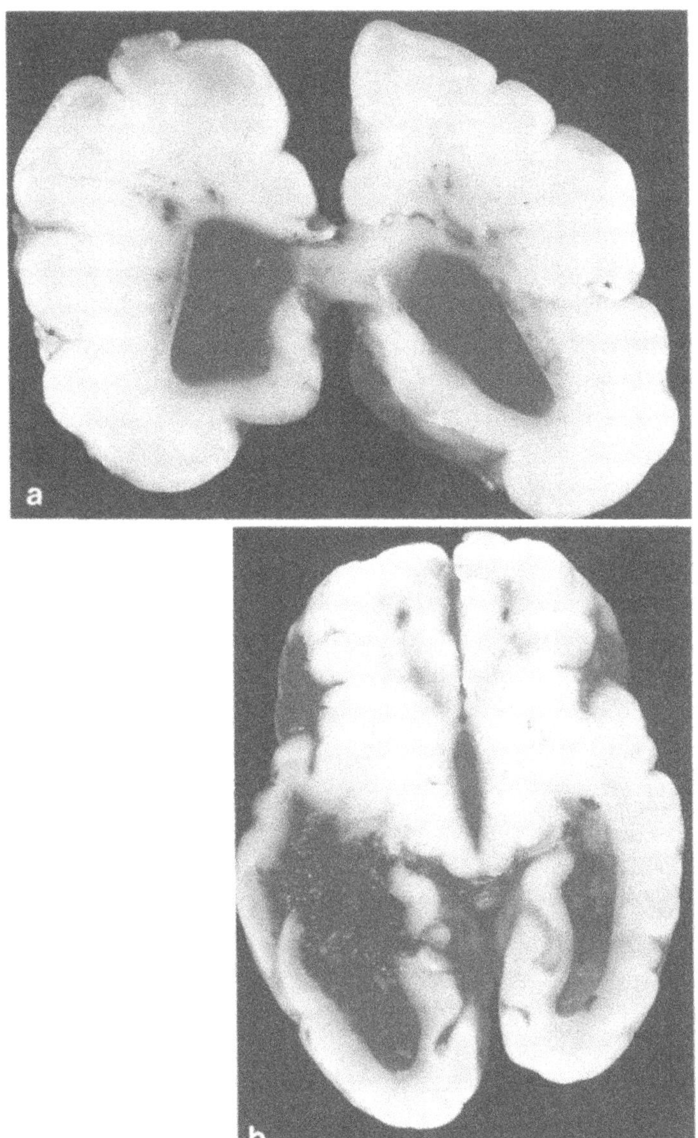

Fig. 2. a Specimen of intraventricular hemorrhage with congested veins in the subependymal germinal layer; b Gross pathology of grade IV: bleeding in the germinal layer has ruptured into the ventricle and into the parenchyma of the left occipital lobe. There are congested vessels in the right posterior horn, blood clots in the third ventricle and in both Sylvian fissures

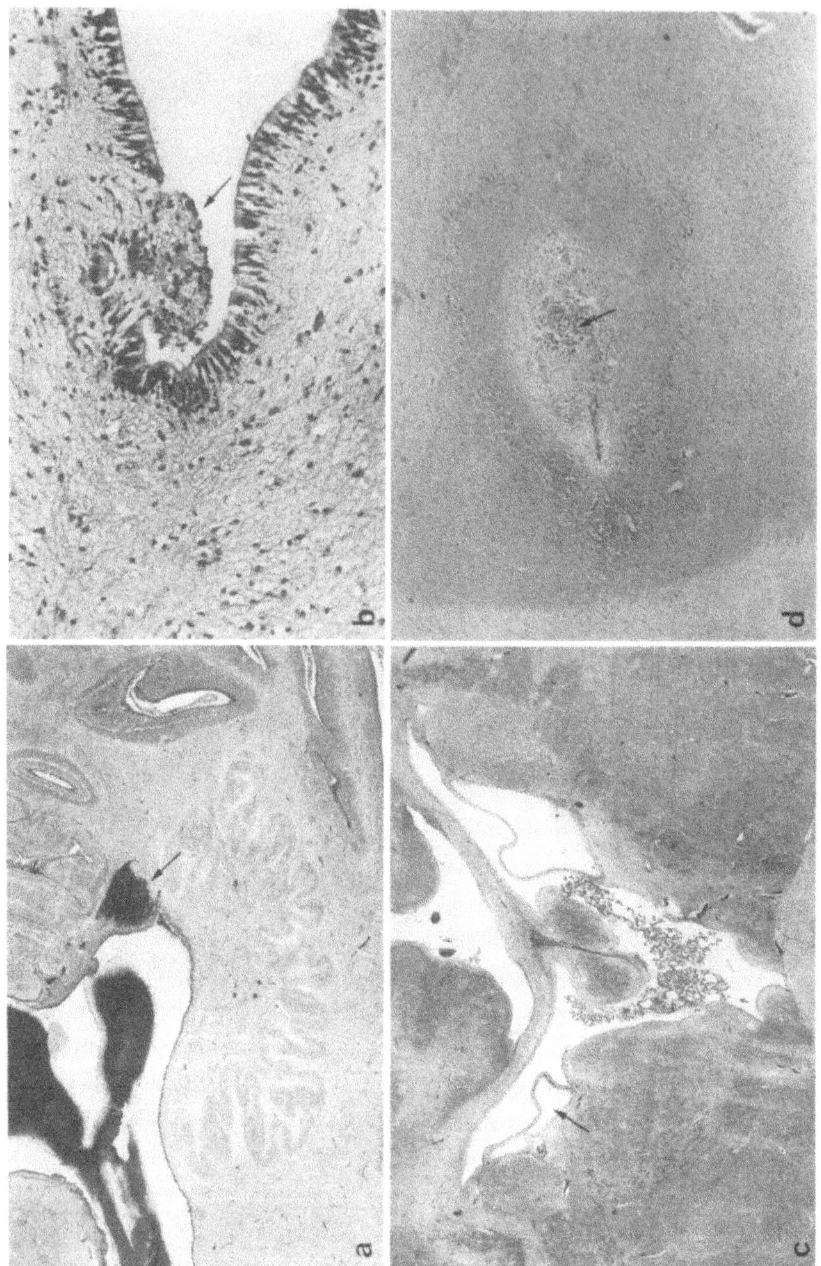

Fig. 3. a Histology of grade II: subependymal hemorrhage (→) near the vermis has ruptured into fourth ventricle. b Sequelae of grade II: → organisation of bleeding, ependymal defect, but persisting lumen of the aqueduct. In this case no signs of raised intracranial pressure, but ventricular dilatation was recorded. c Sequelae of grade I: bicaudate cyst with an intact ependymal layer, which could not be detected in computed tomography. d Occlusion of the aqueduct by the hemorrhage (→). There are only small islets of the ependyma left. In this case signs of raised intracranial pressure occurred and a shunting procedure was performed

Spontaneous Hemorrhages in Intracerebral Brain Tumors and Leukemia

G. Ebhardt[1], W. Müller[2], F. Thun[3], O. Wilche[4], R. Fischer[2]

[1] Forschungsstelle für Hirnkreislaufforschung, Max-Planck-Institut für Hirnforschung, Ostmerheimer Strasse 200, D-5000 Köln 91

[2] Pathologisches Institut der Universität Köln, Joseph-Stelzmann-Strasse 9, D-5000 Köln 41

[3] Radiologisches Institut der Universität Köln, Joseph-Stelzmann-Strasse 9, D-5000 Köln 41

[4] Neurochirurgische Klinik der Universität Köln, Joseph-Stelzmann-Strasse 9, D-5000 Köln 41

During recent years, clinical interest in the stroke syndrome has continued to increase. As a result, more attention has been paid to spontaneous bleeding due to intracranial tumors. Although bleeding in neoplasm was blamed for a sudden deterioration of the patients' condition, CUSHING's opinion, that "severe bleeding in a tumor is an extremely rare occurrence" (3) long remained the prevalent opinion. For autopsy series, the figures given in the relevant literature vary between 0.8% (1, 16, 20) and 10.2% (8, 15). With regard to the types of spontaneous intracranial hemorrhage treated in neurosurgical clinics, the incidence of tumor bleeding lies between 0% (9) and 11% (6, 17). In a series of 236 intracranial tumors, DRAKE and McGEE (3) found 18 cases (7.6%) of tumor bleeding.

There are a number of reasons for these considerable variations in the frequency of spontaneous tumor bleeding. DRAKE and McGEE (3), for instance, reach their figure by relating the 18 cases of tumor bleeding to the total number of intracranial tumors found during operation or autopsies.

In other series, the incidence of tumor bleeding is measured against the total of all intracranial hemorrhages (12).

The aim of this paper is to compare the frequency of *spontaneous intracerebral tumor bleeding* in large series of biopsy specimens and autopsy material.

Material and Method

The diagnoses of all patients operated on or dying from an *intracerebral hematoma* were collected. All primary diseases which had led simply to a subdural hematoma or to subarachnoid hemorrhage were excluded from consideration. In the case of the biopsies, the reference to bleeding in the operation report was accepted as sufficient proof that bleeding had occurred. In the case of the autopsies, the autopsy findings provided the data. In both autopsy and biopsy groups, the frequency figures, that is to say the percentages, are always based on the *total number* of *intracerebral hematomata*.

Advances in Neurosurgery, Vol. 11
Edited by H.-P. Jensen, M. Brock, and M. Klinger

Results

There were 332 deaths from intracerebral hemorrhage, 37 of them (11.1%) from tumor bleeding. In the case of the autopsies, hypertonic massive bleeding (27.8%) and intracerebral hematomas after aneurysm rupture (18.7%) were most frequent. In the case of the biopsies these two causes were shown to have hardly any relevance - six cases (4.0%) and five cases (3.4%) respectively. In this group, tumor bleeding (46 cases = 31%) and hemorrhages without known cause (43 cases = 29%), accounted for over half of the total operation material (Table 1). In the analysis of the various types of tumors with spontaneous bleeding, clear distinctions are to be found between the autopsy findings and the biopsy specimens. In the case of the autopsies, malignant gliomas (13 cases) and metastases (16 cases) are most frequent, whereas fatal hemorrhages are rarer among the better differentiated forms of neuroepithelial tumors. Leukemic hemorrhages resulted in the deaths of 15 patients. In 11 instances an acute form was evident, in four instances the illness was chronic. Amongst the biopsy material there was not one case of leukemia. Of the patients suffering from tumor bleeding which was operable, brain tumors (33 cases) were more frequent than metastases and other tumors. Here too, glioblastomas (12 cases) are to be found in first place, but also a large percentage is made up of oligodendrogliomas (five cases) and unclassified gliomas (five cases) (Table 2). Of the 43 patients with spontaneous intracerebral hemorrhages of unknown etiology 22 were operated on after a clinical diagnosis suggested a possible tumor (19 cases) or tumor, differential diagnosis angioma (three cases). In only five cases could the suspicion of a tumor be proved, or at least traced, histologically. Most of the cases were also impossible to prove morphologically (Table 3a). The situation was different in the group made up of post-mortem cases. An autopsy brought to light malignant tumors in all but one of the 17 patients who had not been subjected to a more thorough diagnostic examination because of their poor general condition (Table 3b). The temporal lobe was found to be the site of the fatal hemorrhages in glioblastoma in nine out of 11 cases (six times right side, three times left). Of the operable patients, the massive bleeding

Table 1. Cause of the intracerebral hematoma in 332 autopsy findings and in 149 biopsy specimens in the period between Jan. 1, 1971 and Feb. 28, 1982[a]

Primary disease	Autopsy case		Biopsy case	
	rate	per cent	rate	per cent
Hypertension	92	27.8	6	4.0
Aneurysm rupture	62	18.7	5	3.4
Angioma	18	5.4	35	23.6
Trauma	37	11.1	9	6.1
Sepsis	28	5.4	∅	
Coagulopathy	3	0.9	∅	
Anticoagulant therapy	15	4.5	3	2.0
Postoperative	17	5.1	2	1.3
Congophile angiopathy	1	0.3	∅	
Unclear etiology	17	5.1	43	29.0
Leukemia	15	4.5	∅	
Tumors	37	11.1	46	31.0
Total	332		149	

a The material comes from the Department of Neurosurgery and from the Institute of Pathology, University of Cologne

Table 2. Analysis of brain tumors and the other tumors with hemorrhages in 37 autopsy cases and 46 biopsy specimens

Brain tumors	Case rate autopsy	biopsy	Other tumors	Case rate autopsy	biopsy
Glioblastoma	11	12	Metastases	12	4
Astroblastoma	Ø	1	Melanoblastoma	4	2
Astrocytoma grade III	1	Ø	Sarcoma	2	3
Astrocytoma grade I	1	2	Malignant lymphoma	1	1
Oligodendroglioma grade III	1	1	Angioblastoma	Ø	2
Oligodendroglioma grade II	1	4	Meningioma with intracerebral		
Mixed glioma	Ø	2	hemorrhage	Ø	1
Unclassified glioma	Ø	5			
Ependymoma	Ø	1	Total	19	13
Plexus papilloma	Ø	1			
Gangliocytoma	Ø	1	Acute myeloid leukemia	11	Ø
Neuroblastoma	2	2	Chronic myeloid leukemia	4	Ø
Medulloblastoma	1	1			
Total	18	33	Total	15	

Table 3a and b. Analysis of the tentative clinical diagnosis and the morphological diagnosis in intracerebral hemorrhages of unknown etiology (17 autopsy cases and 43 biopsy specimens)

Tentative clinical diagnosis (17 autopsy cases)		Morphology Diagnosis	Diagnosis not made
Tumor?	6 cases	2 x glioblastoma 4 x metastases	
Abscess?	1 case	glioblastoma	
Hypertensive intracerebral hemorrhage	2 cases	butterfly glioblastoma metastasis	
Unknown etiology	8 cases	1 x glioblastoma 6 x metastases	1

Tentative clinical diagnosis (43 biopsy specimens)		Morphology Diagnosis	Diagnosis not made
Tumor?	19 cases	1) DD angioma/angiomatous meningioma 2) Melanoblastoma 3) Marginal zone of a pathological lesion 4) Glioma? 5) DD angioma, oligodendroglioma	14 cases
Angioma?	5 cases	1) Marginal zone of an angioma? 2) DD angioma / angioblastoma	3 cases
Angioma or tumor?	3 cases		3 cases
Trauma?	1 case		1 case
Unknown etiology	15 cases		15 cases

Table 4. Analysis of the cause of recurrent hemorrhage in tumors (biopsy specimens)

Case No.	Sex	Age in years I. hem.	Age in years II. hem.	Clinical diagnosis I. hemorr.	Clinical diagnosis II. hemorrh.	Morphological diagnosis I. hemorrh.	Morphological diagnosis II. hemorrh.
1	♀	4	9	Tumor	Tumor? Angioma?	Glioma?	Marginal zone of an old hematoma; glioma? angioma?
2	♀	7	7	Tumor?	Recurrence of a sarcoma	Rhabdomyosarcoma	Ditto
3	♀	54	54	Metastasis of hypernephroma	Ditto	Metastasis of a hypernephroma	Ditto
4	♂	41	45	Astrocytoma right central region	Recurrence of an oligodendroglioma	Oligogendroglioma grade II	Oligodendroglioma and astrocytoma = mixed glioma
5	♀	50	51	Glioma, rich in vessels	Recurrence of an oligodendroglioma	Oligodendroglioma grade II	Oligodendroglioma grade II
6	♂	45	45	Tumor	Recurrence of an oligodendroglioma	Oligodendroglioma grade II/III	Oligodendroglioma grade III
7	♂	44	45	Tumor	Recurrence of an oligodendroglioma	Oligodendroglioma grade II	Oligodendroglioma grade II
8	♀	27	28	Oligodendroglioma?	Recurrence of an oligodendroglioma	Glioma grade II oligodendroglioma?	Oligodendroglioma grade II

in glioblastoma was found to have occurred in the temporal lobe in six out of 12 cases. Among the metastases and the better differentiated gliomas the temporal lobe was not apparent as the source of spontaneous bleeding. After the first instance of tumor bleeding, eight patients experienced further bleeding after an interval of between three weeks and five years. In five cases this manifested itself as hemorrhages in oligodendrogliomas. The ages of these patients (three men and two women) were between 28 and 51 years (Table 4).

Discussion

Intracerebral hemorrhage as a result of a tumor is not as rare as is generally believed. Frequently, tumor bleeding manifests itself clinically as a stroke (12, 23). Trauma is also presumed to be a factor in spontaneous tumor bleeding (4). It is rarely possible to compare reports on frequency. Operation statistics usually base figures for the incidence of tumor bleeding on the total number of tumors operated on (3, 18, 23), and not specifically on the total number of operations for intracerebral hematomas. Other reports are based on investigations of intracranial bleeding in tumors, thus including subarachnoid hemorrhages, for instance in connection with subependymoma (2), meningioma (10) or chiasmal bleeding (11, 19). These observations draw their conclusions from only a very small number of cases (5, 7). The same applies to reports that restrict themselves to intracerebral bleeding in connection with astrocytomas grade I (13, 21, 22) or medulloblastomas (14, 21). The reports that show a high percentage of tumor bleedings are based on extensive post-mortem statistics (8). For both biopsies and autopsies, we have based our figures on the total number of *intracerebral hematomata* in order to be in a position to compare the frequency. As a result, the figure given for patients operated on is high. If we had based the number of tumor hemorrhages on the total number of operations for intracerebral tumors between 1971 and 1982, the percentage would have been much *lower*. Our figure of 11.2% for autopsy cases approximately corresponds to that of KOTHBAUER et al. (8), who, in an autopsy series of 430 spontaneous cerebral hematomas found 44 (10.2%) cases of bleeding in intracerebral neoplasms. Although the frontal lobe is usually considered to be the most frequent seat of glioblastomas, we found the majority of hemorrhages in glioblastomas in the temporal lobe. In our sample, the oligodendrogliomas and the undifferentiated gliomas, together with the glioblastomas, form the majority. Especially in the case of recurring hemorrhages the majority are in oligodendrogliomas. In the interest of producing a differential diagnosis, this tumor should be kept in mind when spontaneous intracerebral hematomas occur in younger patients.

Summary

Between Jan. 1, 1971 and Feb. 28, 1982, 149 patients were operated on for intracerebral hemorrhage. During the same period 332 patients died as a result of intracerebral hematomas. Tumors were the cause of bleeding in 46 (= 31%) of the patients operated on and 37 (= 11.1%) of those that died. Tumor bleeding, then, is not so rare as is generally believed.

Particularly in the case of younger patients and recurring hemorrhages, the possibility of bleeding from better differentiated gliomas, such as oligodendrogliomas, should be given serious consideration.

References

1. Brewer, D.B., Fawcett, F.J., Horsfield, G.I.: A necropsy series of nontraumatic cerebral hemorrhages and softenings, with particular references to heart weight. J. Path. Bact. 96, 311-320 (1968)
2. Changaris, D.G., Powers, J.M., Perot, P.L. Jr., Hungerford, G.D., Neal, G.B.: Subependymoma presenting as subarachnoid hemorrhage. Case report. J. Neurosurg. 55, 643-645 (1981)
3. Drake, C.G., McGee, D.: Apoplexy associated with brain tumors. Can. Med. A. J. 84, 303-305 (1961)
4. Globus, J.H., Sapirstein, M.: Massive hemorrhage into brain tumor. J.A.M.A. 120, 348-352 (1942)
5. Jänisch, W., Schreiber, D., Gerlach, H.: Tumoren des Zentralnervensystems bei Feten und Säuglingen. Jena: VEB Fischer 1980
6. Jellinger, K.: Morphology and aetiology. Pathology and aetiology of ICH. In: Spontaneous intracerebral haematomas. Advances in diagnosis and therapy. Pia, H.W., Langmaid, C., Zierski, J. (eds.), pp. 13-29. Berlin, Heidelberg, New York: Springer 1980
7. Kazimiroff, P.B., Weichsel, M.E. Jr., Grinnell, V., Young, R.F.: Acute cerebellar hemorrhage in childhood: Etiology, diagnosis, and treatment. Neurosurgery 6, 524-528 (1980)
8. Kothbauer, P., Jellinger, K., Flament, H.: Primary brain tumor presenting as spontaneous intracerebral haemorrhage. Acta Neurochirurgica 49, 35-45 (1979)
9. Krayenbühl, H., Siegfried, J.: Der neurochirurgische Beitrag zur Behandlung der intracerebralen Blutung. Wien. klin. Wschr. 76, 401-404 (1964)
10. Lazaro, R.P., Messer, H.D., Brinker, R.A.: Intracerebral hemorrhage associated with meningioma. Neurosurgery 8, 96-101 (1981)
11. Maitland, C.G., Abiko, S., Hoyt, W.F., Wilson, C.B., Okamura, T.: Chiasmal apoplexy. Report of four cases. J. Neurosurg. 56, 118-122 (1982)
12. Mandybur, T.I.: Intracranial hemorrhage caused by metastatic tumors. Neurology 27, 65P.655 (1977)
13. Mauersberger, W., Cuevas-Solorzano, J.A.: Spontaneous intracerebellar hematoma during childhood caused by spongioblastoma of the IVth ventricle. Neuropädiatrie 8, 443-450 (1977)
14. McCormick, W.F., Ugajin, K.: Fatal hemorrhage into a medulloblastoma Case report. J. Neurosurg. 26, 78-81 (1967)
15. McCormick, W.F., Rosenfield, D.B.: Massive brain hemorrhage: A review of 144 cases and an examination of their causes. Stroke 4, 946-954 (1973)
16. Mutlu, N., Berry, R.G., Alpers, B.J.: Massive cerebral hemorrhage. Clinical and pathological correlations. Arch. Neurol. (Chic.) 8, 644-661 (1963)
17. Pia, H.W.: The surgical treatment of intracerebral and intraventricular haematomas. Acta Neurochirurgica 27, 149-164 (1972)
18. Scott, M.: Spontaneous intracerebral hematoma caused by cerebral neoplasms. Report of eight verified cases. J. Neurosurg. 42, 338-342 (1975)
19. Schneider, R.C., Kriss, F.C., Falls, H.F.: Prechiasmal infarction associated with intrachiasmal and suprasellar tumors, J. Neurosurg. 32, 197-208 (1970)

20. Staemmler, M.: Kreislaufstörungen und Gefäßerkrankungen des Zentralnervensystems mit Hirnödem und Hirnschwellung. In: Lehrbuch der speziellen pathologischen Anatomie, Kaufmann, E., Staemmler, M. (eds.), pp. 271-339, Vol. III, Part 1. Berlin: de Gruyter 1958

21. Vaquero, J., Cabezudo, J.M., de Sola, R.G., Nombela, L.: Intratumoral hemorrhage in posterior fossa tumors after ventricular drainage. Report of two cases. J. Neurosurg. 54, 406-408 (1981)

22. Vincent, F.M., Bartone, J.R., Jones, M.Z.: Cerebellar astrocytoma presenting as a cerebellar hemorrhage in a child. Neurology 30, 91-93 (1980)

23. Zuccarello, M., Pardatscher, K., Andrioli, G.C., Fiore, D.L., Lavicoli, R.: Brain tumors presenting as spontaneous intracerebral haemorrhage. Zbl. Neurochirurgie 42, 1-6 (1981)

Follow-Up Observations in Spontaneous Brain Stem Hemorrhage

G. Hopp[1], E. Schneider[1], H. Grau[2], P.-A. Fischer[1], W. I. Steudel[3]

[1] Abteilung für Neurologie, Zentrum der Neurologie und Neurochirurgie, Klinikum der Johann Wolfgang Goethe-Universität, Schleusenweg 2–16, D-6000 Frankfurt 71

[2] Abteilung für Neuroradiologie, Zentrum der Neurologie und Neurochirurgie, Klinikum der Johann Wolfgang Goethe-Universität, Schleusenweg 2–16, D-6000 Frankfurt 71

[3] Abteilung für Neurochirurgie, Zentrum der Neurologie und Neurochirurgie, Klinikum der Johann Wolfgang Goethe-Universität, Schleusenweg 2–16, D-6000 Frankfurt 71

Primary spontaneous brain stem hemorrhage occurs rarely; its frequency is reported to range between 6% and 16% of all intracerebral hemorrhages (1, 5, 7-9, 25). Hypertension as the most frequent cause is to be found in 90 - 96.7% (5, 25). Microangioma, aneurysm, intoxication, acute infection, and tumor-bleeding represent a rare cause of hemorrhage. The fulminating malignant course is characterised by rapid onset of coma, miosis, tetraplegia and respiratory disturbances. Three quarters of all patients die within 24 hours (8, 13, 25). Several authors (19, 21, 22) report on single cases with remission, or a recurrent, benign course. These are mostly cases of younger patients without hypertension. The causes are reported to be arteriovenous malformations.

Case-Material and Approach

From 1979 to 1982, six patients suffering from spontaneous hemorrhage of the brain stem were seen. All cases were verified with CT. The group consisted of three female and three male patients, aged from 28 to 54 years. The case history of three patients showed hypertension. CT examinations were carried out at the onset and during the clinical course of the disease. The mode of treatment was conservative with a symptomatic approach.

Results

Patients without Hypertension

Case 1: A 32-year-old female in the 32nd week of pregnancy had an acute onset of occipital headache, nausea, and vomiting, left hemiparesis with caudal cranial nerve impairment, fixed staring eyes as well as a right hemiataxia and a dissociated left paresthesia. No hypertension and no coma were observed. The CT findings of January 6, 1979 showed a circular space-occupying high density area in the right brain stem with compression of the fourth ventricle (Fig. 1a). The patient was delivered of a male child without complications on January 18, 1979. The condition of the patient worsened on February 3, 1979 with dysphagia, severe vomiting, and fatigue. CT findings showed irregular structure formations in the brain stem, which clearly appeared to be enlarged resulting in an almost complete compression of the fourth ventricle (Fig. 1b); the temporal horns were just visible. No new bleeding occurred but edema developed in the course of resorption. These CT

Advances in Neurosurgery, Vol. 11
Edited by H.-P. Jensen, M. Brock, and M. Klinger

findings might correlate with the symptoms of her worsened condition. The angiograms done by femoral catheter, showed no signs of a vascular malformation and no pathological blood vessels in the brain stem (Fig. 2). Her out-patient follow-up examination in February 1982 showed a left spastic hemiparesis. She was able to carry out her duties as a housewife and care for her baby. The last CT findings showed an irregular high density in the median and paramedian regions of the brain stem (Fig. 1d). Further, there was no increase in density enhancement shown with contrast medium.

Case 2: A 46-year-old female patient was afflicted with recurrent brain stem impairment in January 1981 with dysphagia, dysarthria and right-sided paresthesiae, accompanied at times by nausea and vomiting. The symptoms occurred in September 1981, now with a Horner's syndrome on the left side, rotatory nystagmus on the left, numbness and analgesia in the trigeminal nerve region on the right, paresis of the soft palate and paresis of the hypoglossal nerve on the right side. This resulted in dysphagia and dysarthria. In addition we observed pronounced ataxia. CT findings in September 1981 showed irregular density enhancement with a diameter of up to 2.5 cm from the floor of the fourth ventricle to the foramen magnum. On angiography no signs of vascular malformations or pathological blood vessels were found. The patient improved clinically; only slight right-sided paresthesiae and minimal palsy of the hypoglossal nerve persisted.

Case 3: A 28-year-old patient was affected by slowly progressive brain stem symptoms: phonasthesia, singultus, right-sided paresis of the soft palate, left-sided Horner's syndrome, right-sided hemiparesis, and dissociated paresthesia. Initially the CT findings showed no density enhancement and no signs of intracerebral hemorrhage. The CSF showed no signs of inflammation. The patient died ten days after admission to the hospital from an acute bleeding in the brain stem. The autopsy disclosed extensive bleeding from the medial pontine region extending to the medulla oblongata; histologically M.S. lesions were found.

Patients with Hypertension

Case 4: A 28-year-old female with a fixed hypertension (230/110 mm Hg) with stenosis of the left renal artery and stenosis of the aortic isthmus, suddenly and without warning on August 17, 1981, lost consciousness and this lasted approximately one hour. Thereupon, she showed fixed staring eyes with nystagmus on attempted deviation, palsy of the cranial nerves: VII, VIII, IX, XII combined with dysphagia, dysarthria, and also hemiparesis and hemiataxia on the left side. The blood pressure was found to be 230/150 mm Hg. CT scan showed a high density on the floor and also paramedian on the right side of the fourth ventricle with a diameter of 2.5 to 3 cm. The fourth ventricle was somewhat compressed (Fig. 3a). The clinical disturbances slowly disappeared. After discharge from hospital on October 29, 1981 a residual neurological deficit with dysarthria, right-sided peripheral facial palsy, and a spastic hemiparesis on the contralateral side was observed. The CT showed decrease of the density, which had a cystic appearance, and did not compress the floor of the fourth ventricle (Fig. 3b).

Case 5: This was a 54-year-old male with a history of mild hypertension. He suddenly became unconscious without warning; we observed temporary extension spasms on the right side, pinpoint pupil, and respiratory disturbances. The blood pressure was found to be 300/150 mm Hg. CT showed a 2.5 x 2 cm large high density in the median brain stem area with signs of initial intraventricular rise of pressure. Two days after this acute attack, the patient died.

Case 6: This 52-year-old patient initially experienced dizziness and paresthesiae on the right-side. Within six hours he fell into coma accompanied with respiratory disturbance. The CT showed an extensive high density with a diameter of 2.5 cm in the brain stem, compressing the fourth ventricle. An apallic syndrome developed; the patient died two months after the acute onset.

Discussion

The prognosis of spontaneous brain stem hemorrhage still remains poor and it often occurs as a terminal phase of hypertension (9, 24). Only one of our three patients afflicted with hypertension survived. This patient differed from the others in that she was the youngest and her initial unconsciousness lasted only for a short time. The cases differed from each other neither from the standpoint of duration and peak of the hypertension, nor from the extent of hemorrhage in the CT. The rapid malignant course seems above all to appear in cases of hemorrhage due to hypertension. The fact that one of our patients afflicted with hypertension survived, raises the question of this was due to an extraordinary condition of the individual or can it be explained by the effect of modern intensive medical care? This question can only be answered by further investigation.

In cases with no hypertension, brain stem hemorrhages do not seem to lose consciousness and seem to be recurrent and benign in the follow-up. Arterio-venous malformations have been discussed as a cause of hemorrhage (16, 19, 23). Such microangiomas could not be verified angiographically. In two cases of our group having such negative angiograms, the follow-up CT examinations, however, clearly demonstrated the possibility of an existing cavernous hemangioma. In the cases of hypertensive hemorrhage the CT shows low density after resorption, on the other side; in cases of hemorrhage due to angiomas an isodense phase was observed during resorption, which was accompanied by local brain edema and could result temporarily in a worsening of the condition of the patient. Later we saw irregular zones of high density in the CT, which are probably caused by calcifications. It is possible that before the CT era with recurrent follow-up examinations these were regarded as M.S. lesions or as a glioma of the pons.

Conclusion

Follow-up observations in six patients with hemorrhage, confirm a poor prognosis in cases with vascular hypertension. The hemorrhages not caused by hypertension show recurrences in their follow-up and no loss of consciousness. Probably the cause of hemorrhage is a cavernous hemangioma, which could hardly be detected angiographically. Concerning the differential diagnosis, bleeding into tumors or into inflammatory lesions should be considered.

References

1. Arseni, C., Stanciv, M.: Primary hematomas of the brain stem. Acta Neurochir. 28, 323-330 (1973)

2. Ahwater, H.L.: Pontine hemorrhages. Guy's Hosp. Rep. 65, 339-389 (1911)

3. Brismar, J., Hindfelt, B., Nilsson, O.: Benign brain stem hematoma. Acta Neurol. Scand. 60, 178-182 (1979)

4. Dana, C.L.: Acute bulbar paralysis due to hemorrhage and softening of the pons. Med. J. Rec. 64, 361-374 (1903)

5. Dinsdale, H.B.: Spontaneous hemorrhage in the posterior fossa. Arch. Neurol. 10, 200-217 (1964)

6. Doczi, T., Thomas, G.T.: Successful removal of an intrapontine hematoma. J. Neurol. Neurosurg. Psychiat. 42, 1058-1061 (1979)

7. Epstein, A.N.: Primary massive pontine hemorrhage; clinico-pathological study. J. Neuropath. exp. Neurol. 10, 426-448 (1951)

8. Fang, H.C.H., Foley, J.M.: Hypertensive hemorrhages of the pons and cerebellum. Arch. Neurol. Psychiat. 72, 638-639 (1954)

9. Freytag, E.: Fatal hypertensive intracerebral hematomas: a survey of the pathological anatomy of 393 cases. J. Neurol. Neurosurg. Psychiat. 31, 616-620 (1968)

10. Gowers, W.R.: Diseases of the nervous system, p. 362. London: Churchill 1893

11. Hassler, R.: Erkrankungen der Oblongata, der Brücke und des Mittelhirnes. In: Handbuch der Inneren Medizin, 4. Aufl., Bd. V/3. Berlin, Heidelberg, New York: Springer 1953

12. Kase, C.S., Mauliby, G.O., Mohr, I.P.: Partial pontine hematomas. Neurology 30, 652-655 (1980)

13. Környey, S.T.: Rapidly fatal pontine hemorrhage. Arch. Neurol. 41, 793 (1939)

14. Lavi, E., Rothman, S., Reches, A.: Primary pontine hemorrhage with complete recovery. Arch. Neurol. 38, 320 (1981)

15. Ley, E.P.: Surgical vascular pathology of the brain stem. J. Neurosurg. Sciences 20, 101-108 (1976)

16. Lindheimer, I.H., Barrows, H.S.: Recurrent pontine hemorrhage incident to venous angioma of the pons. Bull. Los Ang. Neurol. Soc. 27, 128-133 (1962)

17. Luce, H.: Zum Kapitel der Pons Hämorrhagien. Dtsch. Z. Nervenheilk. 15, 327-363 (1899)

18. Mastaglia, F.L., Edis, B., Kakulas, B.A.: Medullary hemorrhage. J. Neurol. Neurosurg. Psychiat. 32, 221-225 (1969)

19. McCormick, W.F., Nofzinger, J.D.: "Cryptic" vascular malformation of the central nervous system. J. Neurosurg. 24, 865-875 (1966)

20. McCormick, W.F., Hardman, J.M., Boulter, T.R.: Vascular malformations of the brain with special reference to those occurring in the posterior fossa. J. Neurosurg. 28, 241-251 (1968)

21. Murphy, M.G.: Successful evacuation of acute pontine hematoma; case report. J. Neurosurg. 37, 224-225 (1972)

22. Payne, H.A., Maravilla, K.R., Levinstone, A., Heuter, J., Tindall, R.: Recovery from primary pontine hemorrhage. Annals Neurol. 4, 557-558 (1978)

23. Pia, H.W.: Vasculäre Mißbildungen und Erkrankungen der hinteren Schädelgrube und ihre Behandlung. H.N.O. 17, 165-171 (1969)

24. Schiefer, W.: Klinik der intracerebralen Massenblutungen und spontanen Hämatome. In: Der Hirnkreislauf, pp. 680-709. Stuttgart: Thieme 1972

25. Silverstein, A.: Primary pontine hemorrhage. In: Handbook of Clinical Neurology (P.J. Vinken and G.W. Bruyn), Vol. 12, 37-53

26. Teilmann, K.: Haemangiomas of the pons. Arch. Neurol. Psych. 69, 208-223 (1953)

27. Vaquero, J., Areitio, E., Leunda, G., Bravo, G.: Hematomas of the pons. Surg. Neurol. 14, 115-118 (1980)

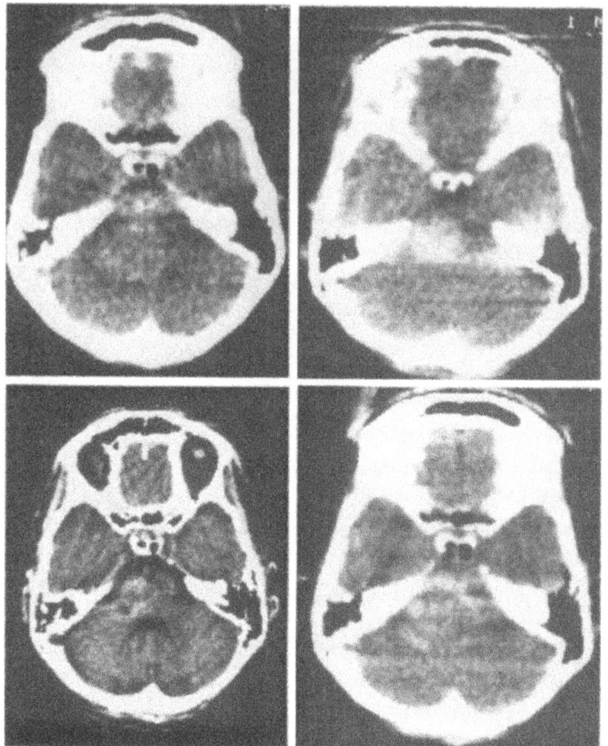

Fig. 1. Follow-up examinations of non-hypertensive hemorrhage. *(Upper left)* CT Jan. 1, 1979: Circular bleeding, non-homogeneous, located in the right side of the pons - extending over the mid-line, reaching to the left side. Compression of the fourth ventricle. Diameter = 2.5 cm; *(Upper right)* CT Feb. 2, 1979: Further extension of the bleeding to the left side as well as dorsally. Diameter increase to 4.5 cm. Increase of the non-homogeneous area; *(Lower left)* CT March 20, 1979: Lysis of the hemorrhage with an almost identical diameter, primarily isodense; *(Lower right)* CT Oct. 8, 1981: Approximately 2.2 cm non-homogeneous, partially high density; in part cystic formation without being space-occupying. Extension into the fourth ventricle and the brain stem cisterns. Possibly cavernous hemangioma

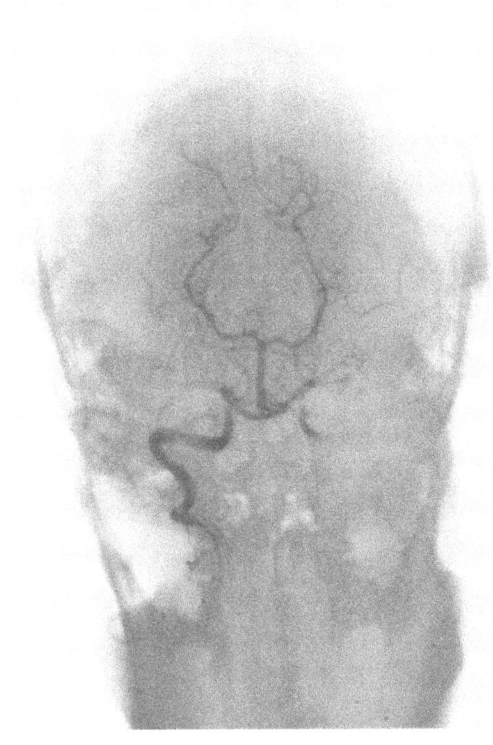

Fig. 2. Angiography of the right vertebral artery (subtraction). Feb. 6, 1979: Increase of volume in the brain stem. No pathological blood vessels

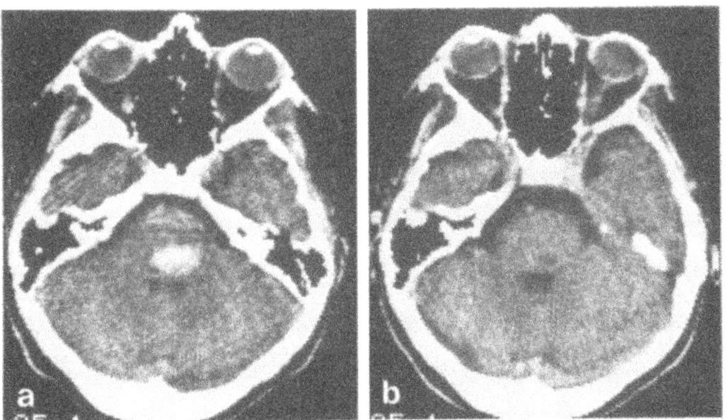

Fig. 3a, b. Follow-up examinations of a hypertensive hemorrhage. a CT Aug. 17, 1981: Bleeding on the floor of the fourth ventricle, with compression of the ventricle; b CT Oct. 2, 1981: Approximate size 0.7 x 0.5 cm blood-cavity

Posterior Fossa Tumors in Infancy

Clinical and Diagnostic Appearance of Tumors of the Posterior Fossa in Childhood, Results of a Retrospective Cooperative Study

B. J. M. Lamers

Neurochirurgische Klinik und Poliklinik, Universitätsklinikum der Gesamthochschule Essen, Hufelandstrasse 55, D-4300 Essen 1

The Co-operative Study was initiated in the summer of 1981 in order to obtain as much information as possible on the subject of posterior fossa tumors in childhood, and to evaluate the relevance of the clinical data for the prognosis of the individual patient as well as to review critically the results of our therapy.

Because of selection procedures the data presented are not representative for the F.R.G., nor do they claim completeness because of problems inherent in collecting such material.

Thirty-one clinics participated in the study (28 neurosurgical clinics) with a total of 1001 patients; 72 cases had to be excluded from further evaluation for various reasons. The follow-up period ranged from over 10 years for 164 patients, between 5 and 10 years for 232 patients and less than 5 years for 533 patients.

Histologically 38.9% of the children suffered from a medulloblastoma, 23.3% from a glioma (astrocytoma I-IV, glioblastoma, oligodendroglioma I-IV) and 17.8% from a spongioblastoma of the posterior fossa. The ependymoma followed with a frequency of 8.3%. All other types of new growths had an incidence of less than 1% and will therefore not be considered (Table 1). Another, more satisfying or more sophisticated classification proved not feasible, mainly because of gaps in information.

The sex distribution was almost equal in cases of spongioblastoma or glioma whereas there was a male preponderance in medulloblastoma or ependymoma (male - female ratio: 2 - 1) (Table 2) (Fig. 1). The age distribution showed uniformly a peak between 2 and 3 years except in glioma. Other increases of incidence were observed between 5 and 6, and 12 and 13 years in medulloblastoma, between 9 and 11 years in spongioblastoma and between 10 and 13 years in ependymoma.

The sex distribution showed no correlation with the age of the children.

Table 1. Histological diagnosis

Medulloblastoma	38.9%
Spongioblastoma	17.8%
Glioma	22.3%
Ependymoma	8.3%

Advances in Neurosurgery, Vol. 11
Edited by H.-P. Jensen, M. Brock, and M. Klinger

Table 2. Sex distribution on admission

	♂		♀
Medulloblastoma	2	:	1
Spongioblastoma	5	:	4
Glioma	6	:	5
Ependymoma	2	:	1

Regarding the clinical state on admission, there was an obvious tendency towards a bad condition, especially in cases of medulloblastoma or ependymoma (according to the Karnowsky-index).

Karnowsky index	0-40%	50-70%	80-100%
Medulloblastoma	52	27	21
Spongioblastoma	33	24	43
Glioma	39	26	35
Ependymoma	51	24	25

Neurologically, almost 80% of the children showed signs of raised ICP, most pronounced in cases of medulloblastoma or ependymoma (Table 3). Papilledema, however, was seen less in ependymoma (64%) than in the other tumor groups (75%). Cerebellar dysfunction was observed in 78% of the patients except in cases of ependymoma (60%). Ataxia was the most frequent symptom here, followed by dysdiadochokinesia and dysmetria. Disturbances of cranial nerve function were also frequent (50%) especially in ependymoma (62%). Here the abducens nerve was usually affected, in contrast to the other groups, where the facial and caudal cranial nerves generally were involved. Nystagmus (Table 4) was the leading symptom in the complex vestibulary malfunctions (50%) except in ependymoma (34%), impairment of consciousness (27%) and respiration (4%) however was diagnosed here twice as often. Psychiatric disturbances, prominent in cases of pyramidal dysfunction with an incidence of 10%, played a minor role in the clinical picture (Table 5).

Table 3. Neurological findings on admission (in%)

	Signs of raised ICP	Cerebellar dysfunction	Cranial nerve malfunction
Medulloblastoma	87	78	49
Spongioblastoma	79	80	46
Glioma	78	77	46
Ependymoma	84	58	69

Table 4. Neurological findings on admission (in %)

	Vestibular dysfunction	Disturbances of consciousness	Psychiatric symptoms
Medulloblastoma	52	17	10
Spongioblastoma	34	12	18
Glioma	50	11	8
Ependymoma	45	27	10

Table 5. Neurological findings on admission (in %)

	Pyramidal motor dysfunction	Disturbances of speech	Respiratory dysfunction
Medulloblastoma	8	7	4
Spongioblastoma	5	6	<1
Glioma	9	<1	2
Ependymoma	5	1	4

Focusing on the extension of tumor growth, it was somewhat depressing to establish that the neoplasms, independent of their histological nature had a respectable size at diagnosis already (Fig. 2).

The preferred location of the diverse tumor types followed a known pattern (Table 6).

The impression arose, that the poor condition at admission was due to the site of the tumor rather than to its size or the presence of hydrocephalus, as ventricular dilatation had an equal incidence (± 70%) in all. In this respect location in the fourth ventricle seemed to play a particular role. Reviewing the radiological diagnostic methods, the plain skull X-ray already showed pathological findings in 55%, mostly signs of raised ICP. Bone erosion was seldom seen and pathological calcification proved even rarer (4 i.e. 1%), and usually seen in cases of ependymoma or spongioblastoma. In 75% a pathological finding in an angiogram, generally vascular displacement, stressed the suspicion of a mass lesion in the posterior fossa. Tumor enhancement however was rare (7%), but seen as a rule in the more rapidly growing tumors.

Ventriculography or cisternography proved to be highly reliable in assessing a space-occupying lesion (95%), although a tumor contour was also seldom visible (14%), but could be expected more often in cases of medulloblastoma or ependymoma. If at all possible CAT scanning was even more accurate in this respect, with a score of 99.9%. Purely indirect signs of a mass lesion were rare (10%) and usually seen in cases of spongioblastoma and glioma. There existed a distinct tendency for medulloblastoma and ependymoma to visualize as a hyperdense structure whereas spongioblastoma and glioma tended to a hypodense imaging.

CAT appearance	Hyperdense (%)	Hypodense (%)
Medulloblastoma	64	22
Spongioblastoma	8	56
Glioma	38	64
Ependymoma	47	19

CAT →	No score	No pathol. finding	Pathol. finding
Ventriculography			
No score ↓	80	0	260
No pathol. finding	2	0	5
Pathol. finding	408	1	173

Table 6. Location of posterior fossa tumor (in %)

	IVth ventricle	Vermis	Cereb. hemisphere	Pons	Medulla obl.
Medulloblastoma	69	72	23	8	6
Spongioblastoma	28	53	27	13	1
Glioma	31	41	48	10	4
Ependymoma	79	16	16	6	12

Diagnostic accuracy in identifying a space-occupying lesion has obviously not been improved since the introduction of the CAT scanner, although the latter method for other reasons will rightly be preferred nowadays. Interestingly the tumor size on admission has not yet changed.

What factors now can be deduced as relevant in assessing the prognosis (for life and function) of the individual patient?

Age proved to be not essential and the same applied for sex, except for children suffering from ependymoma, where the mortality rate amongst boys was twice as high. The tumor type had a striking influence. The mortality in medulloblastoma and ependymoma children at 56% and 62% resp. was more than twice as high as in cases of spongioblastoma or glioma (21% and 26%). The effect on the quality of outcome for survivors was much less obvious however. An extremely poor condition on admission was invariably accompanied by a bad outcome (in both respects) especially in medulloblastoma and spongioblastoma. However an excellent initial clinical state did not necessary mean an excellent outcome, where children suffering from medulloblastoma were concerned. It was interesting finding that it made no difference whether the children were treated before or after 1975 (Table 7).

The neurological state on admission had no significant influence on survival or performance in relation to symptoms of raised ICP, cerebellar dysfunction or vestibular malfunction. Spongioblastoma children had a slightly better survival quality in this respect. Signs of vital dysfunction proved very unfavorable. Cranial nerve distrubances, more pronounced in medulloblastoma, lead to a higher mortality rate, whereas the quality of outcome was decreased in ependymoma. In respect to psychiatric symptoms glioma and spongioblastoma had a more unfavorable prognosis as regards survival. Disturbances of speech generally implied a worse outcome, except perhaps in cases of ependymoma (Table 8).

Table 7. Clinical condition on admission ↔ prognosis

Karnowsky index (%)	✠	Survivors life quality
0 - 40	↑→↑↑↑↑	↓↓
50 - 70	= →↓	↓ = ↑
80 - 100	= →↓	= ↑

Table 8. Neurological findings on admission ↔ prognosis

	✠	Survivors life quality
Signs of raised ICP	= →↑	=
Vital symptoms	↑↑↑↑	?
Cerebellar dysfunction	=	=
Cranial nerve malfunction	= →↑	= →↓
Brain stem dysfunction	↑—↑↑	↓↓

Only excessive tumor growth (5 cm) was accompanied by a distinct rise in fatal outcome or bad performance of the survivors. Small new growths however, did not necessarily mean a better prognosis (Table 9) as seen in cases of spongioblastoma in regard to mortality rate. Focussing on tumor location, the pons as well as the medulla oblongata proved to be often fatal, with a high mortality and an extremely bad proformance. This tendency was less pronounced in spongioblastoma in respect of survival. A medulloblastoma, when situated in the vermis, had a raised mortality rate although this location in ependymoma meant a slightly worse outcome for quality of life. A lesion located in the fourth ventricle did not lead to a raised death rate when a medulloblastoma or a glioma was concerned, but the effect on life quality of the survivors was uncertain (Table 10).

Table 9. Size of tumor at operation ↔ prognosis

Diameter (cm)	✠	Survivors life quality
1	↓ = ↑	= –↓
2 - 3	=	=
3 - 5	=	= –↓
5	↑	↓–↓↓

Table 10. Tumor location involvement ↔ prognosis

	✠	Survivors life quality
IVth ventricle	(↓)=→↑	=–↑
Vermis	↓ –↑	=→↓
Cerebellar hemisphere	=→↓	=→↑
Pons	↑→↑↑	↓
Medulla oblongata	↓→↑↑	↓–↓↓

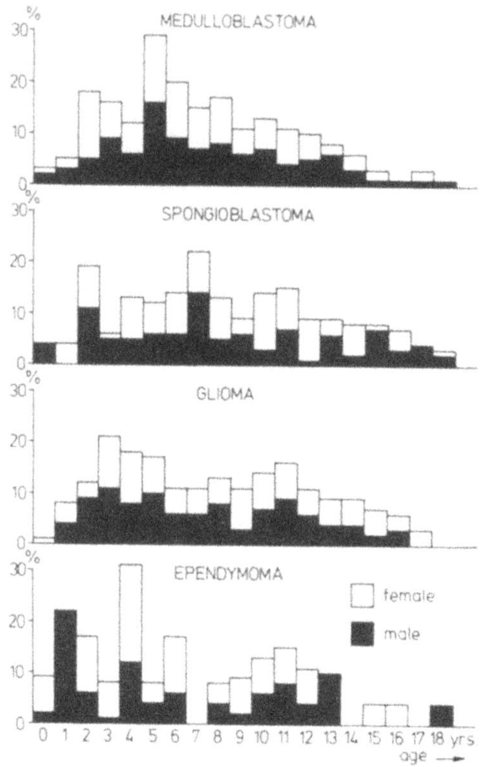

Fig. 1. Incidence and sex ratio of posterior fossa tumors depending on age

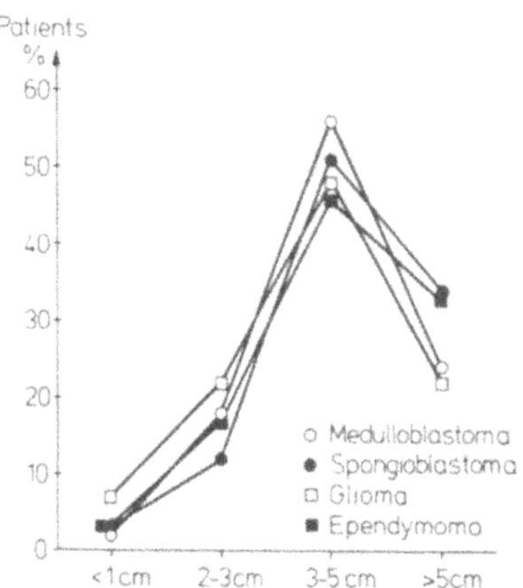

Fig. 2. Tumor size

Posterior Fossa Tumors in Children – Treatment and Prognosis

H.-E. Clar

Neurochirurgische Klinik und Poliklinik, Universitätsklinikum der Gesamthochschule Essen, Hufelandstrasse 55, D-4300 Essen 1

We have investigated a wide variety of factors which could conceivably influence the outcome of tumor treatment.

The subject of our report is the effect which treatment can exert on the prognosis for a patient.

The following factors were investigated:
1. Histology
2. Recurrence of tumor
3. Metastases
4. Recurrence and metastases
5. Operation
6. Radiation treatment
7. Chemotherapy
8. Combined forms of treatment.

The Histology

We were confronted with a fundamental difficulty in our endeavour to analyse morphological findings. Although general statements on the histology (e.g. astrocytoma) were available to us - albeit without any indication of the degree of malignancy - diagnoses such as ependymo-spongioblastoma are recorded only on very rare occasions. It was therefore deemed necessary to form more or less equivalent groups from the purely histological and biological standpoint. Roughly comparable numbers of cases were contained in the two groups.

Table 1

Medulloblastomas	39%	Including desmoplastic medulloblastomas
Spongioblastomas	18%	Spongioblastoma of the brain and pilocytic astrocytoma
Gliomas	25%	Oligodendroglioma Astrocytoma I + II Astrocytoma III + IV Glioblastoma Glioma
Ependymoma	8%	

Advances in Neurosurgery, Vol. 11
Edited by H.-P. Jensen, M. Brock, and M. Klinger

Our analysis of the life expectancy and prognosis was always carried out separately for each of the four categories of tumor. The extent to which the degree of malignancy of a tumor influences the outcome of treatment could be investigated only for the astrocytomas I + II/ III + IV. It was established that astrocytomas of low malignancy represented a better life expectancy and quality than the more malignant types or the outcome of all types of glioma taken together. A differentiation of the remaining types of tumor was not feasible because of diagnoses which were too general in nature.

Recurrence of Tumors

The morphology of a tumor is an important determinant of the recurrence rate in patients. For the two groups we mentioned above, this rate lies between 14% and 27%. However, it has not been possible to determine whether genuine recurrences were involved here or tumor remnants were establishing themselves again. For medulloblastomas and ependymomas the period between the first diagnosis and appearance of a recurrent tumor is <2 years. This is about half as long as for spongioblastomas and astrocytomas (which need around 4 years) (Table 2).

The survival rate of patients experiencing a recurrence falls dramatically. Some 40 - 80% of all children treated for tumors who subsequently suffer a recurrence, die within 6 - 9 months. Those patients who survived for a long period (>5 - 10 years) never experienced any recurrence - with the exception of spongioblastomas.

Accordingly, the goal of the initial treatment must be to undertake all possible measures to prevent any recurrent tumor formation. Once a recurrence has occurred, the situation for the patient becomes therapeutically hopeless (with the exception of spongioblastomas). The period from the first diagnosis during which a recurrence can occur is an extended one, though the average period is

Medulloblastoma	in < 2 years (max. 10 years)
Spongioblastoma	in 4 years (max. 19 years)
Glioma	in 3 years (max. 8 years)
Ependymoma	in < 2 years (max. 6 years)

Metastases

It is primarily tumors of the posterior cranial fossa which undergo metastasis. Whereas the frequency for gliomas and spongioblastomas is well below 1%, the above with 8% in the case of ependymomas represent a substantial risk and the mortality rate is very high. We found deposits mainly in the third and fourth ventricles; diffuse dissemination and spinal clusters were not observed. In the case of medulloblastomas over one third of all patients suffered metastasis. The mortality rate

Table 2. Recurrence

	(%)	† (%)	period (days)
Medulloblastoma	27	85	183
Spongioblastoma	15	40	200
Glioma	14	38	79
Ependymoma	17	70	171

among the latter was 75% with a mean life expectancy of 6 - 7 months (Table 3).

In terms of frequency of occurrence, the spinal canal is the most commonly attacked with some 15% of all the metastases. Basal cisterns and diffuse metastases account for most cases. Only in six cases (1%) were extraneural deposits observed. Metastases were found to have occurred in the bones, viscerally, in the pelvis and in the parotid. Only in one individual case did the extraneural metastases appear to be responsible for the death of the patient.

We were not able to demonstrate any relationship between cell dissemination and shunting operations in our collective study. Only in the case of one child was a valve inserted after metastases had previously been found in the parotid gland.

The extent to which extraneural metastases represent a major threat to the patient is unknown. Our own findings are contrary to those of KLEINMANN et al., who indicated an incidence of 5% and who warned of increased mortality. The proof of the existence of extraneural deposits is frequently obtained by chance or occurs at the same time as generalised metastases in the cerebrospinal fluid system.

Recurrence and Metastasis

Only in 15% of all cases of medulloblastoma were recurrence and metastasis observed together; the patients then died within a period of seven months (Table 4).

These observations support the assertion that as the illness runs its course the medulloblastoma commonly permeates the whole cerebrospinal fluid system. Both invasion of the meninges by the tumor as well as new colonisation of a tumor-free region by tumor cells from metastases after treatment are conceivable.

The fate of the patient is sealed when recurrence and metastasis occur together. In our study *no* child with either of these complications survived for more than five years.

Since the possibilities for treatment are usually restricted in the case of recurrence/metastasis (e.g. a second irradiation may be tried), the goal of the initial therapy must be the elimination of such risks. Where feasible this should be attempted by devising a logical scheme of treatment for the patient. Our study has revealed that only a pri-

Table 3. Metastasis

		✝ (%)	Period (days)
Medulloblastoma	38%	75	<200
Ependymoma	8%	80	

Table 4. Recurrence and metastasis

		✝ (%)	Period (days)
Medulloblastoma	15%	100	<200

mary "systemic treatment" (e.g. irradiation of the whole brain and spinal canal) can be regarded as logical in this context after a "systemic disease" has established itself.

Operation

The indication for a tumor operation arises from the necessity of establishing proof of the histological diagnosis and of removing the proven tumor to a large extent or totally. Only then can further treatment be undertaken.

No Operation

In 6 - 8% of all patients no operation on the tumor was carried out. Most of these patients were in a very poor state when admitted (KARNOWSKI 30%) and practically all of them died shortly after being diagnosed. Altogether, the mortality among these patients was ten times that for the patients operated upon.

One Operation

The overwhelming majority of patients had an operation performed on them to remove their tumor, namely:

	(%)	Operative mortality (%)
Medulloblastoma	91	14
Spongioblastoma	77	12
Glioma	80	7
Ependymoma	86	31

We have endeavoured to investigate the mortality of this operation on the basis of its extent. RAIMONDI has suggested that partial extirpation, e.g. for a medulloblastoma, can yield worse results than total extirpation.

Unfortunately, it proved to be impossible to make any differentiation using the records at our disposal since the reports on the extent of the operation were insufficient.

Removal by operation of individual tumors resulted in a significant improvement in both the life expectancy and quality of the surviving children.

	Rise of quality of life (%)	Period of observation
Medulloblastoma	15	>4 months
Spongioblastoma	20	>3 years
Glioma	30	>2 years
Ependymoma	30	>2 years

The results of the operation in the particular case of medulloblastomas are, however, not so promising.

This fact underlines the necessity of primary operative treatment for tumors of the posterior cranial fossa in children.

More than One Operation

More than one tumor operation was undertaken on:

Medulloblastoma 13% - Spongioblastoma 14% - Glioma 12% - Ependymoma 6%.

Spongioblastomas a quality of life equal to 70% for >8 years
Gliomas a quality of life equal to 74% for 4 - 5 years.

The period of observation was twice as long as the average for patients who had undergone only one operation.

No assessment of the influence of operations for recurrence in patients suffering from medulloblastomas or ependymomas could be made owing to the small number of cases.

This observation leads us to the conclusion that for gliomas and spongioblastomas the indication to operate for a recurrence should be more frequently considered than in the past.

Shunting Operation

Roughly one third of all the children with tumors in the posterior cranial fossa were fitted with a shunt. However, no difference in the outcome either as regards the quality of life or its expectancy could be detected for any of the tumors. Even for the long-term prognosis the manner of relieving the pressure plays no significant role. We did not consider temporary external ventricular drainage.

Radiation Treatment

The irradiation of patients was carried out in a variety of different ways, both with respect to the radiation equipment used and the dose and field size. We have not studied the effects of partial irradiation. Most frequently the therapy was performed by means of a cobalt apparatus, though treatments were also made using accelerators or combined cobalt/accelerator regimens. Twenty patients had been treated with conventional X-rays (ortho voltage).

The total dosage applied to patients was just as variable as the equipment used. This proved to be relatively low both for the individual types of tumor involved (30 - 45 GY) and in regard to the radiation fields (large field, small field, spinal canal 30/40/20 GY). Numerous authors have indicated that during irradiation of the brain area the tumor dosage should lie between 50 - 60 GY and the large field dose including the spinal canal should exceed 36 GY.

In view of the variable values reported, no further attempt to analyse the influence of field size or the focal dosage were made. Only the influence of completed treatment subsequent to an operation was investigated (Table 5).

In the case of medulloblastomas, there was a definite improvement of the quality of life of some 10% as against the non-irradiated patients, as well as an extension of life expectancy of around two years. Only

Table 5. Radiation treatment (mean doses)

	%	GY
Medulloblastoma	74	30 - 40
Spongioblastoma	10	41
Glioma	16	35 - 45
Ependymoma	53	21 - 46

those patients who were irradiated survived for more than five years. The life table also reveals a significant difference in the two curves. Thus, in spite of the varying dosages, the effectiveness of radiation therapy can now be regarded as established; this certainly applies to the patients in our cooperative study.

It would appear that for ependymomas, radiation treatment is also effective, though the individual patient groups here were too small to warrant a definitive statement.

No assessment can be made for either spongioblastomas or gliomas on account of the limited numbers of patients involved.

It is apparent that the focal dosage applied to most of the patients was set too low. We recommend that in future the possibilities are more fully utilized, especially in the treatment of medulloblastomas. In particular, when planning radiation treatment attention should be focussed on a "systemic irradiation" of the whole cerebrospinal fluid system.

The extent to which spongioblastomas or gliomas respond to radiation treatment remains unclear. Even here, however, the therapy must be consistently applied using a dosage between 50 and 60 GY. Orthovoltage therapy is now no longer permitted for the treatment of brain tumors.

Chemotherapy

Because the number of chemotherapeutic agents employed was so large and the individually treated groups of patients so small, it did not prove feasible to form any unified treatment groups. All that can be said here is that, apart from the SIOP study involving 13 patients, virtually every clinic has its own patent brand of therapy.

As though this were not enough, most patients received non-standardised combinations, dosages and courses of treatment. For this reason we find it impossible to make any meaningful statements here. Moreover, we were unable to follow up any effect which chemotherapy may have on the prognosis for patients.

The numbers of patients treated for spongioblastomas, gliomas and ependymomas were too small to enable us to make any statements. It would appear that chemotherapy is not really indicated here.

For medulloblastomas in general - bearing in mind the reservations expressed above - we were able to demonstrate *no* significant difference either in life expectancy or quality even for patients who survived for long periods between

Operation + Irradiation
and
Operation + Irradiation + Chemotherapy (Fig. 1).

This observation is confirmed by our analysis of life tables. Since even the SIOP study (BLOOM) and others have not yet been able to secure any significant increase in survival times, experiments based on chemotherapeutic agents administered to small numbers of patients cannot be justified on ethical grounds.

Combined Forms of Treatment

Analysis of the results of our cooperative study support the following now well-established types of treatment:

Medulloblastoma	Operation + irradiation
Spongioblastomas) Glioma)	Operation, poss. operation for recurrence
Ependymoma	Operation, poss. further irradiation

The treatment of choice for gliomas and spongioblastomas, as tumors not basically sensitive to irradiation, remains operation. Even a tumor which has not been entirely extirpated should be re-extirpated rather than irradiated, provided the patient is in good general health. The decision about operating a second time should be made at the latest when a recurrence occurs. Irradiation can be performed at a later stage.

In the case of ependymomas, irradiation should be considered only after maximal extirpation in the initial operation. It should then be carried out logically using sufficiently large bursts and dosages.

We have already discussed the operative and radiation treatment of medulloblastomas. The irradiation should be performed only with high voltage equipment (cobalt or accelerator) by an experienced doctor fully acquainted with the complicated technology used for irradiating the entire cerebrospinal axis. The best chance for a patient to remain free of recurrence or metastasis lies in a well-designed and logical initial treatment.

Although this treatment has resulted in a clearcut increase in life expectancy and quality - in spite of its limitations - new ways must still be sought to attain further improvements. New agents and combinations of cytostatic drugs *must* be explored. This, however, should not be left to individuals with small numbers of patients to undertake mutually randomised studies (such as the SIOP study).

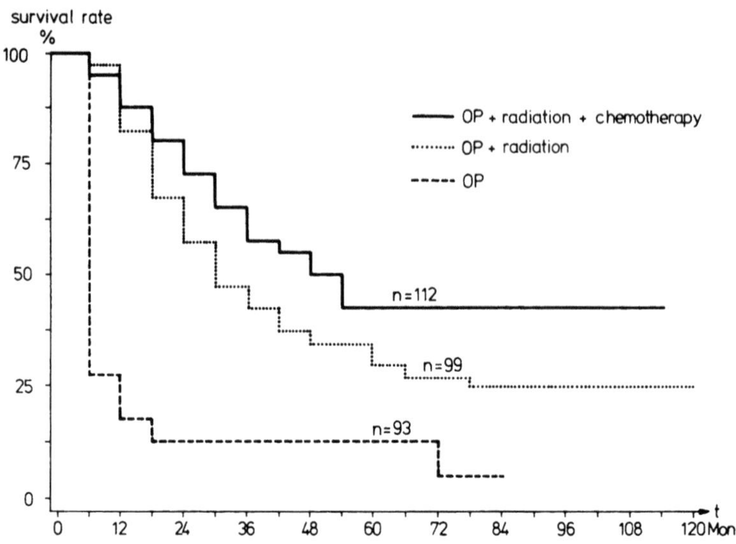

Fig. 1. Medulloblastoma: Types of treatment in relation to survival rate

Critique of the Findings

W. Grote

Neurochirurgische Klinik und Poliklinik, Universitätsklinikum der Gesamthochschule Essen, Hufelandstrasse 55, D-4300 Essen 1

Cooperative studies on particular patterns of disease doubtless present us with some considerable advantages. By such means one is able to study in depth a wide range of patients and thus make more reliable statements. A clearer picture of clinical and diagnostic problems emerges when use is made of extensive sets of data on the development of illnesses, on purely clinical symptomatology and on the results of investigations employing various types of equipment. Moreover, the larger numbers involved enable a fuller assessment to be made of the manifold therapeutic approaches and their associated prognoses.

On the debit side, important single findings, individual experiences and useful leads may be overlooked: they tend to get submerged in the welter of averaged values and trend predictions. We believe that, in addition to comprehensive statements based on statistics, observations and experiences originating from small groups as well as those arising from individuals can be both significant and valuable in shedding light on the total situation.

For these reasons, it would seem advisable to offer the subject for discussion by a broad panel of experts, after the findings on the joint study on children with tumors of the posterior cranial fossa have been delivered. The experiences and problems seen from the point of view of neuropathologists, oncologists, pediatricians and neurosurgeons should be presented and discussed.

At this point it is my pleasant duty to thank all of those involved in this study for their unstinting help. Colleagues from 31 clinics provided us with 1001 case histories or completed questionnaires. We have been able to evaluate 926 of these. The rest had to be discarded as in most instances the upper age limit was exceeded, though in a few cases other reasons obliged us to reject them.

My own colleagues have gone to great pains to read through, to assess and draw conclusions from the great wealth of data made available to us. I should like to extend to them also a special word of thanks.

Perhaps I may be permitted once again at this juncture to outline and summarize those facts arising from the cooperative study which seems to me especially noteworthy and significant.

The observation period for all patients involved in our study extends over 30 years, from 1951 to 1981. Over this period we were able to observe 164 children = 17% for 10 years, another 232 = 35% for a period of 5 to 10 years and most of the patients (533 = 58%) only up to 5 years. In due course an undetermined number of those in the latter two groups will move up into the next more favorable category.

As a result of the histological restriction, medulloblastomas are by far the most common with almost 40%. These are followed by ependymomas and other gliomas with 23%, after which come the spongioblastomas with 18%; all other tumors represent 1% and are thus to be regarded as rarities.

On admission to the clinic about one half of the children with medulloblastomas and ependymomas were in a very poor state of health, corresponding to a value of 10 - 40% on the KARNOWSKY scale - a depressing statistic.

With regard to clinical symptoms, signs of pressure on the brain led with 80%, followed by symptoms of the cerebellum in 70% of the cases. The only exception were the ependymomas in which cranial nerve disturbances predominated.

The size of the tumor in medulloblastomas, spongioblastomas and ependymomas was about the same. Medulloblastomas were located primarily in the cerebellum, vermis and fourth ventricle, spongioblastomas in the hemispheres and ependymomas largely in the region of the fourth ventricle.

The location of these tumors is in accordance with the clinical and neurological symptoms manifested.

All of the above data are well-known and cannot be regarded as representing new knowledge.

With reference to diagnoses made with the aid of apparatus, it is worth mentioning that straight X-ray pictures suggest tumors in 55% of the cases from signs of pressure on the brain, the destruction of bones and calcium deposits.

Angiography provided evidence of spatial expansion in 75% of cases, primarily in the form of displaced arterial and venous vessels.

Ventriculography was successful in identifying a tumor in 95% of the cases.

Computer tomography is able to establish the presence of a tumor in the posterior cranial fossa in 99.9% of the cases.

It seems remarkable that X-ray tumor identification before the "CT era" was not significantly worse. Of course, it should not be forgotten that computer tomography represents a far less invasive diagnostic measure than ventriculography. Not surprisingly, the former has now virtually replaced ventriculography. There has been no reduction in tumor size in our CT age as compared to earlier times.

What factors have an effect on the prognosis?

The sex of the patient has no effect, apart from male ependymoma sufferers.

The age at which the disease manifests itself has no significance for the quality of survival.

Ependymomas have the worst chance of survival with a mortality rate of 62%, followed by medulloblastomas with 56%, spongioblastomas with 26% and other gliomas with 21%.

When neurological examination indicates disturbances of consciousness, this can be expected to have a negative influence on the prognosis. Although signs of pressure on the brain and symptoms associated with the cerebellum are of no significance for the prognosis, disturbances in breathing greatly reduce the chance of survival. Negative effects on the survival are exerted especially by an increasing size of the tumor and understandably a tumor location which involves the caudal brain stem.

Turning now to surgical treatment, we mention here again the mortality rate for the operations:

Medulloblastomas	14%
Spongioblastomas	12%
Ependymomas	over 30%

and for the remaining gliomas around 7%.

These figures appear to be rather high at first sight and in my opinion could be improved upon. In this context it is important to mention that neither complete, subtotal nor partial extirpation of the tumor makes any clear-cut difference to the mortality rate for the operation. We conclude from this that one may and should attempt to operate as radically as possible. Too timid an approach to the removal of a tumor will produce a considerably more unfavorable starting point for any subsequent treatment.

It also became clear that operations for recurrent spongioblastomas and other gliomas should be performed more frequently than hitherto. Not only can one thereby lengthen the life expectancy several times but a good quality of life is to be expected during this period. There seems to me to be an important lesson here. No statement can be made at present on medulloblastoma and ependymoma cases.

The difficulty in making meaningful statements on the efficacy of the various types of operative treatment is compounded when it comes to expressing an opinion on one individual treatment - especially on auxiliary irradiation treatment - and chemotherapy at the present time. Irradiation is carried out in manifold different ways and with very variable dosages. Chemotherapy is still only in its infancy, even though it does appear to be a promising approach.

To conclude, I should now like to present to you and propose for adoption the plan we have devised for the treatment of tumors in the posterior cranial fossa in children based on our survey and the evaluation of data and information provided for the joint study:

1. Medulloblastomas

1.1. Maximal surgical removal of tumor, "pre-planned" on the basis of CT findings
1.2. Ventriculo-atrial shunt (PHD) to relieve pressure on the brain
1.3. Consistent whole brain + cerebrospinal irradiation with high voltage equipment (40 GY + 20 GY focal dose + 36 GY spinal canal) using experienced radiation personnel
1.4. Chemotherapy only in randomized studies

2. Spongioblastomas

2.1. Attempt at complete surgical removal
2.2. PHD for pressure on the brain
2.3. For a recurrence or increase in size of a residual tumor renewed surgical removal

3. Gliomas

3.1. Maximal surgical removal
3.2. For pressure on the brain, PHD
3.3. Irradiation using high voltage therapy (with a focal dose not less than 50 GY) *may* be performed
3.4. For recurrence attempt a further operation

4. Ependymomas

4.1. Attempt at complete surgical removal
4.2. For pressure on the brain, PHD
4.3. Irradiation using high voltage therapy (over 40 GY focal dose). If tumor cells found in cerebrospinal fluid use "systemic irradiation" - spinal canal + whole brain (with 36 GY)
4.4. In the case of recurrence or growth of a residual tumor perform further operation

It is my hope and conviction that the above proposals will form the basis of an efficacious treatment.

Age-Dependent Morphobiological Aspects of "Medulloblastomas"

F. Gullotta
Institut für Neuropathologie der Universität Bonn, Sigmund-Freud-Strasse 25, D-5300 Bonn-Venusberg

The term "medulloblastoma" covers a large group of malignant tumors with common pathogenesis (neural crest) but rather polymorphic patterns (GULLOTTA 1967; 1979). We can find in fact:

1. Isomorphic, "undifferentiated" tumors.

2. Tumors with mixed structures. Here, quantitative differences of both mesenchymal and neuroectodermal components in different parts of the same tumor can be found. This type of medulloblastoma is frequent, appearing mainly in children. An overgrowth of the mesenchymal component can usually be observed, mostly in relation to the patient's age. The final picture is that of a "desmoplastic" medulloblastoma.

3. "Desmoplastic" medulloblastoma, that is, *arachnoidal sarcoma*. This is a common tumor type, especially in adults, i.e. in patients older than 16 years, and represents in our opinion the most differentiated mesenchymal form of the whole medulloblastoma family.

Being aware of the histological polymorphism of these tumors and of their frequent transitional forms, we can explain easily why the discussion concerning the origin of medulloblastoma is still controversal. It must not be forgotten that often the tumor tissue is soft (especially in children) and therefore the surgeon removes most of it by suction. In these cases, samples for histological examination consist mainly of small bits of infiltrated cerebellar tissue, and it is histologically very difficult to determine which of the tissue structures are neoplastic or autochthonous.

"Pure" neuroectodermal types of medulloblastoma (neuroblastoma) are infrequent, the majority of cases reported in literature consisting of mixed tumors or of cerebellar tissue infiltrated by undifferentiated cells. This can easily be demonstrated by studying microscopically the tumor samples by low magnification. GFAP-method can be also very useful in distinguishing true tumor tissue from pre-existing structures (see SCHINDLER and GULLOTTA).

In all these types, the mesenchymal tissue component usually overgrows the neuroectodermal one, and this in relation to age. This age-dependent growth of mesenchymal tissue is connected with other biological factors, every one related to and depending on each other: nodular growth in adults, contrasting with diffuse growth in children, and shifting of

Advances in Neurosurgery, Vol. 11
Edited by H.-P. Jensen, M. Brock, and M. Klinger

tumor location (deep midline in children - superficial, lateral, also extracerebellar in adults).

Age of patient; location of tumor; its manner of growth and its histological structure are four biological factors that can play a very important role in the postoperative course of these neoplasms. Each of these factors, considered singly, does not seem to be important. In fact, children with desmoplastic medulloblastomas have the same poor prognosis as patients with non-desmoplastic tumors. Whereas a desmoplastic tumor, at the same location, but in an adult, can be completely removed, presenting a longer postoperative course until the appearance of recurrence (GULLOTTA and NEUMANN, 1980; GULLOTTA 1981). It is evident that some kind of relation between tumor type and the age of patients does exist, and that radiation treatment or chemotherapy are not solely responsible for the better prognosis of these tumor forms in adults; they act on or together with favorable biological factors. In evaluating the efficacy of any postoperative treatment, the biological factors mentioned above have to be taken into consideration.

Finally a last point has to be emphasized. Sometimes an unusually long survival of a child's "classical" medulloblastoma is reported. It is important to insist on the fact that not every cellular tumor of the posterior fossa has to be a medulloblastoma. Ependymomas and also small-celled astrocytomas may simulate a medulloblastoma, especially if only a few, small tissue fragments are available for histology. The consequences of such an incorrect diagnosis are of course erroneous pathogenetic, prognostic and therapeutic interpretations. We run the risk of recording therapeutic successes on the wrong tumor, whose postoperative course has probably little or nothing to do with the postoperative treatment. The outcome of posterior fossa tumors depends chiefly on their exact location, the modern operation techniques, anesthesia and intensive care as well.

References

Gullotta, F.: Das sogenannte Medulloblastom. Berlin, Heidelberg, New York: Springer 1967

Gullotta, F.: Morphological structure and biological behaviour of medulloblastoma in relation to age. In "Multidisciplinary Aspects of Brain Tumor Therapy; (Paoletti-Walker-Butti-Knerich, eds.). Amsterdam, New York, Oxford: Elsevier / North Holland Biomedical Press 1979

Gullotta, F.: Morphological and Biological Basis for the classification of brain tumors. in "Advances and Technical Standards in Neurosurgery", Krayenbühl, H. (ed.), Vol. 8. Berlin, Heidelberg, New York: Springer 1981

Gullotta, F., Neumann, J.: Medulloblastome und cerebelläre Sarkome. Eine histologisch-katamnestische Untersuchung. Neurochirurgia, Stuttgart, 23, 35-40 (1980)

Immunohistopathology in Diagnostic Neuropathology

H. H. Goebel[1], B. Rama[1], and M. Schlie[2]*

[1] Abteilung für Neuropathologie, Pathologisches Institut der Johannes Gutenberg-Universität, Langenbeckstrasse 1, D-6500 Mainz

[2] Abteilung für Neurochirurgie, Medizinische Einrichtungen der Universität Göttingen, Robert-Koch-Strasse 40, D-3400 Göttingen

Introduction

Within the field of surgical pathology, immunohistochemistry is now frequently applied to the morphological diagnosis of lymphomas (11), certain carcinomas and other selected types of tumor (12). In neuropathology, the study of neuro-oncological and non-neoplastic diseases may also receive diagnostic support from performing immunohistological techniques, which encompass immunofluorescent and immunoperoxidase methods. The following report represents a survey of our experience in this recently developed field of diagnostic neuropathology.

Methods and Material

Conventionally formalin-fixed and paraffin-embedded tissue sections of surgical and autopsy specimens were deparaffinized and submitted to the triple layer peroxidase-antiperoxidase technique (PAP) (10). The material consisted of surgically removed brain tumors, such as astrocytomas, oligodendrogliomas, glioblastomas, gliosarcomas, medulloblastomas, ependymomas, hemangioblastomas, neuroblastomas, pituitary adenomas, and post-mortem autopsy specimens, chiefly affected by herpes simplex infection.

Antisera against the glial fibrillary acidic protein (GFAP), hormones of the anterior pituitary gland, herpes virus antigen and factor VIII were all obtained from either private or commercial sources. Control sections were treated by replacing the specific antiserum by non-immunized serum of the same animal species.

Results

GFAP

GFAP is present in reactive and neoplastic astrocytes as amply documented in reactively gliotic tissues and astrocytomas of various degrees

*We are grateful to Dr. L.F. Eng, Veterans Administration Hospital, Standford (USA), for providing antiserum against GFAP, to the National Pituitary Agency, Baltimore (USA), for supplying antisera against pituitary hormones. These morphological studies were financially supported by the "Büttner-Stiftung" Göttingen

Advances in Neurosurgery, Vol. 11
Edited by H.-P. Jensen, M. Brock, and M. Klinger

of anaplasia (Fig. 1). In glioblastomas, GFAP-positive cells are frequently encountered, especially in large cells with bizarre nuclei. In gliosarcomas, the mesenchymal sarcomatous part does not contain GFAP, whereas the astrocytic part does (Fig. 2). Subependymomas and ependymomas may also contain GFAP. Neoplastic astrocytes may spread beyond their original parenchymatous confines (Fig. 1). GFAP-positive astrocytes may occasionally be entrapped within hemangioblastomas, medulloblastomas, neuroblastomas, pineal tumors or lymphomas. The fact that in glioblastomas, and even more differentiated astrocytomas, neoplastic cells may remain non-reactive to the GFAP antiserum, indicates a variable number of astrocytes, of possible immaturity.

Anterior Pituitary Hormones

Immunohistological demonstration of various hormones such as GH, FSH, LH, ACTH, TSH and prolactin in pituitary adenomas provides more precise classification. Prolactin-containing microadenomas may be detected in autopsy specimens (Fig. 3). More than one hormone may occasionally be encountered in the same pituitary adenoma.

Herpes Simplex Virus Antigen

In biopsy and autopsy specimens of necrotizing encephalitis, antigens of the herpes simplex virus may be demonstrated in neuronal perikarya and processes within oligodendrocytes, particularly inside nuclei, but not in inflammatory infiltrates. Pyramidal cells seem to be more often infected than other cortical neurons, and oligodendrocytes of the cortex are more frequently involved than those of the white matter.

Factor VIII

This endothelial marker (Fig. 4) is present within endothelial cells and neoplasms consisting of capillaries and endothelial cells.

Discussion

Immunohistological techniques are now a well established support in morphological diagnosis, particularly in that of neoplasms. While immunofluorescent methods pertain to those tumors containing intermediate filaments (3), development of the PAP technique applied to paraffin-embedded sections has proved to be a major advance, as tissues do not require special fixation and embedding procedures and also allow retrospective studies. The PAP technique is now a routine method in many neuropathology laboratories throughout the world and is employed in diagnosing astrocytomas and ependymomas and their various forms. The introduction of the PAP technique to demonstrate GFAP has also separated neoplastic entities, such as meningocerebral xanthoastrocytomas (6) or highly lipidized astrocytomas (5). The stromal cell of cerebellar hemangioblastomas has been interpreted in various ways, although it has always been based on the demonstration of GFAP in these cells, either being of astrocyte origin (7) or indicating uptake of GFAP rather than its intrinsic production (2). Extracerebral and extracranial growth of astroglial tumors may be correctly identified on account of their production of GFAP.

The once common classification of pituitary adenomas into chromophobe, basophilic and acidophilic types and another one based on morphological features as sinusoidal, fetal and diffuse adenomas now appears obsolete in view of the feasibility of detecting pituitary hormones within neoplastic cells. Recently, endocrine heterogeneity of individual pituitary adenomas has been shown (4, 8).

The morphological detection of viral antigens in human brain tissue is still in its preliminary stages. When herpes simplex encephalitis is suspected and a biopsy is performed, application of anti-herpes antisera to the tissues provides a less cumbersome and often more accurate investigative approach than does the documentation of viral particles by electron microscopy. Viral antigens in subacute sclerosing panencephalitis, progressive multifocal leucoencephalopathy and rabies may also be immunohistologically detected (1).

More rarely, malignant angioendotheliomatosis or hemangioendotheliosarcoma may require demonstration of factor VIII (9) as an endothelial marker. The presence of lysozyme and alpha-1-antitrypson marks histiocytes and granulocytes in respective lesions. More detailed classification of primary cerebral lymphomas is best left to the general pathologist, who can classify several subentities of the lymphoma group immunohistologically.

Conclusion

Immunohistopathology is an ever increasing subspeciality in diagnostic morphology. This trend has finally also reached neuropathology. The pace of further progress is largely set by the speed, with which new cellular antigens are identified, isolated and purified to procure specific antisera. Those against neuronal and oligodendroglial cells are eagerly awaited to be diagnostically applied to the respective neoplasms. Immunohistology may also aid in identifying the original site of a cerebral metastasis.

Summary

The application of the peroxidase-antiperoxidase technique to conventionally formalin-fixed paraffin-embedded tissue is the most practical approach in immunohistopathological diagnosis. In neuropathology, the glial fibrillary acidic protein is the major target in astrocytomas, ependymomas and their variants. Demonstration of pituitary hormones in pituitary adenomas has led to reclassification of these neoplasms. Immunohistochemical evidence of herpes simplex viral antigen amplifies histological and serological as well as culture techniques in establishing the precise diagnosis of herpes simplex encephalitis. Other immunohistochemical applications pertain to the demonstration of immunoglobulins in lymphomas, of factor VIII as an endothelial marker, and lysozyme and alpha-1-antitrypsin to connote histiocytes.

References

1. Budka, H.: Zur Systembezogenheit von Virusinfektionen des ZNS: Verteilung der Virusantigene bei Rabies, Herpes-simplex-Encephalitis, SSPE und PML. Presentation at "Tagung Deutsche Gesellschaft für Neuropathologie und Neuroanatomie", München (FRG), April 1982

2. Deck, J.H.N., Rubinstein, L.J.: Glial fibrillary acidic protein in stromal cells of some capillary hemangioblastomas: significance and possible implications of an immunoperoxidase study. Acta Neuropathol. (Berl.) 54, 173-181 (1981)

3. Gabbiani, G., Kapanci, Y., Barazzone, P., Franke, W.W.: Immunochemical identification of intermediate-sized filaments in human neoplastic cells. A diagnostic aid for the surgical pathologist. Am. J. Pathol. 104, 206-216 (1981)

4. Heitz, P.U.: Multihormonal pituitary adenomas. Hormone Res. 10, 1-13 (1979)

5. Kepes, J.J., Rubinstein, L.J.: Malignant gliomas with heavily lipidized (foamy) tumor cells: A report of three cases with immunoperoxidase study. Cancer 47, 2451-2459 (1981)

6. Kepes, J.J., Rubinstein, L.J., Eng, L.F.: Pleomorphic xanthoastrocytoma. A distinct meningocerebral glioma of young subjects with relatively favorable prognosis. A study of 12 cases. Cancer 44, 1839-1852 (1979)

7. Kepes, J.J., Rengachary, S.S., Lee, S.H.: Astrocytes in hemangioblastomas of the central nervous system and their relationship to stromal cells. Acta Neuropathol. (Berl.) 47, 99-104 (1979)

8. Martinez, D., Barthe, D.: Heterogeneous pituitary adenomas. Virchows Arch. (Pathol. Anat.) 394, 221-223 (1982)

9. Sehested, M., Hou-Jensen, K.: Factor VIII related antigen as an endothelial cell marker in benign and malignant diseases. Virchows Arch. (Pathol. Anat.) 391, 217-225 (1981)

10. Sternberger, L.A.: Immunocytochemistry, second edition. New York: Wiley & Sons 1979

11. Taylor, C.R.: Immunohistological approach to tumor diagnosis. Oncology 35, 189-197 (1978)

12. Taylor, C.R., Kledzik, G.: Immunohistologic techniques in surgical pathology - a spectrum of "new" special stains. Hum. Pathol. 12, 590-596 (1981)

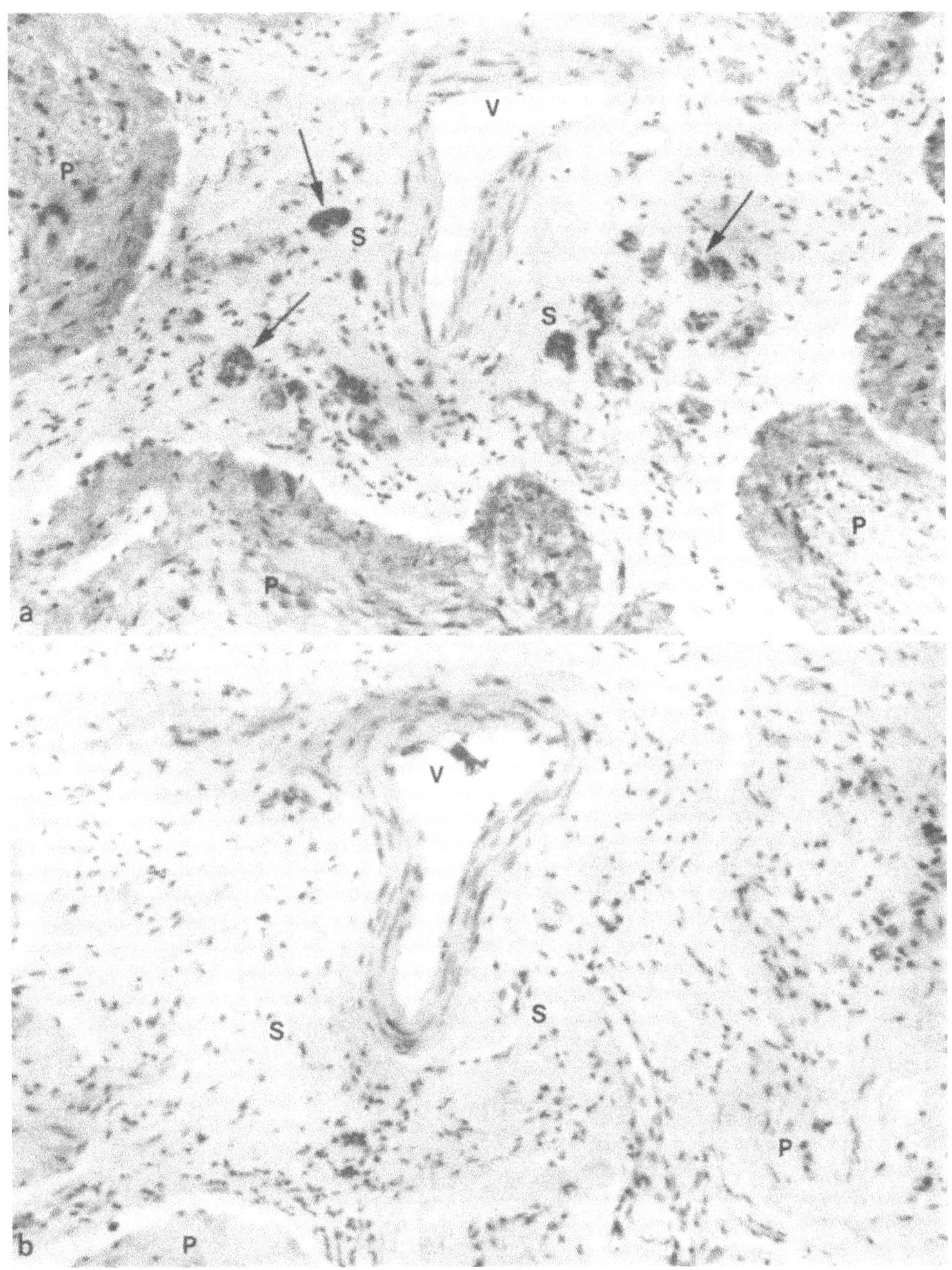

Fig. 1. a GFAP *(arrows)* is found in the optic glioma (pilocytic astrocatoma) both in the optic nerve parenchyma (P) and its vascular (V) septa (S). X 237.5; b control section without specific anti-GFAP antiserum; X 237.5; hematoxylin counterstain for nuclei

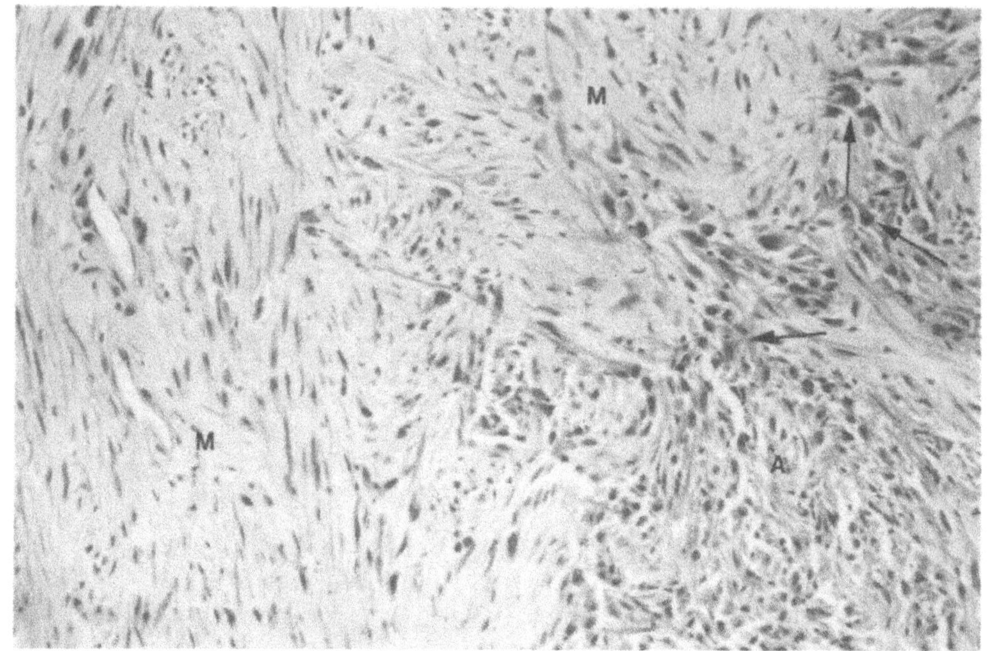

Fig. 2. GFAP *(arrows)* is confined to the astrocytic (A) part *(right)* of a gliosarcoma; the mesenchymal (M) part is on the *left*; hematoxylin counterstain for nuclei. X 475

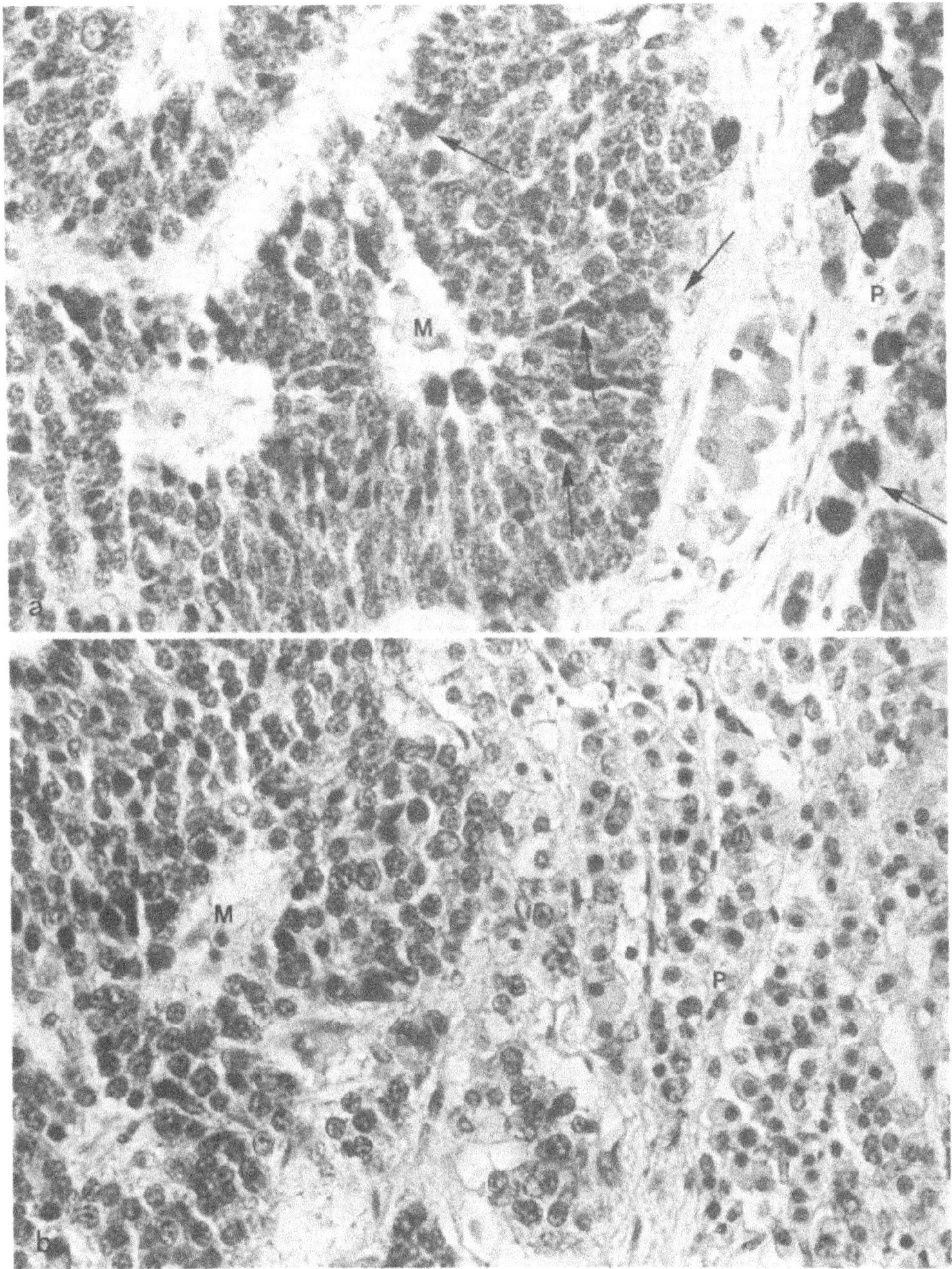

Fig. 3. a Prolactin *(arrows)* is present in regular anterior pituitary (P) epithelial cells *(right)* and in cells of an adjacent microadenoma (M) *(left)*; X 475. b control section without specific anti-prolactin antiserum. X 475; hematoxylin counterstain for nuclei

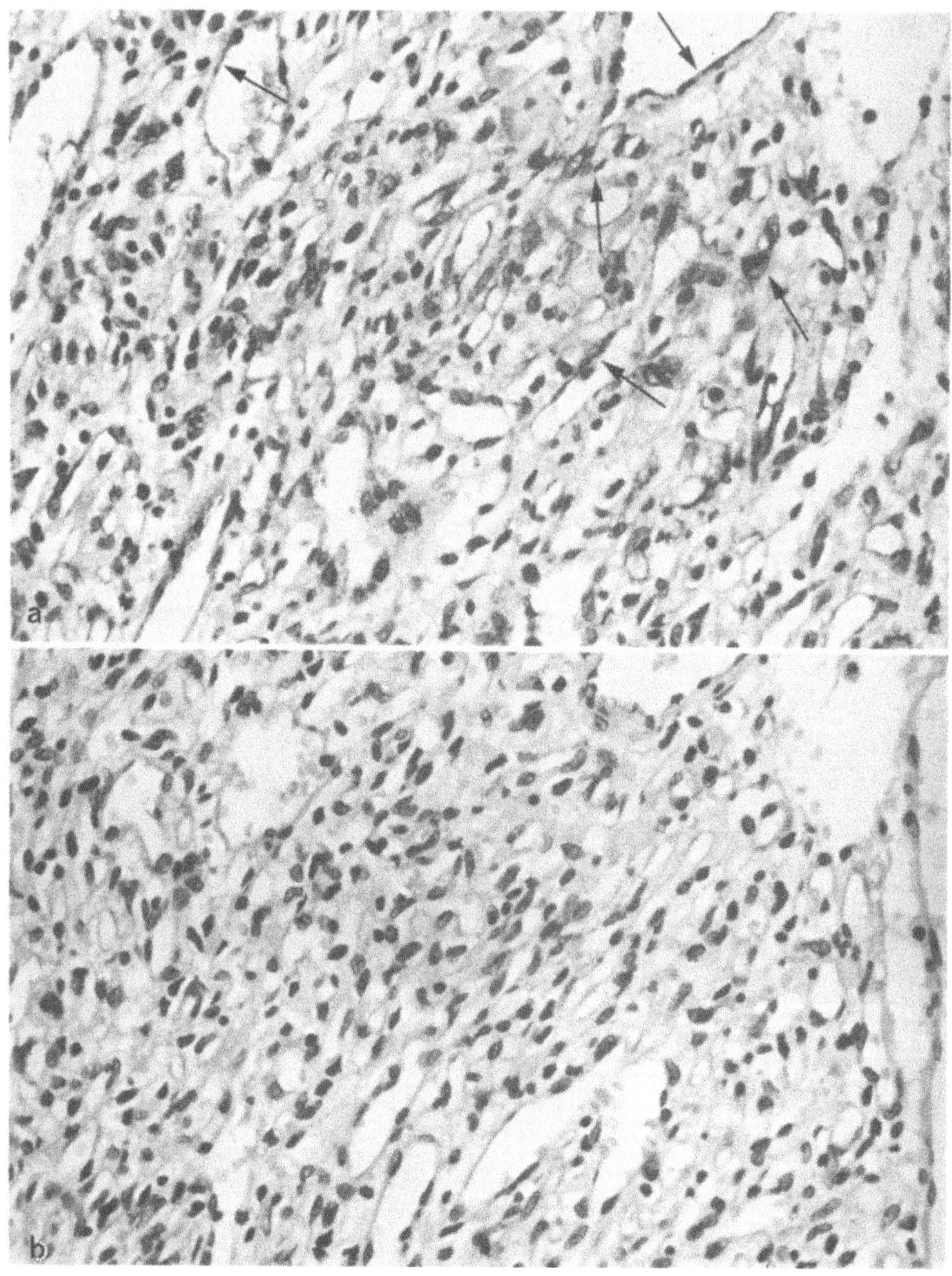

Fig. 4. a Factor VIII *(arrows)* is present in endothelial cells of a cerebellar hemangioblastoma; X 475. b control without specific anti-factor VIII antiserum; X 475; hematoxylin counterstain for nuclei

Glial Fibrillary Acid Protein in Medulloblastomas

E. Schindler and F. Gullotta
Institut für Neuropathologie der Universität Bonn, Sigmund-Freud-Strasse 25, D-5300 Bonn-Venusberg

Glial Fibrillary Acid Protein (GFAP) is regarded as an astroglial-specific component and its presence can therefore be used to control the degree of differentiation of glial elements (BIGNAMI and DAHL, 1974; 1977). This method is being used with increasing frequency in neuro-oncology, with some contradictory results, however (for review see De ARMOND et al., 1980).

We have investigated 52 posterior fossa tumors, histologically diagnosed as "medulloblastomas", with the unlabelled immunoperoxidase method after STERNBERGER, employing a GF-Antiserum kindly supplied by Drs. Doris DAHL and Amico BIGNAMI, Boston, Mass. (USA). The results can be summarized as follows:

Two cases were definitely identified as small-celled malignant gliomas and were therefore discarded from the medulloblastoma group *sensu strictu*. In this group, all three varieties (22 isomorphic; 8 mixed type; 20 desmoplastic tumors - see GULLOTTA, this journal (paper 39)) were present.

17 out of 50 medulloblastomas were GFAP-negative.
23 tumors were classified as "pseudopositives",

because their positivity was due to reactive and/or degenerating astrocytes intermingled with tumor cells. These astrocytes were easily recognized not only by their cellular shape and their strong GFAP-positivity, but also on the basis of their topographical distribution: the exact analysis of the tumor, especially of its manner of growth, permitted us in all these cases to distinguish "pure" tumor parts (GFAP-negative) from pseudopositive areas, i.e. infiltrated cerebellar tissue. Some positive astrocytes, reactive type, were also detected in a few arachnoidal sarcomas (i.e., "desmoplastic" medulloblastomas). A proliferation of glial cells from the molecular layer into the leptomeninges, in consequence of a local chronic irritation with breakage of pia-glia-barrier is well known since SPIELMEYER (1922).

In ten tumors, intermingled with neoplastic cells and bearing morphological similarities to them, some GFAP-positive cells were found. These cells were mostly scattered among tumor elements, but occasionally they also appeared as small clusters. Taking into account the theoretical possibility of a glial development from undifferentiated medulloblastoma cells, we did classify these cases as "positives". In five of them, however, a careful investigation of many additional tumor samples, also revealed pseudo-positive areas. It cannot therefore be

Advances in Neurosurgery, Vol. 11
Edited by H.-P. Jensen, M. Brock, and M. Klinger

excluded with certainty that these "positive" cells were also in reality, pre-existing reactive glial cells - or tumor cells bearing phagocytized GFAP. It has in fact to be stressed that in some meningiomas and cerebral epithelial metastases as well, we detected sporadically some GFAP-positive elements undoubtly neoplastic in origin. In these cases, GFAP-positivity can only be explained on the basis of phagocytosis (occasionally macrophages are found to be GFAP-positive), or of unspecifity of immunochemical reaction.

Our results do in part confirm those presented by others (PALMER et al.; MANNOJI et al.), although our interpretation is quite different, however:

1. An assured glial differentiation in medulloblastomas is lacking or is extremely rare. For an exact interpretation of the immunocytochemical results, the analysis of the whole tumor, of its manner of growth, and of further histological factors as well are of paramount importance. The positivity of a single cell, or the investigation on only a few small tumor samples are inadequate for a correct evaluation of the findings.

2. This immunocytochemical method is useful in differentiating medulloblastomas from small-celled astrocytomas and, in part, ependymomas. In these cases too, large tissue samples have to be examined, however.

3. In our material, no correlation between presence or quantity of GFAP and biological behaviour of the tumor, i.e. course of disease, could be demonstrated.

References

1. Bignami, A., Dahl, D.: Astrocyte-specific protein and radial glia in the cerebral cortex of newborn rat. Nature 252, 55-56 (1974)
2. Bignami, A., Dahl, D.: Specificity of glial fibrillary acid protein for astroglia. J. Histochem. Cytochem. 25, 466-469 (1977)
3. De Armond S.J., Eng, L.F., Rubinstein, L.J.: The application of glial fibrillary acidic (GFA) protein immunohistochemistry in neurooncology. Path. Res. Pract. 168, 374-394 (1980)
4. Mannoji, H., Takeshita, I., Fukui, M., Ohta, M., Kitamura, K.: Glial fibrillary acidic protein in medulloblastoma. Acta neuropath. 55, 63-69 (1981)
5. Palmer, J.O., Kasselberg, A.G., Netsky, M.G.: Differentiation of medulloblastoma. Studies including immuno-histochemical localization of glial fibrillary acid protein. J. Neurosurg. 55, 161-169 (1981)
6. Spielmeyer, W.: Histopathologie des Nervensystems. Berlin: Springer 1922
7. Sternberger, L.A.: Immunocytochemistry. 2nd. ed. New York, Toronto: John Wiley and Sons, 1979

The Value of Intraoperative Histological Examination in Order to Decide on the Surgical Procedure in Medulloblastomas and Astrocytomas of the Posterior Fossa

R. Kalff[1], B. J. M. Lamers[1], M. Maksoud[1], H. Wiedemayer[1], and C. Bayindir[2]

[1] Neurochirurgische Klinik und Poliklinik, Universitätsklinikum der Gesamthochschule Essen, Hufelandstrasse 55, D-4300 Essen 1

[2] Neuropathologische Klinik, Universitätsklinikum der Gesamthochschule Essen, Hufelandstrasse 55, D-4300 Essen 1

The removal of tumors in the posterior fossa in childhood depends on the type and quality of the tumor on the one hand and on the anatomical location on the other.

In spite of the pre-operative diagnosis as well as of the intraoperative finding, the surgeon is not always able to decide what kind of tumor he is dealing with.

In comparison with cerebellar astrocytomas in which the surgical treatment is the most important, in patients with medulloblastomas postoperative radiotherapy will follow. Therefore the surgeon should strive for a radical removal of cerebellar astrocytomas even if neurological deficits must be taken into account, while a radical removal in the case of a medulloblastoma is not so necessary because of good possibilities in postoperative treatment.

Because of this we have checked our patients and we laid great emphasis on how far the intraoperative macroscopical aspect agreed with the histological examination and further to compare the opinion about the intraoperative histological diagnosis with the results of the later histological examination.

We would like to present 25 patients with medulloblastomas and 24 patients with cerebellar astrocytomas who have been treated in our hospital since 1974.

The sex and age distribution corresponds with that in the literature (Table 1 and Fig. 1).

The patients were divided into four groups according to the Bloom classification when they entered the hospital (Table 2).

Table 1. Sex distribution

	Medulloblastoma	Astrocytoma
♂	16	14
♀	9	10

Advances in Neurosurgery, Vol. 11
Edited by H.-P. Jensen, M. Brock, and M. Klinger

Table 2. Bloom's classification (1971)

	Medulloblastoma	Astrocytoma
I	1	5
II	9	6
III	13	12
IV	2	1

After the initial X-ray examination the tumor was operated on in all cases. In all patients with medulloblastomas and in 15 patients with astrocytomas a Pudenz-Heyer drainage was inserted.

The anatomical localisation is shown in Table 3. In 11 cases with medulloblastomas and in 12 cases of astrocytoma we found the tumor was limited to the vermis and/or the cerebellar hemisphere. In 14 cases of medulloblastoma and in 12 cases of astrocytoma infiltration into the brain stem was seen.

In ten patients with medulloblastoma the tumor was removed totally, in nine cases there was a subtotal removal and an excision only for histological examination was done in two (Table 4).

In 12 cases of cerebellar astrocytoma we had success with a total removal of the tumor. A subtotal removal was done in ten and only a biopsy in two cases.

For patients with infiltrating tumors a total removal is dangerous to life and often serious neurological deficits must be taken into account. Because of the different methods of treatment needed for each kind of tumor an intraoperative diagnosis is desirable.

During the operation we established that there was infiltration of the tumor in eight patients with medulloblastoma. In all cases the diagnosis was confirmed by the intraoperative histological examination and therefore a radical removal of the tumor was carried out. Because of the pre-operative diagnosis infiltration was confirmed in six patients, and therefore in these cases only a subtotal removal of the tumor or an excision for histological examination was planned. The intraoperative diagnosis agreed in all cases with the final histological opinion.

Table 3. Anatomical location

	Medulloblastoma	Astrocytoma
Vermis and/or cerebellar hemisphere	11	12
Infiltration into the brain stem	14	12

Table 4. Type of surgical procedure

	Medulloblastoma	Astrocytoma
Total	10	12
Subtotal	9	10
PE	6	2

In case of 12 patients with infiltrating cerebellar astrocytomas an intraoperative histological examination confirmed the diagnosis in eight cases. Here we tried to remove the tumor as radically as possible. Finally we had success in three cases. In the other patients with infiltrating tumors a radical removal was not possible because of the extension of the tumor and because of the patients' poor condition. The intraoperative histological diagnosis was confirmed in all cases by the final histological result.

I would like to comment on six cases of cerebellar astrocytomas in which the surgeon diagnosed a medulloblastoma because of the macroscopic appearance, whereas in all cases of medulloblastoma the macroscopical appearance agreed with the histological diagnosis.

Thirteen patients of 25 with medulloblastomas died, whereas only six of 24 with cerebellar astrocytomas died. The patients were checked up in our clinic and were divided according to the classification of Bloom (Tables 5, 6).

Eleven patients have had a good chance to continue their lives and avoid any recurrence of the tumor because the astrocytoma was totally removed (Table 6).

In spite of computerized tomography and the intraoperative macroscopic appearance, even today it is impossible to be absolutely definite about the true nature of a tumor. Therefore we think that an intraoperative histological diagnosis is very helpful in planning the operation in case of infiltrating medulloblastomas and cerebellar astrocytomas.

Table 5. Bloom's classification (1971)

	Medulloblastoma		
	Preoperative	✝	Postoperative
I	1	-	5
II	9	4	4
III	13	8	8
IV	2	1	1

Table 6. Bloom's classification (1971)

	Astrocytoma		
	Preoperative	✝	Postoperative
I	5	2	2 (2 t)
II	6	-	10 (6 t)
III	12	3	6 (3 t)
IV	1	1 (t)	-

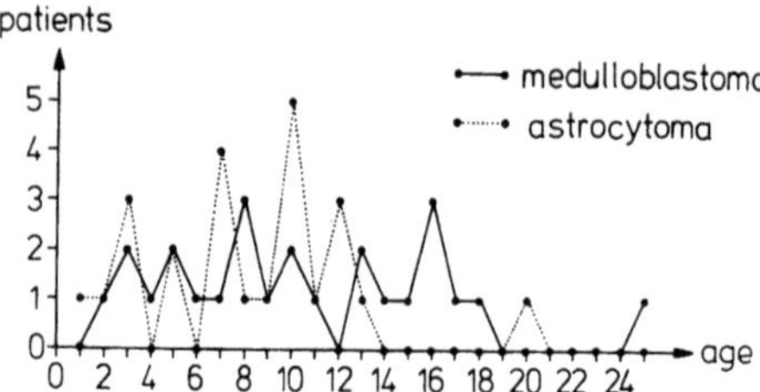

Fig. 1. Age distribution of patients with medulloblastoma and cerebellar astrocytomas

Tumors of the Posterior Cranial Fossa in Childhood: Role of Radiotherapy and Chemotherapy

H. J. G. Bloom
Department of Radiotherapy and Oncology, Royal Marsden Hospital and Institute of Cancer Research, GB-London SW3

Introduction

Although the infratentorial brain volume is only one tenth of that in the supratentorial region, some 55% of all intracranial tumors in children are found in the posterior cranial fossa (Table 1). This is in sharp contrast to the situation in adults where most tumors occur in the larger volume of the supratentorial region and only 10% arise in the posterior fossa.

In children under age 16, tumors occurring in the posterior fossa have a wide range of biological behavior. Here are found the most favorable tumors which are curable by surgery alone and also the most lethal lesions requiring multi-modal therapeutic endeavours to try and achieve maximum tumor control. The prognostically favorable low grade astrocytomas of the cerebellum make up about one third of all posterior fossa tumors, whilst the highly malignant medulloblastomas constitute some 30%, brain stem gliomas about 17% and ependymomas 11% (Table 2). Thus,

Table 1. Intracranial tumors in children; 2720 cases (5 series)[a]

	Cases	%
Supratentorial	1202	44
Infratentorial	1518	56

a MATSON (1969), Boston; KOOS (1971), Vienna; KOOPER (1975), Melbourne; HEISKANEN (1977), Finland; GJERRIS (1978), Denmark

Table 2. Posterior fossa tumors in children. From total 2720 cases (5 series)

	Cases	%
Cerebellar astro.	491	32
Medulloblastoma	469	31
Brain stem	254	17
Ependymoma	174	11
Others	130	9
Total	1518	100

Advances in Neurosurgery, Vol. 11
Edited by H.-P. Jensen, M. Brock, and M. Klinger

two thirds of all posterior fossa tumors in children are cerebellar astrocytomas or medulloblastomas in roughly equal proportions. Among the additional rare tumors found in this region for which radiotherapy may have a role are lepto-meningeal sarcomas, choroid plexus tumors, hemangioblastomas and chordomas.

Roayl Marsden Hospital Series (1952 - 76)

Reference will be made to 267 children with posterior fossa tumors referred to the Royal Marsden Hospital for radiotherapy. These constitute 58% of our total series of 461 patients under age 16 with primary intracranial tumors seen at this hospital between 1952 - 1976 (Table 3). The proportion of posterior fossa cases in this series is almost identical to that found in the collected general or neurosurgical series (Table 1). Survival of the children with infratentorial tumors is compared with that of 192 children with supratentorial tumors in Fig. 1.

The precise composition of our infratentorial group of patients is shown in Table 4. The high proportion of medulloblastoma cases (51%) and the low incidence of those with cerebellar astrocytomas (11%), compared with the collected series (Table 2), merely reflects the practice of a radiotherapy department which has a special interest in medulloblastoma and is rarely called upon to treat children with low grade cerebellar astrocytomas for which operation is, of course, the principal treatment.

Medulloblastoma

Medulloblastoma is the most malignant and most frequent intracranial type of cancer found in childhood, constituting about 20 - 25% of all intracranial tumors occurring before the age of 16. The first treatment is the only one which carries any hope of success: after recurrence, further measures by operation, radiotherapy or chemotherapy can only be palliative.

Table 3. Royal Marsden Hospital (1952-76). Intracranial tumors in children (461 cases)

Site	Cases	%
Supratentorial	192	41.7
Infratentorial	267	57.9
Multiple	2	0.4
Total	461	100.0

Table 4. Royal Marsden Hospital (1952-76). Post. fossa tumors in children

	Cases	%
Medulloblastoma	135	51
Cerebellar astrocytoma	30	11
Ependymoma	31	12
Brain stem glioma	57	21
Others	14	5
Total	267	100

Surgical excision and re-excision cannot cure these tumors. They are extremely radiosensitive and with post-operative irradiation to the whole neuraxis using precise megavoltage techniques and high doses compatible with reasonable safety, particularly to the posterior fossa, survival rates of approximately 40% at five years and 30% at ten years are to be expected (Table 5). This is in contrast to a five-year survival rate of less than 10% associated with operation and inadequate radiotherapy. The survival of 118 children completing radiotherapy at the Royal Marsden Hospital between 1952 and 1976 is shown in Fig. 2 (BLOOM, 1982[a]).

Children with medulloblastoma die essentially because of failure to eradicate the primary tumor: recurrence in the posterior fossa, with or without spread elsewhere, is found in some 75% of treatment failures (BLOOM et al., 1969; CASTRO-VITA et al., 1978; NÜCHEL and ANDERSON, 1978). The incidence of local disease is even greater among patients subjected to autopsy - 13 of 14 (93%) cases reported by BLOOM et al. (1969), and 13 of 15 (87%) reported by NÜCHEL and ANDERSON (1978). The risk to life and to cerebral function associated with radical surgery for medulloblastoma, a tumor which not infrequently involves the brain stem, imposes serious limitations on attempts at total tumor excision. Furthermore, the maximum dose of irradiation which can be delivered without serious risk to normal brain tissue, especially in very young children, is also restricted. Even after doses of 50 - 60 Gy, which surely must approach the upper limits of normal brain tolerance in children, local failure still occurs in 50 - 60% of patients. Thus, further advances in the treatment of medulloblastoma cannot be expected from present conventional methods of surgery and radiotherapy.

There are several possible therapeutic approaches to the problem of residual radioresistant medulloblastoma. These include the use of chemical radiosensitizers, particle beams (fast neutrons) or chemotherapy (BLOOM, 1979[a]). From a review of these options, the most reasonable and most hopeful choice at present would seem to be chemotherapy (BLOOM, 1982[b]).

Table 5. Medulloblastoma in children 1970 - 82. (Series of > 30 patients)

		Cases	5 Year Survival (%)
KOOS & MILLER (1971)		120	30
SMITH et al. (1973)		36	32
PEARSON (1974)		84	37
HENDRICK et al. (1975)		40	42
HARISIADIS & CHANG (1977)		58	40
MEALY & HALL (1977)		32	41
NÜCHEL & ANDERSON (1978)		38	33
HARDY et al. (1978)		42	40
McINTOSH (1979)		66	21
SCHWEISGUTH (1979)		82	50
BERRY et al. (1981)		122[a]	56
BLOOM (1982)	1952-70	87	32
	1970-80	37	71

a Includes 24 adults

Chemotherapy

There are certain problems to be considered in developing possible cytotoxic drug regimes for children with medulloblastoma. First and foremost, we need effective agents against the tumor. Second, given the availability of such agents, they must be able to pass the blood-brain barrier. Third, the treatment must be feasible since marrow reserve in such cases is inevitably seriously compromised by cerebro-spinal axis irradiation. Fourthly, a combination of chemotherapy and radiotherapy must be reasonably safe, since enhanced changes may occur in normal brain tissue leading to late sequelae, especially in young children in whom the developing brain is known to be particularly vulnerable to toxic injury. Finally, any potentially useful drug regimen must not jeopardise the administration of radiotherapy which, at least for the present, remains the most effective and reliable adjuvant treatment for this disease.

Over the past decade there have been only sporadic studies of chemotherapy for recurrent medulloblastomas. Numbers of cases have generally been small, additional treatment other than chemotherapy often used, and criteria for response not always clearly defined. Nevertheless, it seems that this tumor may respond to the nitrosourea drugs, procarbazone, vincristine and also methotrexate.

Royal Marsden Hospital Medulloblastoma Study

In 1970 we started a pilot study at the Royal Marsden Hospital to assess the feasibility and possible value of adjuvant chemotherapy following radical surgery and megavoltage radiotherapy (BLOOM, 1975). We have now treated 40 children using mainly vincristine during radiotherapy, followed by intermittent maintenance courses of vincristine and CCNU cycled every six weeks over a period of approximately one year after completion of radiotherapy (BLOOM, 1982[b]). The survival for this group is greater than that of an earlier historical series of 87 patients from the same centre in which chemotherapy was not used (Fig. 3). The difference between the two groups at five years is highly significant ($p < 0.001$).

Chemotherapy was reasonably well tolerated but, as expected in children undergoing prior cranio-spinal irradiation, marked myelo-suppression occurred in some patients which made a reduction in dose of CCNU or delays in drug administration inevitable. Serious neurotoxicity with vincristine was rarely seen and full dosage of this agent throughout was generally achieved. There were not drug-related deaths.

Historical controls, of course, are unsatisfactory and in recent years modern megavoltage techniques alone have improved results in children with medulloblastoma without the addition of chemotherapy. Furthermore, advances in neurosurgery, especially the advent of the operating microscope, may permit more complete tumor excision, a factor which is related to a more favorable prognosis. Clearly, for these reasons a controlled study was required to assess the role and value of adjuvant chemotherapy for children with medulloblastoma.

SIOP Medulloblastoma Study (1975 - 79)

In 1975 the Brain Tumor Committee of the International Society of Pediatric Oncology (SIOP) started a multi-centre prospective randomised trial. Fourty-four centres from 15 countries entered 287 children with medulloblastoma into the study which was closed in September, 1979. The treatment protocol has been given elsewhere (BLOOM, 1979[a]).

The results, as reported to the SIOP Annual Conference in Marseilles, September, 1981, revealed a difference in disease-free survival between the two principal arms of the trial in favor of adjuvant chemotherapy (Fig. 4): The difference, however, was not statistically significant ($p = .056$) (BLOOM et al., 1982). In virtually all sub-group correlations the survival of patients receiving chemotherapy was superior to that of the control groups, although again the difference often did not reach the generally accepted level of statistical significance. However, significant results were obtained in favor of chemotherapy in certain high risk sub-groups, namely, children aged less than two years ($p = .04$), those having only partial or sub-total tumor excision as opposed to macroscopically "total" removal ($p = .005$), and those with brain stem involvement ($p = .004$) (Fig. 5).

In the American CCSG/RTOG Medulloblastoma Study (EVANS et al., 1979) there also has been no statistically significant overall advantage for chemotherapy but recently, as in the SIOP study, such benefit may be emerging in certain high risk groups (Dr. Audrey EVANS, personal communication, 1982). More advanced results from both sides of the Atlantic are awaited with interest.

Further Studies

On the basis of the present SIOP data, it seems that adjuvant chemotherapy must improve the survival of children in certain high risk groups and that further work in this direction is fully warranted. Various pediatric teams are searching for more effective chemotherapy regimens. Bearing in mind the need to avoid compromising marrow function which may interfere with subsequent radical radiotherapy, care will be required in devising any chemotherapy schedule which is to be additional to the vincristine/CCNU combination already used. Methotrexate with leucovorin rescue would appear to be one possible option. Following the lead by ROSEN et al. (1977) in using high dose systemic methotrexate with leucovorin rescue for recurrent brain tumors in children, this approach is now being applied to primary adjuvant chemotherapy for medulloblastoma by Dr. VOUTE in Amsterdam (personal communication, 1980), in combination with other drugs by RIEHM et al. (1981) and as an extension to the SIOP protocol at the Royal Marsden Hospital (BLOOM, 1982[b]).

A successful adjuvant chemotherapy regimen in children with medulloblastoma may ultimately permit a reduction in the dose of prophylactic irradiation to the cerebral hemispheres and also to the spinal cord. At present, however, it would be premature to reduce the dose of irradiation, especially to the posterior fossa where tumor recurrence generally first appears. Furthermore, the value of prophylactic spinal irradiation is evident when we consider the past high failure rate when treatment was limited to the cranium (JENKIN, 1969; HOPE-STONE, 1970; GULLOTTA, 1979; LANDBERG et al., 1980). Thus it seems that spinal irradiation can eradicate occult metastases and this is also supported by the recent report by DEUTSCH and REIGEL (1980) who found positive myelograms, one to four weeks after craniotomy for medulloblastoma, in no less than seven of 16 children (43%) in whom spinal spread was not suspected.

Cerebellar Astrocytomas

The low grade astrocytoma of the cerebellum is one of the most important tumors met with in children. It is among the commonest intracranial lesions, and is also one of the most curable. The majority are slow growing, well-circumscribed, often cystic lesions, frequently amenable to complete excision. In such circumstances 100% cures are expected. However, in about 30% of children total removal may not be possible. Based on histological criteria, there appear to be two distinct types of cerebellar astrocytoma in children. There is the more frequent juvenile piloid lesion characterised by elongated unipolar and bi-polar cells with oval nuclei, scanty cytoplasm, rare mitoses, prominent fibrillary processes and Rosenthal fibres: tumors of this type are associated with a 25-year survival rate of more than 90%. There is also a less common more diffuse type of cerebellar astrocytoma with which only some 30 - 40% of patients are still alive at 25 years (GJERRIS and KLINKEN, 1978): it is in this variety that radiotherapy may have a role to play.

Since incomplete removal of cerebellar astrocytoma may be followed by a prolonged symptom-free period, perhaps extending over many years (GEISSENGER and BUCY, 1971; GRIFFEN et al., 1979), the value of postoperative radiotherapy in such cases is obviously difficult to assess. However, from our experience with low grade astrocytomas in the cerebral hemispheres, particularly the hypothalamic region, and in the optic pathways, it appears that such tumors may respond remarkably well to irradiation (BLOOM, 1982[a]). It would therefore seems reasonable to recommend post-operative irradiation for cases of cerebellar astrocytoma with known residual disease, particularly if the tumor is of grade II or higher malignancy and especially if there is invasion of the brain stem. GRIFFEN et al. (1979) and also LEIBEL et al. (1975) report results which suggest that post-operative irradiation following partial tumor removal increases survival of patients with low grade astrocytoma. Treatment results in children with cerebellar astrocytomas referred to radiotherapy departments obviously cannot be compared with purely neurosurgical series, in view of the selection of only unfavorable cases for post-operative irradiation.

Royal Marsden Hospital Series of Cerebellar Astrocytomas

In 30 children with cerebellar astrocytoma referred for post-operative radiotherapy, 70% were alive at five years and 58% at 10 - 20 years (Fig. 6). Prognosis appears to be influenced by age, the youngest children having the lowest survival (Fig. 7).

High Grade Cerebellar Astrocytomas

Cerebellar astrocytomas of high grade malignancy are rare in children. They may develop *de novo* (FRESH et al., 1976; KEPES et al., 1980) or as recurrence of a previously treated low grade lesion after a latent period, sometimes as long as 30 - 40 years (SCOTT and BALLANTINE, 1973; BUDKA, 1975). Unlike the situation in the cerebral hemispheres of adults, progression with time to less differentiated and histologically more malignant looking astrocytomas is very rare in childhood cerebellar tumors. Incidentally, the histological appearance may not always reflect the inherent biological potential of these tumors: a perfectly indolent-looking cerebellar tumor can behave in a malignant fashion (AUER et al., 1981): the reverse may also occur.

It has been suggested that therapeutic irradiation may be a factor in the progression of a tumor of low grade malignancy to one of higher grade. We know that this change often occurs in the cerebrum without prior radiotherapy (MÜLLER et al., 1977). Examples of malignant transformation in cerebellar astrocytomas of children after a long latent period following treatment by radiotherapy have been reported by BUDKA (1975), SCOTT and BALLANTINE (1973) and by KLEINMAN et al. (1979), but such a change has also been observed in patients treated solely by operation (BERNELL et al., 1972).

SHAPIRO and SHULMAN (1976) report three extraordinary cases of low grade juvenile cerebellar astrocytoma which seeded to the spinal subarachnoid space. This is a very rare event and prophylactic irradiation of the spine in children with juvenile cerebellar astrocytomas is certainly not indicated. If such spread occurs it seems that long-term control of the spinal disease may be achieved by operation and radiotherapy.

Chemotherapy

There is very little information concerning the use of chemotherapy for cerebellar astrocytomas in children. In a recent report EDWARDS et al. (1980) referred to two or three patients with recurrent solid low grade astrocytomas that responded well to nitrosourea agents. From our experience with recurrent cerebral hemisphere astrocytomas in adults, it is the *low* grade lesions which are more likely to show a good response to treatment with the nitrosourea agents (BLOOM, 1982[a]).

Ependymoma

Ependymomas are mainly tumors of children and young adults with the peak incidence during the first decade. They constitute approximately 10% of all CNS tumors in childhood: one-third occur above, and two thirds below the tentorium.

Past experience has shown that attempts to treat ependymomas solely by operation, although somewhat more successful than with medulloblastoma, have produced five-year survival rates of only 15 - 20%. To improve on these figures post-operative radiotherapy is essential and to obtain a maximum number of cures the principle of large volume, high dose treatment is essential. After such postoperative radiotherapy the five-year survival rate has increased to about 50% (SALAZAR et al., 1975).

Over 90% of posterior fossa ependymomas arise from the floor of the fourth ventricle. They may fill the ventricle, grow out of the foramina to extend laterally to the cerebello-pontine angle and also inferiorly through the foramen magnum along the cervical cord, sometimes as far down as the fifth cervical vertebra. They may seed along the CSF pathways with the formation of widespread satellite nodules and this risk will influence radiotherapy policy. However, opinions vary regarding the frequency of this complication. The average incidence among 598 cases collected from the literature was 12% with a range of between 0% and 60% (BLOOM, 1982[a]). The risk seems to depend upon primary tumor site, histological grade and, of course, whether the diagnosis of seeding is based on clinical, radiological, cytological or autopsy findings.

In a series of 39 patients of all ages seen at the Royal Marsden Hospital with histologically graded ependymomas, clinical evidence of dissemination developed in seven (19%): this complication occurred more frequently in patients with tumors situated in the posterior fossa (21%), tumors of high grade malignancy (26%), and especially when both these features were present (29%) (BLOOM, 1982[c]). Unlike the situation with medulloblastoma, ventricular and subarachnoid deposits from ependymomas often do not seem to progress and may remain in a sub-clinical phase for a considerable time, perhaps during the entire life of the host.

For all patients with high grade tumors and for all those with tumors of any grade situated in the posterior fossa we recommend irradiation of the whole cerebro-spinal axis. The risk of dissemination in supratentorial low grade ependymomas seems to be quite small and for these lesions treatment should be given to the primary site with a substantial margin of clearance without the need to irradiate the whole brain and spinal cord.

Royal Marsden Hospital Ependymoma Series

Of 47 children with histologically verified ependymoma referred for radiotherapy between 1952 and 1976, 53% were alive at five years, 37% at ten years, and 32% at 20 years (BLOOM, 1982[b]) (Fig. 8).

Because of their surgically more dangerous anatomical site, posterior fossa ependymomas carry a greater operative mortality than supratentorial lesions. On the other hand, supratentorial ependymomas tend to be more infiltrative and of higher grade malignancy than those found in the fourth ventricle: ultimately, the prognosis is better for patients with posterior fossa lesions. The survival rates for children with infratentorial and supratentorial ependymomas were comparable at five years (50% and 54%, respectively): after this time, however, whilst the level of survival was maintained up to 20 years for patients with infratentorial tumors (45% - 38%), it declined rapidly to 18% for those with supratentorial lesions (Fig. 9). The better outlook for posterior fossa ependymomas was also found by MARKS and ADLER (1982) and by GJERRIS et al. (1978).

Prognosis in patients with ependymoma is largely determined by histological grade: for 13 children with grade I and II infratentorial tumors the five year survival rate was 69%, compared with only 33% for 14 with grade III and IV lesions. Age also influences the outcome, the highest survival being associated with older children.

Adjuvant Chemotherapy

Because of the poor results in children with high grade ependymoma after operation and radiotherapy we decided to test the value of adjuvant chemotherapy in a pilot study at the Royal Marsden Hospital using the same programme as for children with medulloblastoma. Although our early results were encouraging, improvement with adjuvant chemotherapy was not maintained beyond six years: life appears to have been prolonged without an increase in cure rate (Fig. 10).

The International Society for Paediatric Oncology have also included high grade ependymomas as a separate group in their brain tumor study. So far, there is no difference between the survival rates for those receiving adjuvant chemotherapy and the controls (BLOOM et al., 1982).

It appears that ependymomas are more resistant than medulloblastoma to the same chemotherapy: a different type of cytotoxic drug combination is required to achieve a greater effect in patients with ependymomas.

Brain Stem Tumors

Brain stem gliomas are intrinsic neoplasms of the medulla and pons and constitute about 10% of all intracranial tumors in childhood. In most patients the diagnosis is based on clinical and radiological features, since operation, even when limited to biopsy, is hazardous. The correlation between cases histologically verified by operation or at postmortem and the clinico-radiological diagnosis is high and it is therefore considered justifiable to treat patients with brain stem tumors solely on the basis of radiological and clinical findings.

Radiotherapists naturally feel uneasy when patients are referred for treatment with unverified tumors. However, in the case of the brain, they have had to come to terms with the situation and fully appreciate that for some patients the risks of treatment are often less than those associated with biopsy. The results for patients with verified brain stem tumors are similar to those in which histology is not available (KIM et al., 1980).

Irradiation is the treatment of choice, after which there is usually a good initial clinical response rate in about 70% of cases (Table 6). Responses appear to be greater in children than in adults (LASSITER et al., 1971). Cranial nerve palsies, long tract signs, speech difficulties and mental disturbance may all improve substanitally after irradiation: in some cases resolution may be complete. Regrettably, these responses are generally not maintained and tumor recurrence occurs in the majority of patients, resulting in progressive distressing disability and early death.

Royal Marsden Hospital Series of Brain Stem Tumors

Fifty-seven children with tumors of the brain stem were referred to the Royal Marsden Hospital for radiotherapy between 1952 and 1976. Ten of the 57 failed to complete the full course of treatment but are

Table 6. Brain stem tumors response to radiotherapy. Various reported series

Authors	Cases	Clinical improvement (%)
BRAY et al. (1958)	24	15 (63)
LASSMAN & ARJONA (1967)	15	12 (80)
PANITCH & BERG (1970)	28[a]	21 (75)
LASSITER et al. (1971)	16[a]	15 (94)
	9[b]	2 (22)
MARSA et al. (1973)	15[a]	13 (87)
SHELINE (1975)	27	19 (70)
GREENBERGER et al. (1977)	26	19 (73)
KIM et al. (1980)	63	35 (56)
Total	223	151 (68)

a Children only; b Adults only

not excluded. At five years, 17% of the entire series were still alive and this figure was maintained to 20 years (Fig. 11). Failures were rapidly fatal: 60% of the children were dead at 12 months. Similar results, reported by other authors have been summarised elsewhere (BLOOM, 1982a). A few authors have reported surprisingly high survival rates of up to 30% or 40% at five years (WHYTE et al., 1969; SHELINE, 1977; VILLANI et al., 1975; KIM et al., 1980). Different results from various centres, using similar treatment, is most likely explained by the incidence of different tumor types, age distribution, and especially inclusion of other anatomical sites adjacent to the brain stem such as the fourth ventricle, mid-brain, hypothalamus and third ventricle, all of which have a better prognosis than those arising in the brain stem proper (pons and medulla).

Operation in Brain Stem Cases. Posterior fossa surgical exploration in patients with suspected brain stem tumors may be of value for clinical diagnosis, for biopsy of exophytic lesions, for cyst aspiration and for making a by-pass. Removal of a mural nodule from a cystic brain stem glioma has been reported by LASSITER et al. (1971). Recently, HOFFMAN et al. (1980) have sought to identify a small group of circumscribed, non-infiltrating brain stem glioma, which compress the aqueduct and bulge into the fourth ventricle, for which surgical excision appears to be feasible. However, this concept was applicable to only ten of their 121 cases. The possible benefit of surgery must always be weighed against the operative risk. An alternative approach is to use radiotherapy initially and to undertake exploration only if there has been no reduction in tumor mass.

Gliomas of the brain stem represent a great challenge to the radiotherapist. The tumor burden is relatively small, a good initial response to irradiation is obtained but there is a high recurrence rate without dissemination. Little further progress can be expected from megavoltage radiotherapy alone and, so far, adjuvant chemotherapy appears to be only of limited value. The situation is ideal for exploring the value of chemical radiosensitizers in the hope of rendering residual hypoxic tumor tissue, after the initial good response to irradiation, more vulnerable to treatment. We are conducting a pilot study in which misonidazole is used as a radiosensitizer in the primary treatment of children with brain stem tumors. This agent is given in daily oral doses of 600 mg/m^2 during the *last* four weeks of a six-week course of radical megavoltage radiotherapy during which time a total tumor dose of 50 Gy is delivered. The sensitizer is administered four hours before irradiation and this is followed by blood levels of between 30 and 40 micrograms/ml. It is too early for results but, so far, it can be said that misonidazole has been extremely well tolerated and gastro-intestinal disturbance and peripheral neuropathy have been slight (BLOOM and BUGDEN, 1982).

Other Posterior Fossa Tumors

Among the rarer tumors occurring in the posterior fossa of children, radiotherapy may be of value in the treatment of choroid plexus tumors, hemangioblastoma, chordoma and lepto-meningeal sarcoma.

Radiotherapy is recommended for *choroid plexus tumors* if the lesion is incompletely excised or shows histological features suggestive of malignancy. Such cases are treated by whole cerebro-spinal axis irradiation. We have had two cases of choroid plexus of the fourth ventricle: in one the spine was not treated and the child died within one year with spinal metastases, whilst the second, who received whole

CNS irradiation, is alive and well 15 years after treatment. From our experience in adults, recurrent *hemangioblastomas* may respond well to radiotherapy with prolonged survival of 10 - 20 years or more. Postoperative irradiation is recommended for primary cases. Although the dose of irradiation required to eradicate *chordomas* of the clivus is in excess of normal brain tolerance, control over several years may be achieved by doses within the acceptable range after surgical reduction of the tumor mass.

Poorly differentiated *lepto-meningeal sarcomas* generally of the spindle cell or polymorphic cell type, may occur in infants and older children. They probably arise from the surface meninges or from perivascular pial extensions within the brain substance. The tumor may form circumscribed masses or infiltrate diffusely over a wide area. Postoperative radiotherapy may achieve limited control. Very rarely, an intracranial sarcoma may develop many years later at the site of previous cranial radiotherapy, but most of the reported cases have occurred in the region of the sella turcica after treatment of pituitary tumors.

Quality of Survival

In no other condition can there be more concern over possible late sequelae as following the successful treatment of a child with a brain tumor, especially as the survival rates for such patients have increased in recent years. Approximately 45% of all 461 children aged under 16 in our series with all types of intracranial tumor who completed radiotherapy, given either postoperatively or as the sole treatment, survived more than ten years. Furthermore, 36% of those under the age of three, in whom the rapidly developing central nervous system may be particularly vulnerable to the effect of disease and of treatment, were alive after ten years. The spectrum of survival over 20 years for the different types of posterior fossa tumor is shown in Fig. 12.

The dose/time factors we have used in radiotherapy appear to be relatively safe in that they are not associated with gross irradiation damage. On the other hand, such treatment may produce more subtle changes which could lead to disturbance of higher mental function, emotional problems and endocrine deficiency, notably of growth hormone. It is of course difficult to separate out the precise cause of such complications which, at least in some cases, may be due more to the direct and indirect effects of the tumor, particularly hydrocephalus and perhaps even to the act of surgery itself, rather than to radiotherapy. Mental impairment may be found in patients treated for brain tumors who have never received radiotherapy. One should also not forget the impact on a child's mind of prolonged illness, physical disability, loss of schooling and parental anxiety (BLOOM, 1982[d]).

Age at the time of treatment is an important factor in determining the risk of late changes, the youngest children being the most vulnerable. Treatment given in a shorter total time than that generally recommended, using greater individual fractions of irradiation, carry considerably greater risk to the young central nervous system. Our treatment policy ensures that the individual daily maximum tumor doses never exceed 1.6 Gy (i.e. 8 Gy per week), and for the youngest children under age two, 1.4 Gy (i.e. 7 Gy per week). In children with infratentorial tumors the maximum dose of irradiation is directed to the posterior fossa and risk of late intellectual and endocrine sequelae is probably less than after treatment of primary tumor situated in the cerebral hemispheres.

Elsewhere, reference has been made to a number of authors who have reported favorably on the overall functional results in children treated for brain tumors by irradiation, some 70 to 80% of all surviving cases leading apparently "active, useful lives" (BLOOM, 1979[b]). The criteria for these assessments were based on general observation and enquiry and clearly a more detailed and more precise investigation of intellectual capabilities and socio-economic status are needed by specially trained personnel. What at first appears to be "an active, useful life" with little or no neurological disability may, in fact, turn out to be one that is significantly altered by changes in preexisting intellect and behavior which may not be obvious on general examination.

When all treatment has been completed, regular follow-up and supervision is necessary to deal with general rehabilitation and any problems that may arise from lost time at school, reduced vision and impaired physical and mental function, endocrine disturbance and loss of scalp hair. It requires considereable team-work and resources to provide adequate and comprehensive care for children who have been through the disturbing experience of treatment for an intracranial tumor. To try and deal with these problems we have set up a Brain Tumor Board and also an Endocrine and Growth Clinic consisting of medical and paramedical staff representing the relevant disciplines.

In Conclusion

Treatment results for intracranial gliomas in children by current methods of conventional surgery and megavoltage radiotherapy, alone or in combination, appear to have reached a plateau. The extent of operation and of current radiotherapy practice cannot be increased without additional risks to life and cerebral function. For further progress we must try to gain additional tumor control with adjuvant cytotoxic agents, render residual radio-resistant tumor tissue more vulnerable to irradiation by chemical radiosensitizers, and achieve greater biological effects more safely with particle beam therapy. Ultimately, such measures in combination may decrease the tumor burden to such an extent that the residue can be eradicated by immunological maneuvers aimed at enhancing natural host resistance. Finally, in our aspiration to achieve greater destruction of the tumor we must constantly bear in mind that combination therapy may not only exert an enhanced effect on tumor cells, but also on normal tissues. We must try and ensure that any significant gain in response and survival, especially in young children, is not offset by an increase in serious early or late complications. Above all we need to define high-risk cases for which more intensive and perhaps more hazardous treatment is justified and from which lower risk cases can be spared.

References

1. Auer, R.N., Rice, G.P.A., Hinton, G.G., Amacher, A.L., Gilbert, J.J.: Cerebellar astrocytoma with benign histology and malignant clinical course. Case report. J. Neurosurg. 54, 128-132 (1981)
2. Berry, M.P., Jenkin, R.D.T., Keen, C.W., Nair, B.D., Simpson, W.J.: Radiation treatment for medulloblastoma. A 21 year review. J. Neurosurg. 55, 43-51 (1981)
3. Bernell, W.R., Kepes, J.J., Seitz, E.P.: Late malignant recurrence of childhood cerebellar astrocytoma: Report of two cases. J. Neurosurg. 37, 470-474 (1972)

4. Bloom, H.J.G.: Combined modality treatment for intracranial tumors. Cancer 35, 111-120 (1975)

5. Bloom, H.J.G.: In: Recent result- in cancer research. G. Bonadonna, G., Mathe, S.E., Salmon (eds.), Vol. 68, pp. 412-422. Berlin, Heidelberg, New York: Springer 1979[a]

6. Bloom, H.J.G.: Recent concepts in the conservative treatment of intracranial tumors in children. Acta Neurochirurgica 50, 103-116 (1979[b])

7. Bloom, H.J.G.: Intracranial tumors: Response and resistance to therapeutic endeavors (227 references). Int. J. Radiat. Oncol. Biol. Phys. 8, 1083-1113 (1982[a])

8. Bloom, H.J.G.: In: Controversies in pediatric oncology. A.J. Freeman (ed.). New York: Masson Publishing USA Inc., 1982[b], in press

9. Bloom, H.J.G.: In: Tumors of the central nervous system: Modern radiotherapy in multi-disciplinary management. Chang, C.H., Housepian, E.M. (eds.). New York: Masson Publishing 1982[c]

10. Bloom, H.J.G.: Brain gliomas in children. Treatment policy and prognosis. In: Childhood cancer: Triumph over tragedy, Vaeth, J.M. (ed.), Front. Radiat. Ther. Onc. 16, 90-104 (1982[d])

11. Bloom, H.J.G., Budgen, R.D.: In: Progress in radio-oncology. Karcher, K.H., Kogelnik, H.D., Reinartz, G. (eds.). New York: Raven Press 1982

12. Bloom H.J.G., Thornton, H., Schweisgutz, O.: In: Pediatric oncology. Proceedings 13th Meeting of International Society of Pediatric Oncology, Marseilles, September 1981. Raybaud, C., Clement, R., Lebreuil, G., Bernard, J.L. (eds.), pp. 309-322. Amsterdam: Excerpta Medica 1982

13. Bloom, H.J.G., Wallace, E.N.K., Henk, J.M.: The treatment and prognosis of medulloblastoma in children: A study of 82 verified cases. Am. J. Roentgenol. 105, 43-62 (1969)

14. Bray, P.F., Carter, S., Taveras, J.M.: Brain stem tumors in children. Neurology (Minneap.) 8, 1-7 (1958)

15. Budka, H.: Partially resected and irradiated cerebellar astrocytoma of childhood: malignant evolution after 28 years. Acta Neurochirurgica 32, 139-146 (1975)

16. Castro Vita, H., Salazar, O.M., Cova, M., Rubin, P.: Cerebellar medulloblastoma: Spread and failure patterns following irradiation (Papers presented at ASTR). Int. J. Radiat. Oncol. Biol. Phys. Suppl. 2, 211, 1978 (Abstract 192)

17. Deutsch, M., Reigel, D.H.: The value of myelography in the management of childhood medulloblastoma. Cancer 45, 2194-2197 (1980)

18. Edwards, M.S., Levin, V.A., Wilson, C.B.: Chemotherapy of pediatric posterior fossa tumors. Child's Brain 7, 252-260 (1980)

19. Evans, A.E., Anderson, J., Chang, C., Jenkin, R.D.T., Kramer, S., Schoenfeld, D., Wilson, C.: Adjuvant chemotherapy for medulloblastoma and ependymoma. In: Multidisciplinary Aspects of Brain Tumor Therapy. Paoletti, P., Walker, M.D., Butti, G., Knerich, R. (eds.), pp. 219-222. Amsterdam: Elsevier-North Holland Biomedical Press 1979

20. Fresh, C.B., Takel, Y., O'Brien, M.S.: Cerebellar glioblastoma in childhood. Case report. J. Neurosurg. 45, 705-708 (1976)

21. Geissenger, J.D., Bucy, P.S.: Astrocytoma of the cerebellum in children: long term study. Arch. Neurol. 24, 125-135 (1971)

22. Gjerris, F., Harmsen, A., Klinken, L., Reske-Nielsen, E.: Incidence and long-term survival of children with intracranial tumors treated in Denmark 1935 - 1959. Brit. J. Cancer 38, 442-451 (1978)

23. Gjerris, F., Klinken, L.: Long term prognosis in children with benign cerebellar astrocytoma. J. Neurosurg. 49, 179-184 (1978)

24. Greenberger, J.S., Cassady, J.R., Levene, M.B.: Radiation therapy of thalamic, midbrain and brain stem gliomas. Radiology 122, 463-468 (1977)

25. Griffin, T.W., Beaufait, D., Blasko, J.C.: Cystic cerebellar astrocytoma in childhood. Cancer 44, 276-280 (1979)

26. Gullotta, F.: In: Multidisciplinary aspects of brain tumor therapy. Paoletti, P., Walker, M.D., Butti, G., Knerich, R. (eds.), pp. 69-76. Amsterdam: Elsevier-North Holland Biomedical Press 1979

27. Hardy, D.G., Hope-Stone, H.F., McKenzie, C.G., Sholtz, C.L.: Recurrence of medulloblastoma after homogeneous field radiotherapy. J. Neurosurg. 49, 434-440 (1978)

28. Harisiadis, L., Chang, C.H.: Medulloblastoma in children: A correlation between staging and results of treatment. Int. J. Radiat. Oncol. Biol. Phys. 2, 833-841 (1977)

29. Heiskanen, O.: Intracranial tumors of children. Child's Brain 3, 69-78 (1977)

30. Hendrick, E.B., Hoffman, H.J., Humphreys, R.P.: Treatment of infratentorial gliomas in childhood. In: Gliomas: Current concepts in biology, diagnosis and therapy. Hekmatpanah, H., (ed.), pp. 102-106. Berlin, Heidelberg, New York: Springer 1975

31. Hoffman, H.J., Becker, L., Craven, M.A.: A clinically and pathologically distinct group of benign brain stem gliomas. Neurosurgery 7, 243-248 (1980)

32. Hooper, R.: Intracranial tumors in children. Child's Brain 1, 136-140 (1975)

33. Hope-Stone, H.F.: Results of treatment of medulloblastomas. J. Neurosurg. 32, 83-88 (1970)

34. Jenkin, R.D.T.: Medulloblastoma in childhood: radiation therapy. Canad. Med. Ass. J. 100, 51-53 (1969)

35. Kepes, J.J., Lewis, R.C., Vergara, G.G.: Cerebellar astrocytoma invading the musculature and soft tissues of the neck. Case report. J. Neurosurg. 52, 414-418 (1980)

36. Kim, T.H., Chin, H.W., Pollan, S., Hazel, J.H., Webster, J.H.: Radiotherapy of primary brain stem tumors. Int. J. Radiat. Oncol. Biol. Phys. 6, 51-57 (1980)

37. Kleinman, G.M., Schoene, W.C., Walshe, T.M., Richardson, E.P.: Malignant transformation in benign cerebellar astrocytoma. Case report. J. Neurosurg. 49, 111-118 (1978)

38. Koos, W.Th., Miller, M.H.: Intracranial Tumors of Infants and Children, pp. 9-27. Stuttgart: Thieme 1971

39. Landberg, T.G., Lindgren, M.L., Cavallin-Stahl, E.K., Svahn Tapper, G.O.: Improvement in the radiotherapy of medulloblastoma 1946 - 1975. Cancer 45, 670-678 (1980)

40. Lassiter, K.R., Alexander, E., Davis, C.H., Kelly, D.L.: Surgical treatment of brain stem glioma. J. Neurosurg. 34, 719-725 (1971)

41. Lassman, L.P., Arjona, V.E.: Pontine gliomas of childhood. Lancet I, 913-915 (1967)

42. Leibel, S.A., Sheline, G.E., Wara, W.M., Boldrey, E.B., Nielson, S.L.: The role of radiation therapy in the treatment of astrocytoma. Cancer 35, 1551-1557 (1975)

43. McIntosh, N.: Medulloblastoma - a changing prognosis? Arch. Dis. Childhood 54, 200-203 (1979)

44. Marks, J.E., Adler, S.J.: A comparative study of ependymomas by site of origin. Int. J. Radiol. Oncol., Biol. Phys. 8, 37-43

45. Marsa, G.W., Probert, J.C., Rubinstein, L.J., Bagshaw, M.A.: Radiation therapy in the treatment of childhood astrocytic gliomas. Cancer 32, 646-655 (1973)

46. Matson, D.D.: Neurosurgery of Infancy and Childhood, 2nd edition, pp. 410-479. Springfield, Ill.: Thomas 1969

47. Mealey, J., Hall, P.V.: Medulloblastoma in children. Survival and treatment. J. Neurosurg. 46, 56-64 (1977)

48. Müller, W., Afra, D., Schroeder, R.: Supratentorial recurrences of gliomas. Morphological studies in relation to time intervals with astrocytoma. Acta Neurochirurg. 37, 75-91 (1977)

49. Nüchel, B., Anderson, A.P.: Medulloblastoma: Treatment results. Acta Radiol. 17, 305-311 (1978)

50. Panitch, H.A., Berg, B.O.: Brain stem tumors of childhood and adolescence. Amer. J. Dis. Child. 119, 465-472 (1970)

51. Pearson, D.: In: Modern radiotherapy and oncology: Malignant diseases in children. Deeley, T.J. (ed.), pp. 98-119. London: Butterworths 1974

52. Riehm, H., Neidhardt, M.K., Janka, G.: Aktuelle Neuropädiatrie 2, Neuropediatrics Supplement, pp. 80-85. Stuttgart: Hippokrates-Verlag 1981

53. Salazar, I.M., Rubin, P., Bassano, D., Marcial, V.A.: Improved survival of patients with intracranial ependymoma by irradiation: Dose selection and field extension. Cancer 35, 1563-1573 (1975)

54. Schweisgutz, O.: Tumours Solides de l'Enfant, pp. 191-208. Paris: Flammarion Medicine-Sciences 1979

55. Scott, R.M., Ballantine, H.T.: Cerebellar astrocytoma: malignant recurrence after prolonged postoperative survival. Case report. J. Neurosurg. 39, 777-779 (1973)

56. Shapiro, K., Shulman, K.: Spinal cord seeding from cerebellar astrocytomas. Child's Brain 2, 177-186 (1976)

57. Sheline, G.E.: Radiation therapy of tumors of the central nervous system in childhood. Cancer 35, 957-964 (1975)

58. Sheline, G.E.: Radiation therapy of brain tumors. Cancer 39, 873-881 (1977)

59. Smith, C.E., Long, D.M., Jones, T.K., Levitt, S.H.: Medulloblastoma: an analysis of time-dose relationships and recurrence patterns. Cancer 32, 722-728 (1973)

60. Villani, R., Gaina, S.M., Tomei, G.: Follow-up study of brain stem tumors in children. Child's Brain 1, 126-135 (1975)

61. Whyte, T.R., Colby, M.Y., Layton, D.D.: Radiation therapy of brain stem tumors. Radiology 93, 413-416 (1969)

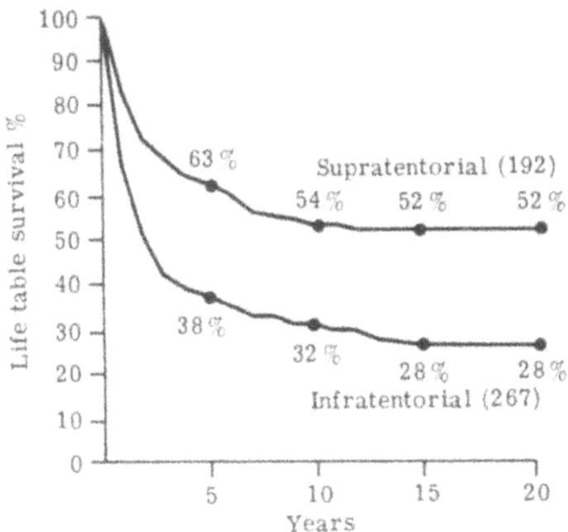

Fig. 1. Royal Marsden Hospital (1952 - 76). Supratentorial and infratentorial tumors (459 children)

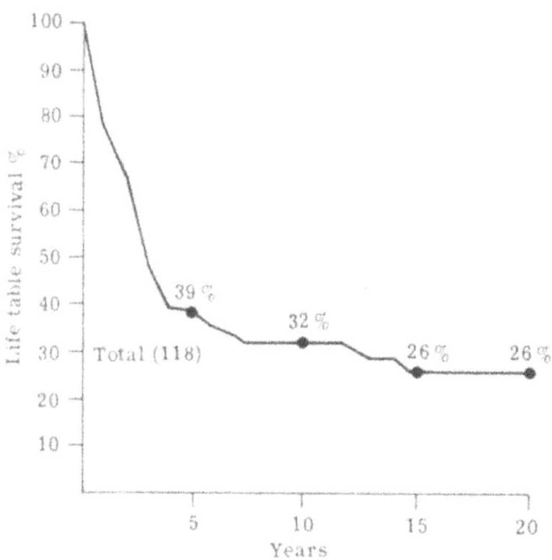

Fig. 2. Royal Marsden Hospital (1952 - 76). Medulloblastoma 118/135 children completing radiotherapy

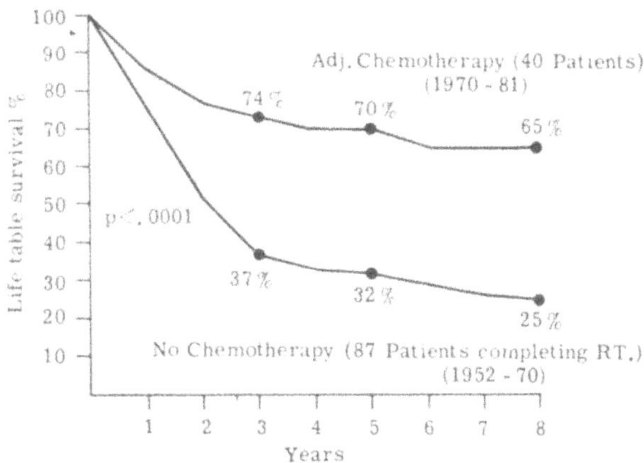

Fig. 3. Royal Marsden Hospital. Medulloblastoma in children (<16) ± adjuvant chemotherapy (1952 - 1970)

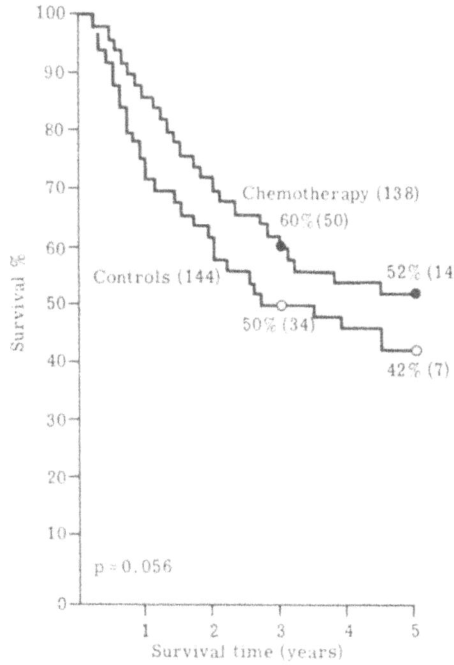

Fig. 4. SIOP study medulloblastoma by treatment

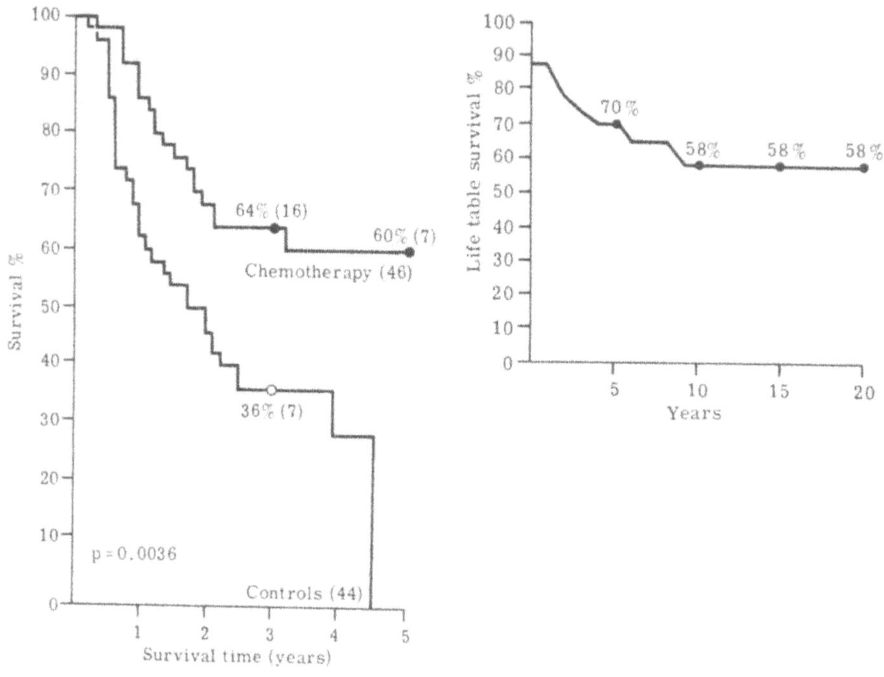

Fig. 5. *Left:* SIOP study medulloblastoma brain stem involved

Fig. 6. *Right:* Royal Marsden Hospital (1952 - 76). Cerebellar astrocytoma (30 children) (RT not completed in 2)

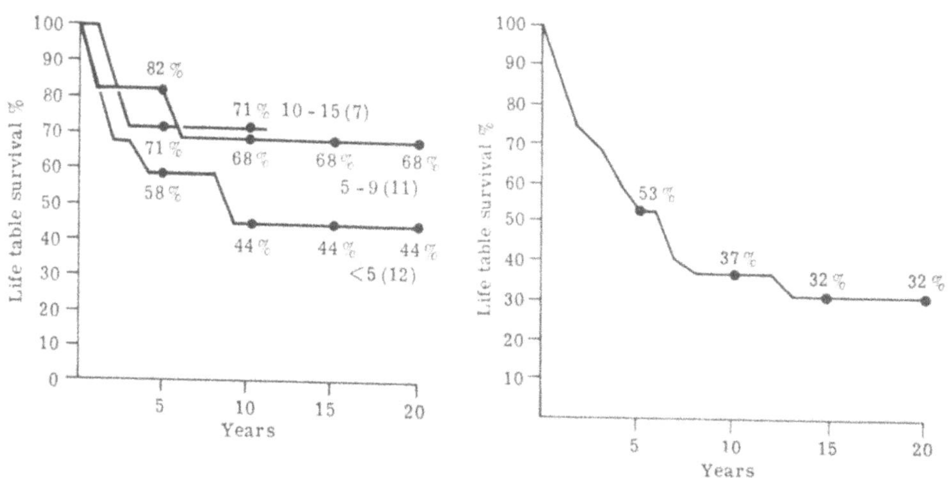

Fig. 7. *Left:* Royal Marsden Hospital (1952 - 76). Cerebellar astrocytoma in children; survival by age (30 patients)

Fig. 8. *Right:* Royal Marsden Hospital. Ependymoma 47 children (1952 - 76)

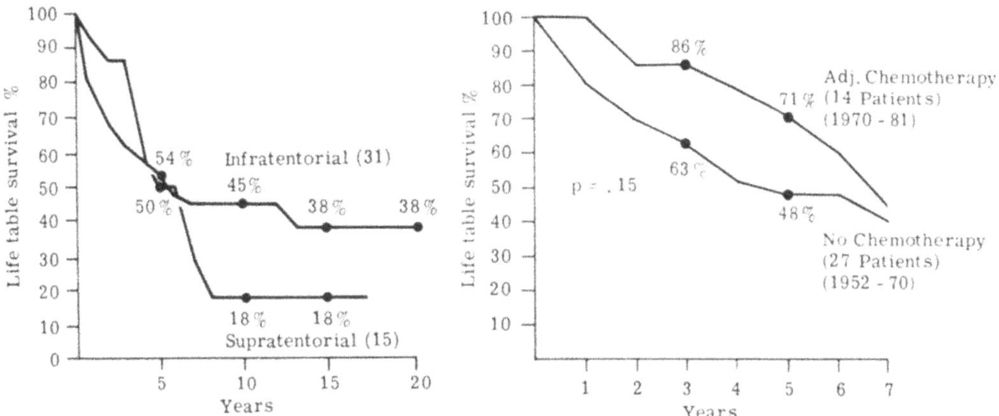

Fig. 9. *Left:* Royal Marsden Hospital. Ependymoma in children; survival by site (46 patients)

Fig. 10. *Right:* Royal Marsden Hospital. Ependymoma in children (<16) + adjuvant chemotherapy (1952 - 1981)

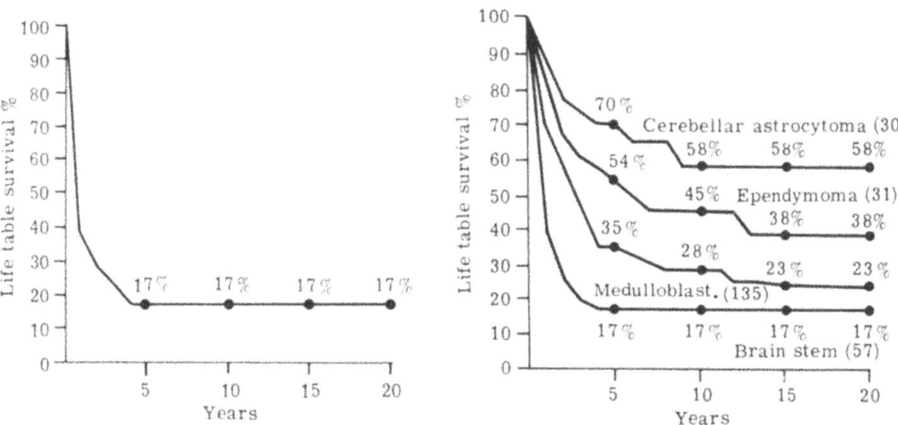

Fig. 11. *Left:* Royal Marsden Hospital (1952 - 76) Total brain stem tumors (57 children)

Fig. 12. *Right:* Royal Marsden Hospital (1952 - 76) Infratentorial tumors (253 children)

The Importance of Fixation Supression in the Detection of Posterior Fossa Tumors

B. Hofferberth[1] and H. E. Nau[2]

[1] Abt. Klinik für Neurologie, Psychiatrische und Nervenklinik der Westfälischen Wilhelms-Universität Münster, Albert-Schweitzer-Strasse 131, D-4400 Münster
[2] Neurochirurgische Klinik im Universitätsklinikum der Gesamthochschule Essen, Hufelandstrasse 55, D-4300 Essen

Summary

We investigated 12 children (6 boys, 6 girls, average age 14.6 years) operated on for a posterior fossa tumor, by means of EEG, ENG, blink reflex and CT. Whereas CT scans show the anatomical situation (tumor, defect), the blink reflex can give hints as to the site of disturbances in brain stem pathways and ENG of those in cerebello-vestibular ones.

Introduction

Disorders of eye movements, especially the occurrence of nystagmus are early and typical symptoms of space-occupying lesions in the posterior fossa. In healthy subjects physiological vestibular induced nystagmus will be suppressed by fixing a target with the eyes. This is called fixation suppression of vestibular nystagmus.

Inhibitory impulses are mediated from the cerebellum to the vestibular nuclei via cerebellar-vestibular pathways. In cases of cerebellar and brain stem disease the inhibitory impulses to the vestibular nuclei disappear causing disturbances of fixation suppression (1, 3, 4). Whereas in healthy persons the visual system dominates the vestibular, in cases of disorders in the posterior fossa vestibular phenomena appear in spite of fixation because of the disinhibition of the vestibular system.

The clinician can prove fixation suppression by a simple test. This test can be quantified by means of electronystagmography (ENG) (2).

In recent years the pathways and generators of R_1 and R_2 components of the blink reflex could be localized in the pontine and medullary brain stem. So it is that the blink reflex can differentiate between pontine and medullary disorders in general. A further analysis of reflex components can give hints about the site and extent of brain stem lesions (Fig. 1).

Material and Methods

We analysed 50 healthy persons and 12 children and adolescents operated on because of posterior fossa tumors in the neurosurgical clinic of Essen University, between 1975 and 1981.

Advances in Neurosurgery, Vol. 11
Edited by H.-P. Jensen, M. Brock, and M. Klinger

We examined 6 boys and 6 girls with an average age of 14.6 years at the time of investigation. The histology of the underlying lesions was cerebellar astrocytoma in six, medulloblastoma in five cases and ependymoma in one case. All patients underwent EEG, blink reflex, ENG and fixation suppression by the simple test, as well as CT investigation.

The EEG was recorded with the usual ten-twenty-system using silver chloride electrodes.

Blink reflexes were analysed by means of the EMG.

Fixation suppression was examined by means of a motor revolving chair. The period of sinusoidal movements was 20 sec, the angular velocity 60° (max.). During two half periods the frequency and the total amplitude of the left or right beating nystagmus with eyes open or closed were determined.

Fixation index means the quotient of these values with eyes opened and closed. An index "1" means absence of fixation suppression, "0" a completely undisturbed fixation suppression. Besides the large-scale electronystagmographic performance, a simple test on a rotating chair with fixing the hands stretched forward was carried out. This test can only show whether fixation suppression fixation exists. Quantification is not possible by this test.

Results

In five out of 12 children we found normal results. A high fixation index corresponded with the CT findings of subtotal tumor removal. The simple clinical test corresponded in all cases to the ENG. In healthy subjects the fixation index never exceeds 0.1.

With EEG we found in two patients a mild general slowing, in one case theta rhythm. In all other cases the EEG was normal corresponding to the age of the patient.

CT demonstrated the state after tumor removal in eight cases, cerebellar or brain stem tumor after subtotal tumor removal in four cases (Table 1, Fig. 2).

Table 1. Synopsis of patients' data and results of investigations FSI, Fixation suppression index (ENG); FS, Fixation suppression (simple clinical test)

Name	Age	Sex	CT	EEG	FSI	FS
B.O.	19	m	Tumor	Slowing	0.8	+
B.H.	2	m	Tumor	Normal	∅	∅
C.M.	21	f	Defect	Normal	0.2	+
F.A.	22	f	Defect	Normal	-	-
G.A.	9	m	Defect	Normal	-	-
K.S.	20	f	Defect	Normal	-	-
K.I.	10	f	Tumor	Slowing	0.5	+
O.S.	15	f	Tumor	Slowing	0.8	+
P.C.	13	m	Defect	Normal	0.4	+
R.C.	4	m	Defect	Normal	-	-
S.J.	13	m	Defect	Normal	0.3	+
S.U.	23	f	Defect	Normal	0.1	-

Discussion

Whereas the blink reflex and other usual neurophysiological investigations can give hints about location in cases of posterior fossa lesions, we want to stress the role of fixation suppression as a neurophysiological method in brain stem diagnosis as well as a simple test in the rotating chair for detection of lesions in the cerebellar-vestibular pathways.

The results of the large-scale ENG test and the simple rotating test in a chair corresponded very well.

Therefore pathological findings should be the indication for further clinical investigations and investigations with apparatus.

References

1. Alpert, J.N.: Failure of fixation suppression: A pathological effect of vision on caloric nystagmus. Neurology (Minneap.) 24, 891-896 (1974)
2. Hofferberth, B.: The use of computer in the equilibrium laboratory. The Royal Society of Medicine, International Congress and Symposium Series 33, 13-22 (1980)
3. Takemori, S.: Visual suppression test. Ann. Otol. 86, 80-85 (1977)
4. Takemori, S.: Visual suppression of vestibular nystagmus after cerebellar lesions. Ann. Otol. Rhinol. Laryngol. 84, 318-326 (1975)
5. Trontelj, M.A., Trontelj, J.V.: Reflex arc of the first component of the human blink reflex: a single motoneuron study. J. Neurol. Neurosurg. Psychiat. 41, 538-547 (1968)

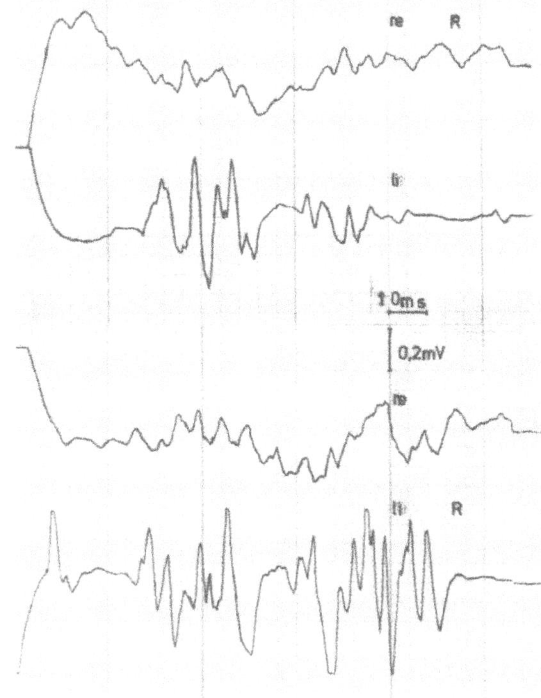

Fig. 1. Blink reflex: O.S., 15-year-old, female. Astrocytoma in the right cerebello-pontine angle and brain stem. On the right side is no R_1-component

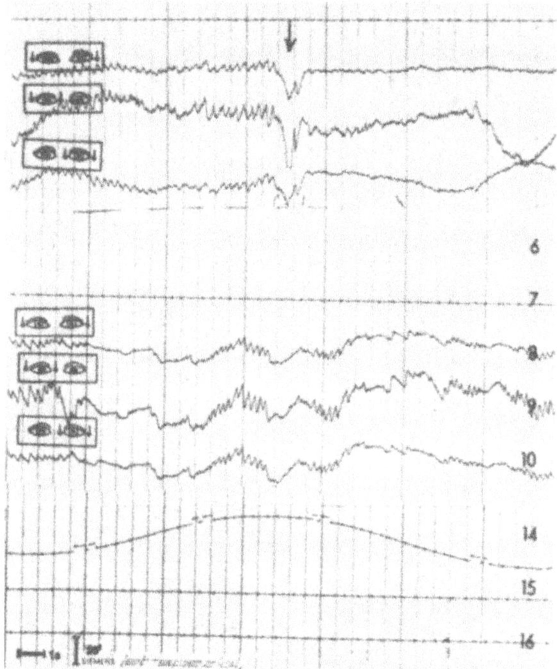

Fig. 2. ENG: examination of fixation suppression with eyes open (narrow) and fixing a target during sinusoidal movement. Upper trace normal fixation suppression, lower trace disturbed fixation suppression. O.S., 15-year-old, female

The Importance of Neurophysiological Examinations for Confirmation of the Diagnosis in Brain Stem Tumors in Children

D. Kountouris[1], A. Servet[2], and W. Gehlen[1]

[1] Neurologische Universitätsklinik, Knappschafts-Krankenhaus Bochum-Langendreer, In der Schornau 23/25, D-4630 Bochum 7

[2] Neurochirurgische Universitätsklinik, Knappschafts-Krankenhaus Bochum-Langendreer, In der Schornau 23/25, D-4630 Bochum 7

Summary

The blink reflex was examined in six children with brain stem tumors.

In only two children could the blink reflex be proved before and after an operation on a brain stem tumor.

Thus, it became possible to estimate the extent of the lesion before and after the operation.

The neurological state was compared with the neuroradiological and blink reflex examinations. In this way the great value of the blink reflex could be seen.

Introduction

Painful examinations such as cerebral arteriography or ventriculography, are often indispensable for localizing a brain stem tumor precisely.

In many cases it is necessary to perform these diagnostic procedures under general anesthesia, especially in children who generally cannot cooperate. This means additional stress for the patients.

Compared with computed tomography, the blink reflex examination is relatively uncomplicated and can be easily performed even in children. It also provides enough information on the functional state of the brain stem and of the trigeminal and facial nerves (1, 6, 10). Furthermore it can be easily repeated so that it becomes possible to follow the progress of the lesion in children with a brain stem tumor. Thus, the blink reflex seems to be a most convenient method for localizing lesions in this area.

Patients and Methods

In six children with a brain stem tumor, blink reflex examinations were performed in addition to neurological examination, computed tomography, cerebral arteriography and ventriculography. In two of these children the blink reflex was recorded before and after resection of the tumor. The blink reflex examination was performed as described by

Advances in Neurosurgery, Vol. 11
Edited by H.-P. Jensen, M. Brock, and M. Klinger

KUGELBERG (1953) (10) and others (1, 6, 13, 15, 16). After electric stimulation of the supraorbital nerve on one side, it is possible to register responses from the orbicularis oculi muscles on both sides. In healthy subjects one early component R_1 (ipsilateral) and two late components, R_2 (ipsilateral) and R_{2x} (contralateral) can be recorded (Fig. 1).

A recording of the blink reflex in a healthy subject is given in Fig. 1. Measurements were referred to normal data established in our own laboratory where the latencies of the R_1 component according to experimental studies are ascribed to its transmission via the pons and via the medulla oblongata (1, 5-7, 10, 11, 14). The latencies of the R_1 and R_2-components after ipsilateral stimulation were 10.9 ± 9 ms and 33.9 ± 4.2 respectively. The latency of the R_{2x} component (contralateral) to the stimulus was 33.8 ± 5.5

Results

In all six children, a pathological blink reflex was recorded. In two of them, the first component R_1 was lost ipsilateral to the tumor, while the other two responses (R_2, R_{2x}) were normal; confirming the neuroradiological localization of the tumor in the pons area. After stimulation of the supraorbital nerve on the opposite side all components appeared normal. This indicated normal function of the brain stem on this side.

In three other children, the latencies of the R_1 and R_2 response were delayed on the side of the tumor, as also was the R_{2x} component after contralateral stimulation, confirming lesions involving both pons and medulla oblongata (Fig. 2).

In one case, where the lesion was found to be in the medulla oblongata, the late response R_2 appeared delayed and reduced in amplitude after stimulation of both sides, while the other two responses were normal.

In two children, whose blink reflex was examined after the operation, the responses were found even more delayed and reduced in amplitude, compared to the pre-operative state. In all children, the neuroradiological examinations confirmed the diagnosis but did not distinctly define the extent of the lesion.

Discussion

As already reported by many authors in various studies, it is possible to localize peripheral lesions of the trigeminal and facial nerves as well as lesions in the brain stem area by means of the blink reflex examination (1, 3-14, 16).

As shown in our patients, it was possible to localize lesions of the pons and of the medulla oblongata caused by brain stem tumors in children and the extent of the lesion could not be defined so well by means of cranial computed tomography or cerebral arteriography (8, 9, 15).

Brain stem lesions seem to be the ideal situation for using the blink reflex examination. The blink reflex offers the further advantage of being repeatable without complication and stress to the patients.

References

1. Dengler, R., Struppler, A.: Beurteilung der Lokalisation und Ausdehnung von Hirnstammaffektionen mit Hilfe des Orbicularis-oculi-Reflexes. Z. EEG-EMG 12, 50-55 (1981)
2. Fischer, M.A., Shahani, B.T., Young, R.R.: Assessing segmental excitability after acute rostral lesions. II. The blink reflex. Neurology 29, 40-45 (1979)
3. Kilimov, N., Linke, D.: Study of blink reflex after trigeminal neurotomy. Electromyogr. Clin. Neurophysiol. 17, 463-491 (1977)
4. Kilimov, N., Linke, D.: Alteration in blink reflex after chordotomy. Electromyogr. Clin. Neurophysiol. 18, 3-10 (1978)
5. Kimura, J.: Alteration of the orbicularis oculi reflex by pontine lesions. Study in multiple sclerosis. Arch. Neurol. 22, 156-161 (1970)
6. Kimura, J.: The blink reflex as test for brain stem and higher central nervous system. function. In: New development in electromyography and clinical neurophysiology. Desmedt (ed.), Vol. 3, pp. 682-691. Basel: Karger 1971
7. Kimura, J.: Electrodiagnostic study of brain stem strokes. Stroke 2, 576-586 (1971)
8. Kountouris, D.: Blinkreflexe bei kindlichen Hirnstammtumoren. 7 Tagung der Gesellschaft für Neuropädiatrie, Frankfurt 8-10, 10-10 (1981)
9. Kountouris, D., Gehlen, W.: Blink reflex of a child brain stem tumor. Neurologia et Psychiatria (in press)
10. Kugelberg, F.: Facial reflex. Brain 75, 385-396 (1953)
11. Lyon, L.W., Kimura, J., McCormick, W.F.: Orbicularis oculi reflex in coma: clinical, electrophysiological and pathological correlations. J. Neurol. Neurosurg. Psychiat. 35, 582-588 (1972)
12. Malin, J.P., Stölzel, R., Freund, G.: Veränderungen des Blinkreflexes (Orbicularis-oculi-Reflexes) im Koma: Lokalisatorische und prognostische Bedeutung. Z. EEG-EMG 11, 12-18 (1980)
13. Namerov, N.J.: The orbicularis oculi reflex in multiple sclerosis. Neurology 20, 1200-1203 (1970)
14. Ongerboer de Visser, B.W., Kuypers, H.G.: An electrophysiological and neuro-anatomical study of Wallenberg's syndrome. Brain 101, 285-294 (1978)
15. Straschill, M.: Orbicularis oculi Reflex mit fehlender früher Komponente und normaler später Reaktion bei einem Patienten mit intrapontinen Tumor. Z. EEG-EMG 11, 19-20 (1980)
16. Struppler, A., Dobbelstein, H.: Elektromyographische Untersuchungen des Glabella-Reflexes bei verschiedenen neurologischen Störungen. Nervenarzt 34, 347-352 (1963)

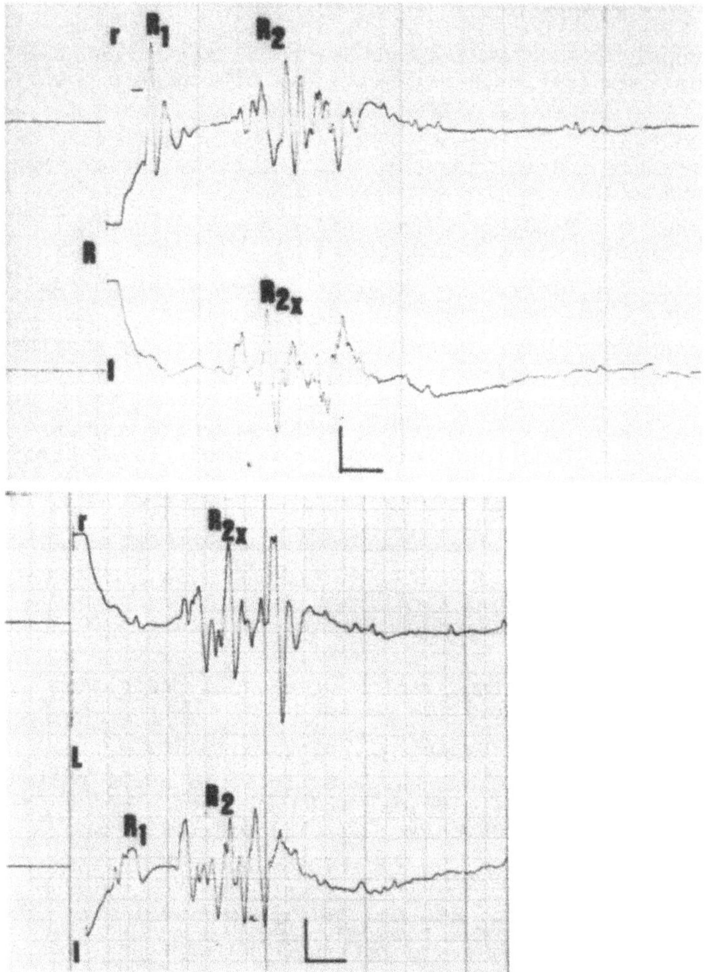

Fig. 1. Recording of the blink reflex responses in healthy subjects. The R_1, R_2 and R_{2x} components can be seen after the simulation of the supraorbital nerve above and below (200 µV/cm, 10 ms/cm).
R, Stimulation of the right supraorbital nerve; L, Stimulation of the left supraorbital nerve; R, Recording of the right orbicularis oculi; L, Recording of the left orbicularis oculi

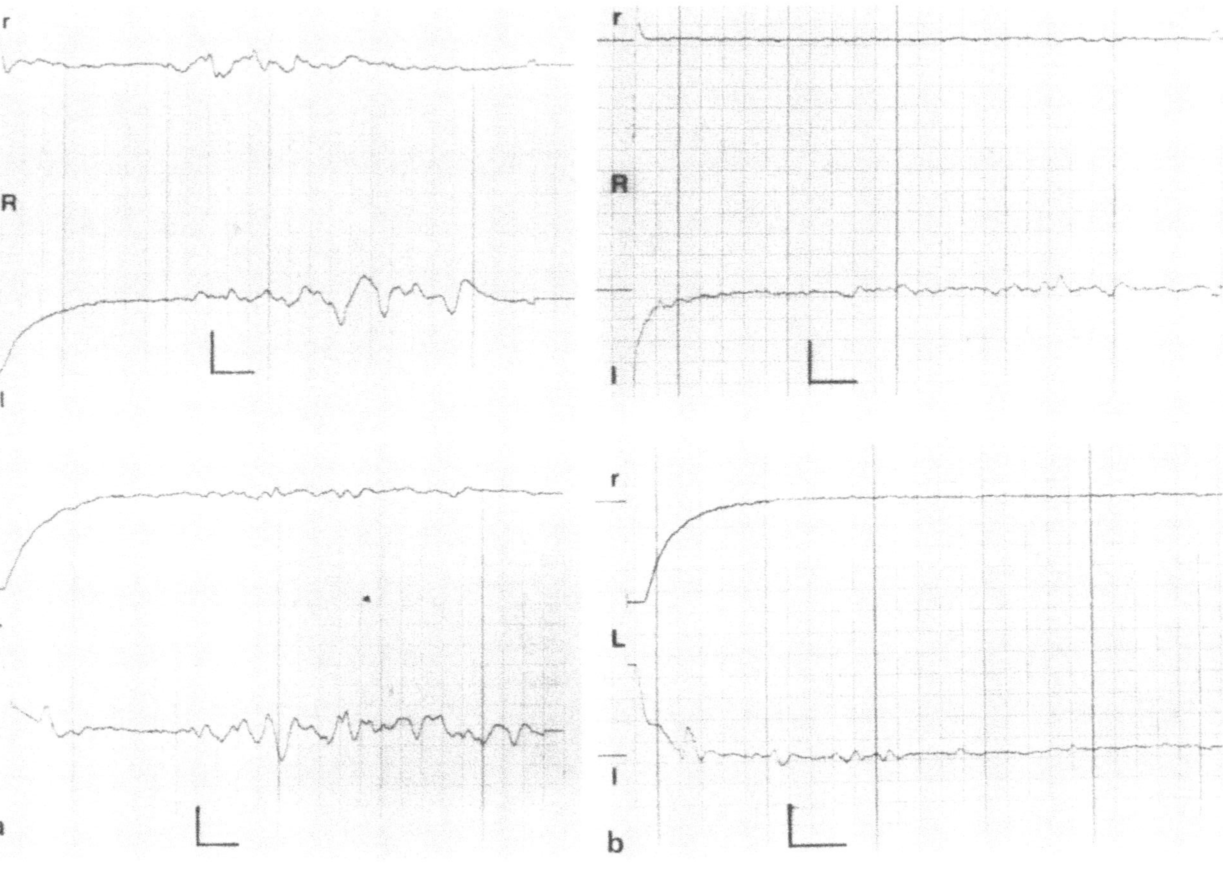

<u>Fig. 2 a,b.</u> The blink reflex examinations of a 12-year-old girl with medulloblastoma before (a) and after (b) an operation. Reduction of the responses can be well seen

Early Auditory Evoked Potentials (EAEP) in Neurosurgery – A New Method for Diagnosis and Localization of Posterior Fossa Tumors in Childhood

K. Maurer[1], M. Rochel[2], P. Gutjahr[2], and D. Voth[3]

[1] Psychiatrische Klinik, Klinikum der Johannes-Gutenberg-Universität Mainz, Langenbeckstrasse 1, D-6500 Mainz
[2] Abteilung für Pädiatrie, Klinikum der Johannes-Gutenberg-Universität Mainz, Langenbeckstrasse 1, D-6500 Mainz
[3] Abteilung für Neurochirurgie, Klinikum der Johannes-Gutenberg-Universität Mainz, Langenbeckstrasse 1, D-6500 Mainz

Introduction

Auditory stimuli of suprathreshold intensity (above 60 dBHL) evoke about 15 waves: an early series (EAEP) during the initial 10 milliseconds (ms), a middle latency sequence (8 to 50 ms) and the longer latency cortical potentials (50 - 300 ms). PICTON et al. (1974) made a survey of all three potential groups. Only the EAEP (waves I to IV) are generated in the infratentorial part of the brain and reflect progressive activation of the auditory tracts and nuclei (Fig. 1): Wave I is assumed to originate at the distal part of the acoustic nerve, wave II in the medulla, wave III in the caudal and wave IV in the rostral pons and wave V in the midbrain (STARR and ACHOR, 1975; STOCKARD and ROSSITER, 1977; MAURER et al., 1979).

Age is one of the most important factors affecting latency and amplitude of the EAEP (HECOX and GALAMBOS, 1974). In the present study, therefore, control values were obtained in three age groups: newborn (birth to 4 weeks), infants (4 weeks to 3 years) and children (3 years to adulthood). Seventeen children with posterior fossa tumors were then investigated. Abnormalities of the waves I to V could be correlated with the extent and location of the neoplastic lesions.

Methods

The methods applied have been already published elsewhere (MAURER et al., 1980). The patients were examined in the supine position. Skin electrodes were applied to the vertex (C_z) and the mastoid ipsilateral to the stimulated ear. The inter-electrode impedances were kept below 1500 ohms. As stimulus we used electrically a sine half wave (duration 250 μs); the acoustic result at the earphone was a tone-pip with a duration of about 1 ms. The acoustic waveform and EAEP recorded from a ten-year-old boy are shown in Fig. 2. The stimuli were presented monaurally by Beyer DT 48 earphones at a randomized rate of 10 per second and an intensity of 70 and 80 d BHL (decibels above the hearing threshold for normals). Masking white noise was presented to the contralateral ear. The amplified and filtered EEG-signal (bandpass of the system: 300 - 3200 Hz) recorded in the first 20 ms after stimulation was summated with a signal averager (1024 sampling points). According to the earlier reports (MAURER et al., 1980) latencies were measured between the start of the sine-wave and the upward positive peak of each wave; the interpeak conduction times were determined between the

Advances in Neurosurgery, Vol. 11
Edited by H.-P. Jensen, M. Brock, and M. Klinger

different peaks. Amplitudes were measured from peak to troughs and vice versa. A latency delay or an amplitude reduction by more than 2.5 standard deviations was considered to be significant. There were at least two runs for each ear and intensity and only congruent components were accepted for evaluation. An automatic artifact rejection monitored the output of the amplifier continuously and averaging was not performed, when the signal was contaminated with muscle activity and artifacts. An otoscopy was done in all newborn and puretone audiometry in cooperative children.

Results

1. *Normal values in newborn, infants and adults.*
   At birth the latencies are prolonged and the amplitudes reduces (Table 1). To determine the maturational changes, at least five infants in the period between 6 months and 3 years were measured. The adult values were obtained from 50 adult subjects. Fig. 3 shows the decrease of latencies of waves II, III and V in the first three years of life.

2. *Wave abnormalities in children with brain stem tumors.*
   There were seven children with medulloblastoma, five with a cerebellar astrocytoma and five patients with a brain stem tumor. Two cases with a ponto-mesencephalic lesion will be described in more detail. The first case was a 15-year-old girl complaining of dysarthria, diplopia and a bilateral hearing loss of varying intensity. Neurological examination revealed horizontal eye movement paresis as a sign of a pontine lesion. The computerized tomography (CT) showed a tumor in the pons. EAEP are shown in Fig. 4 in a child of the same age with normal brain stem function (upper trace) and in the patient with the pontine tumor (lower trace). Waves 1 and II are normal, waves III and IV reduced in amplitude. It was not possible to record a reliable wave V. The EAEP indicated a ponto-mesencephalic lesion.

The second case was a four-year-old boy with a pontine tumor. He complained of diplopia. Neurological examination showed a paresis of the sixth cranial nerve and a right sided hemiparesis. Further symptoms such as dysarthria and unsteadiness in walking were suggestive of an infratentorial lesion.

Table 1. Latencies (ms), interwave latencies (ms) and amplitudes (nV) obtained in 12 newborn

Absolute latencies			Interwave latencies			Amplitudes		
Wave	Mean	SD	Wave	Mean	SD	Wave	Mean	SD
I	1.85	0.26	I-II (pct)	1.1	0.25	I	120	50
II	2.95	0.22	I-III	2.8	0.26	II	64	45
III	4.65	0.28	I-IV	4.0	0.28	III	235	45
IV	5.85	0.38	I-V	5.15	0.21	IV	75	30
V	7.00	0.26	II-V (CCT)	4.05	0.22	V	160	55

EAEP were indicative of a midbrain lesion (Fig. 5). By stimulating the right ear (upper part) waves I to IV were normal, wave V was definitely delayed with a latency of 6.46 ms (normal: 5.6 ms); the I-V interwave latency was prolonged (4.85 ms; normal 4.15 ms). The left ear (lower part) showed a late wave I (2.2 ms) caused by a conductive hearing loss. The interwave latencies, however, still allowed a diagnosis in the central part of the brain stem; the I-V conduction time was 4.7 ms and showed a delay similar to that on the opposite side. This delay was caused by a late wave V.

Conclusion

The first description of wave alterations of EAEP in neuro-otological disorders (SOHMER et al., 1974) and a report concerning mainly maturational changes of waves I to V (HECOX and GALAMBOS, 1974) appeared at the same time. There are few reports about wave abnormalities caused by space-occupying lesions in the posterior fossa (MAURER and ROCHEL, 1982).

For evaluations of brain stem function interwave latencies are of higher significance since they allow a localisation in the presence of a hearing disturbance (Fig. 5). This is of special importance in children with middle ear disease combined with a conductive hearing loss. The peripheral conduction time (pct), i.e. the interwave latency between I and II allows a diagnosis of tumors in the cerebello-pontine angle. The central conduction time (cct), i.e. the interwave latency between II and V is able to test the dorsolateral part of the brain stem.

There are minor differences between the pct in newborn and adults, whereas the cct is significantly longer at birth. This confirms a nearly complete maturation of the inner ear in the newborn. It is more likely to be in the brain stem, rather than in the cochlea itself, that changes occur which lead to a decrease in latency of brain stem potentials.

It was possible to correlate abnormalities of waves II to V with well defined neurological defects in the brain stem. Waves II and III were altered in medullo-pontine and waves IV and V in ponto-mesencephalic lesions. Besides the brain stem level the side of the lesion, i.e. left or right side of the dorsolateral part of the brain stem is indicated. Strictly unilateral wave abnormalities prove the generation of waves II to V at different levels of the auditory pathway of only one side. Bilateral abnormalities at the same or different levels may be caused by a large tumor which crosses the midline.

In most of the patients the tumor was confirmed by computerized tomography. There are, however, a few cases with a normal CT in presence of a brain stem tumor indicating, that EAEP are able to detect clinically "silent" lesions. This may facilitate an early diagnosis in juvenile posterior fossa tumors.

In a few cases with a midbrain syndrome, i.e. compression of the inferior colliculi in the tentorial hiatus, wave V was singularly affected, whereas impulse conduction remained normal from the cochlea to the rostral pons. This can be interpreted as a further indication of a genesis of wave V in the midbrain and different generator loci of waves IV and V.

In summary, EAEP are a non-invasive test of high reliability, which allow one to diagnose functional disturbances in the cerebello-pontine angle and in the dorsolateral part of the brain stem where the auditory pathway has its nuclei and fiber systems. The test is recommended in children suspected of having a tumor in the posterior fossa.

References

1. Hecox, K., Galambos, R.: Brain stem auditory evoked responses in human infants and adults. Arch. Otolaryngol. 99, 30-33 (1974)
2. Maurer, K., Leitner, H., Schäfer, E., Hopf, H.C.: Frühe akustisch evozierte Potentiale, ausgelöst durch einen sinusförmigen Reiz. Dtsch. med. Wschr. 104, 546-550 (1979)
3. Maurer, K., Schäfer, E., Hopf, H.C., Leitner, H.: The location by early auditory evoked potentials (EAEP) of acoustic nerve and brain stem demyelination in multiple sclerosis (MS). J. Neurol. 223, 43-58 (1980)
4. Maurer, K., Rochel, M., Early auditory evoked potentials (EAEP) in children with neoplastic lesions in the brain stem. In: Voth, D., Gutjahr, P., Langmaid, C. (eds.): Central nervous system tumors in childhood. Berlin, Heidelberg, New York: Springer 1982
5. Sohmer, H., Feinmesser, M., Szabo, G.: Sources of electrocochleographic responses as studied in patients with brain damage. Electroencephal. clin. Neurophysiol. 37, 663-669 (1974)
6. Starr, A., Achor, L.J.: Auditory brain stem responses in neurological disease. Arch. Neurol. 32, 761-768 (1975)
7. Stockard, J.J., Rossiter, U.S.: Clinical and pathological correlates of brain stem auditory response abnormalities. Neurology 27, 316-325 (1977)

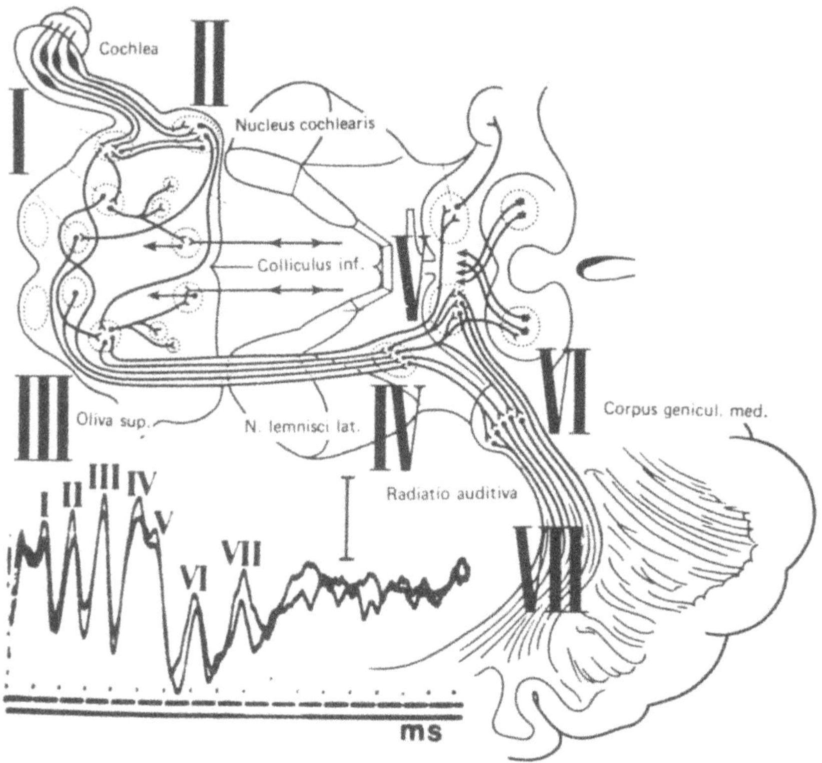

Fig. 1. Schematic drawing of the brain stem and the course of the auditory pathway. Roman numerals indicate the assumed generator sites of EAEP

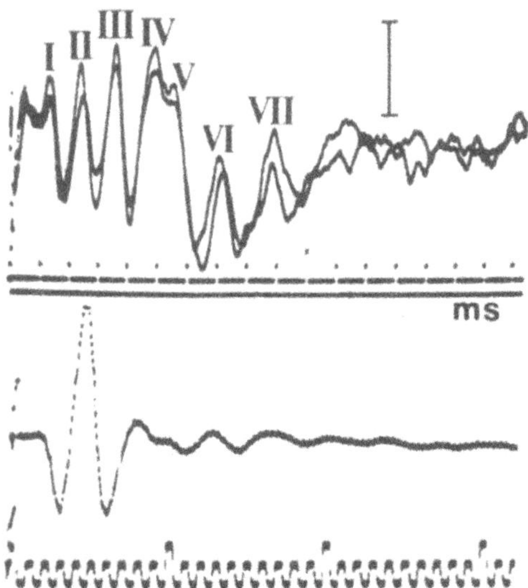

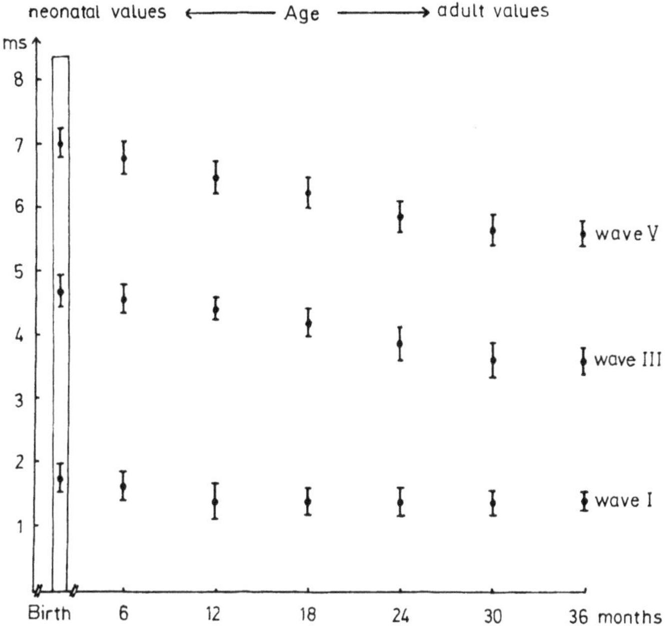

Fig. 3. Latencies of waves I, III and V as a function of age. Vertical bars indicate mean and standard deviations (SD). The column contain values at birth (n = 12). In the period between 6 and 30 months at least nine infants were tested. The adult values were obtained in 50 persons. Age in the figure means gestational age

◄

Fig. 2. Early auditory evoked potentials *(upper part):* The wave of the acoustic nerve (I), waves from the brain stem (II to V) and supratentorial waves (VI and VII) in a normal child. Calibration 200 nanovolts. *Lower part:* Acoustic waveform of the stimulus as produced by a Beyer DT 48 earphone

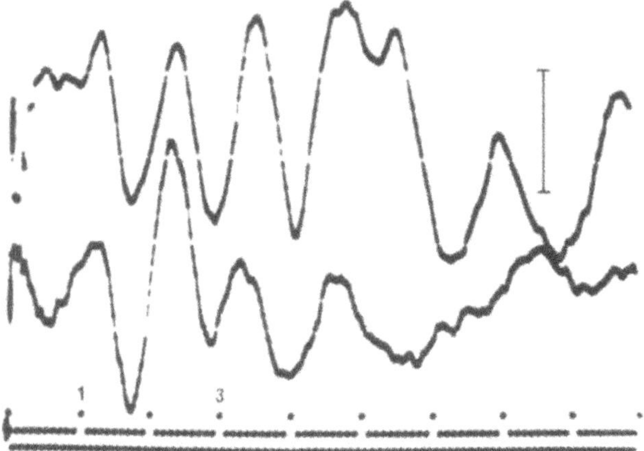

Fig. 4. *Upper trace:* EAEP in a normal child. *Lower trace:* EAEP in a pontine tumor. Calibration 200 nV. Time scale in milliseconds

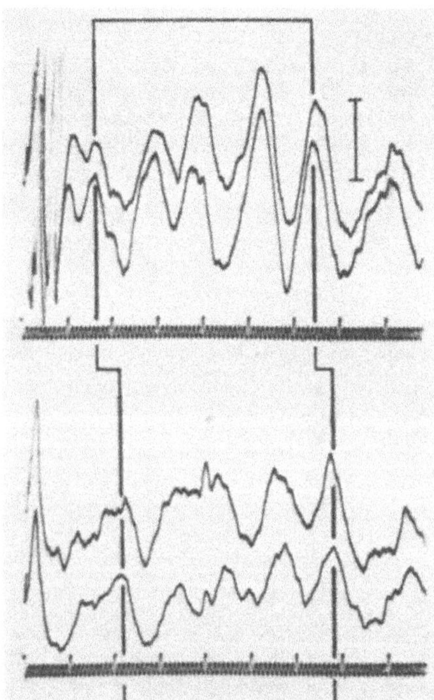

Fig. 5. *Upper part:* EAEP from the right ear; wave V was delayed (6.46 ms), the I-V interwave latency prolonged (4.85 ms). *Lower part:* EAEP from the left ear; the conductive hearing loss produced a late wave I (2.2 ms); the I-V interwave latency was prolonged (4.7 ms). Calibration 200 nV. Time both in milliseconds

Diagnostic Methods, Length of History and Survival Rate of Brain Tumor Patients in Childhood Before and After the Introduction of CT

J. Holldack, C. Roosen, and W. Havers

Neurochirurgische Kinderklinik, Universitätsklinikum der Gesamthochschule Essen, Hufelandstrasse 55, D-4300 Essen 1

Introduction

Cranial computerized axial tomography (CT) is a method of investigation, that has proved its usefulness in a short period of time. Its reliability and the relatively slow invasiveness for the patient were probably the cause, in spite of the high cost. With its help many patients should be spared other more invasive methods of investigation.

With this new diagnostic method the question is whether the prognosis could be improved and most of all, if the period, until the diagnosis is reached, could be shortened. The latter, at least, seemed a possibility.

Material

Included in the study were all patients, that were admitted to the Department of Pediatrics of the University-Hospital Essen from 1967 - 1981 because of brain tumor.

The time until the diagnosis was established was taken from the patients records, while the time of survival was obtained from the records as well as through information from the registration office. The computerized axial tomography was introduced in the Essen Clinic in 1976. That is why group I in the study includes patients from the year 1967 - 75, while group II includes those from 1976 - 81.

The average was about ten patients a year.

Results

The total number of patients was 146 (see Table 1) group I included 97 patients, 49 of whom had a tumor of the posterior cranial fossa. Group II included 49 patients, of whom 27 had posterior cranial fossa tumors.

In group I, the histories of 88 patients could be assessed to determine the time before the diagnosis was established. The average was 8.4 months for the whole group, while for patients with posterior cranial fossa tumors it was only 4.5 months.

Advances in Neurosurgery, Vol. 11
Edited by H.-P. Jensen, M. Brock, and M. Klinger

Table 1. General detail about the patients

	All	Tumors of posterior fossa	Other locations
Group I	97	49	48
Group II	49	27	22
Totals	146	76	70

In group II, this time interval before reaching the diagnosis averaged 8.3 months, for the whole group. Calculating it for those patients with posterior fossa tumors the mean was 3.7 months.

In both groups, that period ranged between a few days and several years (Fig. 1).

Figure 1 makes it clear, that computerized tomography did not influence the time before diagnosis, since both groups showed nearly identical values.

In patients of group II who had posterior fossa tumors, this time was only shorter by 0.8 months.

The time of survival of the patients was analysed according to the method suggested by S.I. CUTTLER and F. EDERER.

It was obvious by analysing both groups, that there was no statistically significant difference at any time during the period of observation. Although this period of observation was longer for patients in group I, it is quite clear that after the sixth year no great changes were to be expected.

At the end of the respective periods of observation, 36% of group I patients and 32% of group II patients were surviving (Fig. 3).

In Fig. 3 the time of survival of patients with posterior fossa tumors in both groups, does not show any relevant difference during the first four years of the period of observation.

In the fifth and sixth years of observation, the life-table analysis has no relevance for group II patients, since only 2 of them were observed for so long because for those patients from years 1976 - 1981 31/12/81 was chosen as a dead-line to end the study. Of this group of patients 33% and 38% respectively survived.

Summary

The patients were divided into two groups as the only criterion whether the diagnosis was achieved with or without the CT scan, not taking into consideration the different treatment used or the different results of histological examinations.

Thus the question if using that new diagnostic method we can reach our final diagnosis earlier, or if the prognosis of these patients can be improved by it, can be clearly answered by: No!

Neither the time between the appearance of the first symptoms and reaching the final diagnosis can be shortened, nor can the period of survival be increased.

A new diagnostic method cannot shorten the prediagnostic period and only helps those, who take the necessary steps to find the diagnosis. It was expected, that the prognosis would improve after introducing the CT, but the fact that the prognosis of the second group has not improved compared to the first, is not satisfactory. It means 15 years without any progress.

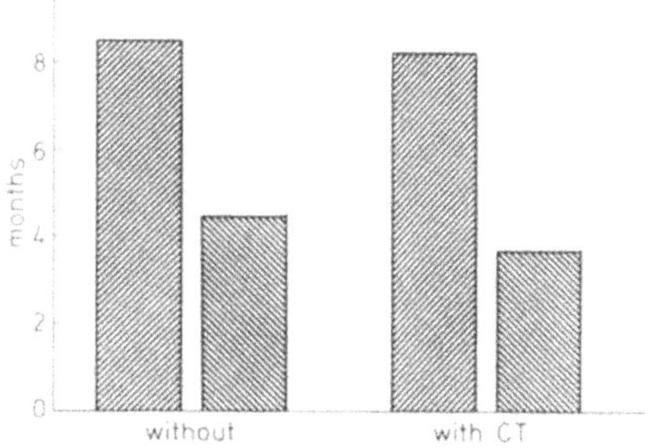

Fig. 1. Time which passed before reaching the diagnosis in both groups in general and selectively for patients with posterior fossa tumors

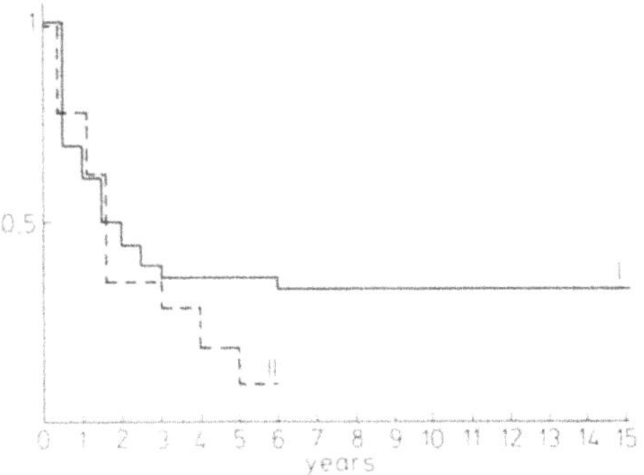

Fig. 2. Curve I shows time of survival of patients before introducing CT, Curve II after introducing CT

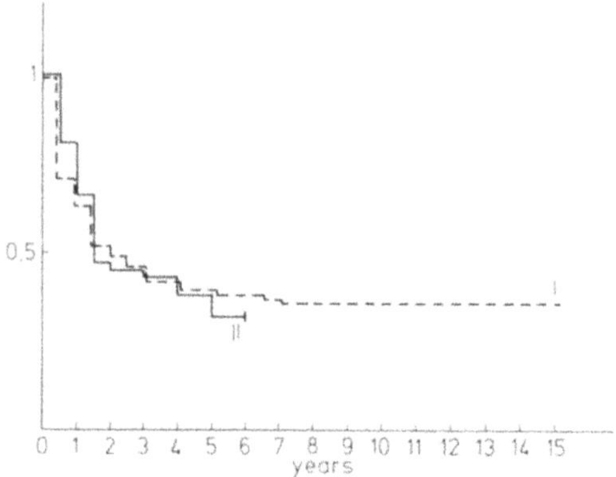

Fig. 3. Time of survival of patients with posterior fossa tumors before (I) introducing CT and after (II)

Internal Shunt or Peri-Operative Pressure-Controlled Ventricular-Fluid Drainage (C-VFD) in Children and Juveniles with Infratentorial Tumors

K. E. Richard, R. Heller, and R. A. Frowein

Neurochirurgische Universitätsklinik Köln, Joseph-Stelzmann-Strasse 25
D-5300 Bonn-Venusberg

In the last two decades opinions have varied concerning the technique of prophylaxis of postoperative disturbances of CSF circulation in patients with infratentorial tumors. The advocates of temporary perioperative external ventricular fluid drainage (C-VFD) (1, 8, 9, 11-13, 15, 17) are opposed to those authors who demand the preoperative implantation of an internal shunt system (1-3, 5, 6, 10).

Recently ALBRIGHT (2) pointed out that so far there have been no investigations regarding the question whether C-VFD or internal shunt is the better method of treatment.

Material

From 1973 - 1981 we treated 91 children and juveniles with an infratentorial space-occupying lesion. There were 32 patients with an astrocytoma, 23 with a medulloblastoma, five with cystic lesions, five with neurinomas and 24 with brain stem tumors of unknown histological diagnosis.

Twenty patients remained without C-VFD or internal shunt, 36 patients had a C-VFD, either before, after or during the operation. In the latter cases the ventricular fluid pressure (VFP) was monitored simultaneously.

In 35 patients an internal shunt was inserted either pre- or postoperatively.

Results

Survival, mortality rates and frequency of complications were compared (Table 1) in the three treatment groups - i.e., patients with neither drainage nor shunt, patients with C-VFD and patients with internal shunt.

In the first group of patients without drainage or shunt the postoperative or intra-operative *mortality*[1] *rate* was strikingly high: 55%. By contrast, mortality in patients with a C-VFD was 6%, in pa-

1 Mortality, fatal outcome during stay in hospital

Advances in Neurosurgery, Vol. 11
Edited by H.-P. Jensen, M. Brock, and M. Klinger

Table 1. Posterior fossa tumors < children + juveniles > n= 91

MODE OF TREATMENT	Nº	†	Intra-op. Early p.op. DEATH	↑	FREE of COMPL.	COMPLICATIONS: CSF FISTULA	CSF-INFECTION / SEPSIS	Post-op. INTRACRAN. HYPOTEN.	TUMOR HEMORRH.	SUBDURAL HYGROMA (HEMATOMA)	COLLAPSE of VENTRICLES	MISCELL COMPLIC
NO DRAINAGE / NO SHUNT	20	11 (55%)	10 (50%)	9 (45%)	8 (89%)	1 (11%)	—	—	—	—	—	—
EXTERNAL DRAINAGE												
PRE-OP	5	—	—	5	2	—	—	—	—	—	—	—
POST-OP	10	—	—	10	8	—	—	—	—	—	—	1
PERI-OP	21	2	1	19	16	2 (SH)	2	—	—	—	—	2
S	36	2 (6%)	1 (3%)	34 94%	26 (76%)	2 (6%)	2 (6%)	—	—	—	—	3 (9%)
INTERNAL SHUNT [V-A / V-P]												
PRE-OP	10	1	—	9	5	—	1	4	—	2+(1)	—	1
POST-OP	11	3	—	8	3	1	3	2	—	2+(1)	3	1
BRAINSTEM TUMORS	14	8	6	6	—	—	1	2	2	—	—	2
S	35	12 (34%)	6 (17%)	23 66%	8 (35%)	1 (4%)	5 (22%)	8 (35%)	2 (9%)	6 (26%)	3 (13%)	4 (17%)
TOTAL	91	25 (27%)	17 (19%)	66 73%	42 (64%)	4 (6%)	7 (11%)	8 (12%)	2 (3%)	6 (9%)	3 (5%)	7 (11%)

tients with an internal shunt 34%. The relatively high mortality rate of the third group is due to the high proportion of inoperable brain stem lesions with a post-operative mortality rate of 57%.

Ten out of 11 patients of the first group died in the early post-operative period in the first week, frequently after an acute respiratory paralysis, or even intra-operatively (four patients) (Table 2).

In the first group five out of six patients with cystic brain tumors survived without complications, but only three of eight patients with solid tumors in the fourth ventricle (Table 2).

The few children with tumors in the cerebello-pontine angle and in the brain stem who received no drainage all died post-operatively (Table 2).

Continuous VFP-monitoring showed acute pressure increases both pre-operatively, during induction of anesthesia, and in the early post-operative period of the first week, if C-VFD was interrupted, especially in cases with difficulties in respiration (Fig. 1).

In one patient, there were also long periods of B-waves after implantation of a ventriculo-atrial shunt system which pointed to a malfunction of the system.

With C-VFD the postoperative course was more frequently free of *complications* than in patients with an internal shunt (Table 1). Two of the externally drained patients developed a *CSF fistula* at the site of operation and therefore the insertion of an internal shunt was necessary. A fistula which continued in spite of the internal shunt only closed after lumbar drainage for several days with drastic lowering of VFP down to negative pressure values.

All patients of group 2 received antibiotic prophylaxis with chloramphenicol. *Infections* of the CSF pathways were seen in two out of 34 (6%) of the patients with C-VFD (Table 3). In both cases of infection, the duration of drainage was more than 14 days.

Table 2. Prognosis of infratentorial tumors in children and juveniles without external drainage or an internal shunt (group I): Survivors; Deaths; "Sudden Death" = term used by H. CUSHING in 1930/31. It means acute deterioration with fatal outcome in the early postoperative period

No drainage / No shunt				
Site of tumor	Number	↑	✠	"Sudden death"
Hemisphere (Cyst. tumor)	6 (4)	5 (4)	1	(1)
Vermis	9	3	5	(4)
4th Ventricle	3	1	2	(2)
Cerebello-pontine angle	1	-	1	(1)
Brain stem	2	-	2	(2)
Total	20	9	11	(10)

Table 3. CSF infection / sepsis

External drainage	No.	Infection (No. of patients)	Pleocytosis (No. of patients) $> 300/3$	Duration of drainage (days) $\bar{x}$	mx
pre-operative	5	-	2	2.8	4
post-operative	10	-	4	4.8	8
peri-operative	19	2	5	8.4	25
Total	34	2 (6%)	11 (35%)	6.7	25
Internal shunt				Manifestation (post-operative day)	
pre-operative	9	1	-	21	
post-operative	8	3	-	10, 13, 300	
brain stem	6	1	4	1	
Total	23	5 (22%)	4 (19%)		

There was also a relation between the duration of drainage and the frequency of CSF-pleocytosis (Table 3). After insertion of an internal shunt, CSF infection or sepsis developed in 22% of the patients. This necessitated a removal of the shunt system, most frequently in cases where the system had been inserted postoperatively (Tables 1 and 3). In all these five patients there had been an earlier ventriculography with positive contrast medium. In only two out of five cases antibiotic protection had been given after examination.

After implantation of an internal shunt, a *CSF hypotension* syndrome developed in 35%, accompanied by headache and vomiting. With C-VFD this syndrome did not occur.

The effects of CSF hypotension were *tumor hemorrhages* immediately after shunt implantation in two patients, and development of uni- or bilateral *hygromas* or *hematomas* in six patients (26%). The treatment used was temporary closing of the internal shunt with simultaneous external drainage of the subdural fluid accumulation. VFP was monitored until reformation of the collapsed brain mantle.

In a three-year-old boy, large subdural abscesses (bitemporo-occipital) with thick membranes had to be removed surgically 20 months after resection of a cerebellar astrocytoma (Fig. 2).

In three children, the ventricles collapsed with the consequence of life-threatening dysfunction of the internal shunt.

The remaining complications consisted of obstructions or dislocations of the internal shunts in four patients, and postoperative hemorrhage in two patients. The latter was detected quite early in one case by the VFP-monitoring.

Comment

There is a basic agreement in modern literature that an infratentorial space-occupying lesion should not be operated on without preliminary correction of raised VFP and that postoperative VFP increases as a result of disturbances of CSF resorption have to be prevented.

For this purpose selected methods of a temporary C-VFD or internal shunts each have their specific advantages, and also their risks which should not be overlooked.

The *temporary C-VFD* offers the possibility of continuous peri-operative VFP-monitoring (11-13). An obstruction of the drainage system can be detected early on by the increase in intracranial pressure.

After the return of the CSF circulation to normal the C-VFD can generally be removed between the third and the seventh post-operative day (12, 13).

This method carries a risk of infection of between 2% and 10% (8, 9, 11, 12, 15, 16, 18). Complications such as "upward herniation" (6, 10) or tumor bleeding (17) as a result of an abrupt lowering of the VFP are probably avoidable by C-VFD. This occurred in two patients after insertion of an internal shunt.

The *internal shunt* has the advantage of quick mobilisation of the patient. Specific dangers include the CSF hypotension syndrome (4, 10) with collapse of the ventricles and development of subdural hygromas or subdural hematomas (10), which have to be operated on. None of the systems used nowadays offers sufficient protection against this complication (4), which amounts to 8% in RAIMONDI's large material (10). As a cause of the high infection rate in our patients with internal shunts, in all of these one has to consider the preceding ventriculography with positive contrast medium without sufficient antibiotic protection. The percentages given by other authors range from 2 to 11% (1, 13) and are therefore no lower than in patients with C-VFD.

With such a low incidence of complications and with the possibility of simultaneous pressure regulation, a C-VFD is therefore preferable. In the choice of the various drainage methods, the type and the position of the space-occupying lesion can be decisive. C-VFD is suitable for solid, removable lesions of the cerebellum and the vermis, requiring a period of drainage of no more than seven days. With cystic lesions, no drainage is generally required.

The implantation of an internal shunt system is indicated in the following cases: inoperable tumors of the brain stem with hydrocephalus; only partially removable tumors of the fourth ventricle and cerebellopontine angle; rare cases of continued disturbance of CSF circulation after operations on tumors of the cerebellum or brain stem.

References

1. Abraham, J., Chandy, J.: Ventriculo-atrial shunt in the management of posterior fossa tumors. J. Neurosurg. 20, 252-253 (1963)
2. Albright, L., Reigel, D.H.: Management of hydrocephalus secondary to posterior fossa tumors. J. Neurosurg. 46, 52-55 (1977)
3. Böhm, B., Mohadjer, M., Hemmer, R.: Preoperative continuous measurements of ventricular pressure in hydrocephalus occlusus with tumors of the posterior fossa: the value of ventricular-auricular shunt. Adv. Neurosurg. 5, 194-198 (1978)
4. Gruss, Pl, Gaab, M., Knoblich, O.E.: Disorders of CSF circulation after interventions in the area of the posterior cranial fossa with prior shunt operation. Adv. Neurosurg. 5, 199-202 (1978)

5. Hekmatpanah, J., Mullan, S.: Ventricular caval shunt in the management of posterior fossa tumors. J. Neurosurg. 26, 609-613 (1967)

6. Hoffman, H.J., Hendrick, E.B., Humphreys, R.P.: Metastasis via ventriculoperitoneal shunt in patients with medulloblastoma. J. Neurosurg. 44, 562-566 (1976)

7. Kessler, L.A., Dugan, P., Concannon, J.P.: Systemic metastases of medulloblastoma promoted by shunting. Surg. Neurol. 3, 147-152 (1975)

8. Lundberg, N., Kjällquist, A., Kullberg, G., Pontén, U., Sundbärg, G.: Non-operative management of intracranial hypertension. In: Advances and Technical Standards in Neurosurgery. Krayenbühl, H. et al. (eds.), pp. 1-60. Wien, New York: Springer 1974

9. Pertuiset, B., Van Effenterre, R., Horn, Y.: Temporary external valve drainage in hydrocephalus with increased ventricular fluid pressure. Acta Neurochir. 33, 173-181 (1976)

10. Raimondi, A.J., Tadaneri, T.: Hydrocephalus and infratentorial tumors. Incidence, clinical picture, and treatment. J. Neurosurg. 55, 174-182 (1981)

11. Richard, K.E.: Liquorventrikeldruckmessung mit Mikrokatheter und druckkontrollierte externe Liquordrainage. Acta Neurochir. 38, 73-87 (1977)

12. Richard, K.E.: Longterm measuring of ventricular fluid pressure with tumors of the posterior fossa. Adv. Neurosurg. 5, 179-185 (1978)

13. Richard, K.E., Frowein, R.A., Vanner, K.G.: Ventricular fluid pressure in posterior fossa tumors of childhood. Mod. Probl. Paediat. 18, 35-39 (1977)

14. Sayers, M.P.: Shunt complications. Clin. Neurosurg. 28, 393-400 (1976)

15. Shalit, M.N., Ari, Y.B., Eynan, N.: The management of obstructive hydrocephalus by use of external continuous ventricular drainage. Acta Neurochir. 47, 161-172 (1979)

16. Smith, R.W., Alksne, J.F.: Infections complicating the use of external ventriculostomy. J. Neurosurg. 44, 567-570 (1976)

17. Vaquero, J., Cabezudo, J.M., De Sola, R.G., Nombela, L.: Intratumoral hemorrhage in posterior fossa tumors after ventricular drainage. Report of two cases. J. Neurosurg. 55, 390-396 (1981)

18. Wyler, A.R., Kelly, W.A.: Use of antibiotics with external ventriculostomies. J. Neurosurg. 37, 185-187 (1972)

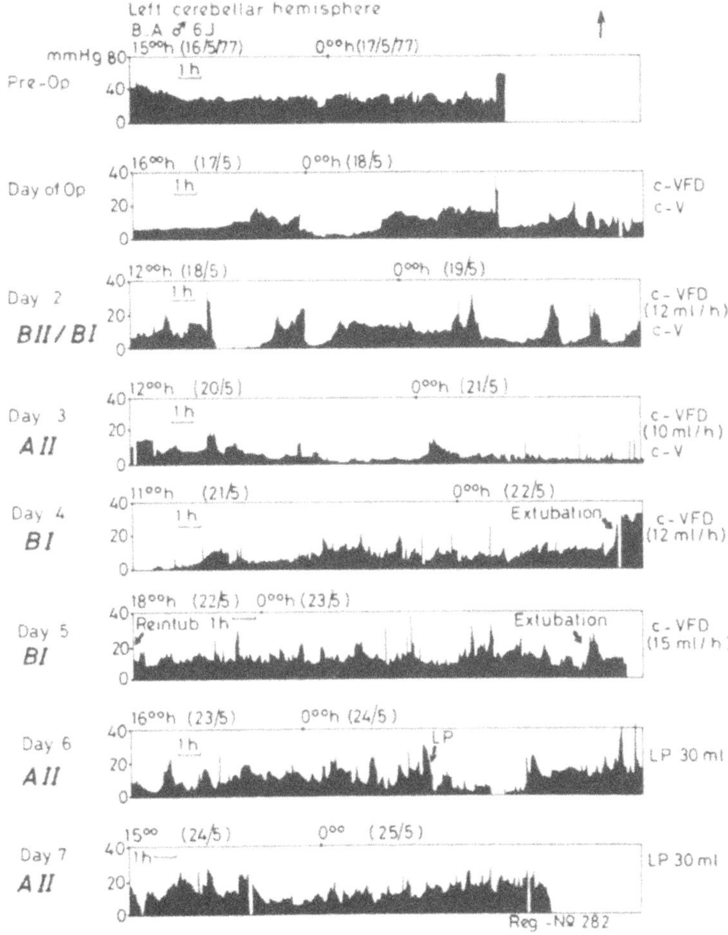

Fig. 2. Three-year-old boy with large subdural abscesses after resection of a cerebellar astrocytoma 20 months earlier

Schedule of Treatment for Pineal Tumors in Children

H. E. Clar[1], M. Bamberg[2], and H. C. Nahser[1]

[1] Neurochirurgische Klinik und Poliklinik, Universitätsklinikum der Gesamthochschule Essen, Hufelandstrasse 55, D-4300 Essen 1

[2] Radiologische Klinik, Universitätsklinikum der Gesamthochschule Essen, Hufelandstrasse 55, D-4300 Essen 1

Tumors of the pineal region are rare and histological diagnosis is seldom possible because of their deep midline location. A wide variety of different tumors can occur in the pineal region. The most common groups are those of germ cell origin, germinomas and teratomas. The germinoma (pinealoma) accounts for over 50% of all pineal neoplasms and is regarded as highly radiosensitive (8). Since the results of operative treatment are not adequate, we established a technique of a surgical approach combined with high voltage irradiation (1).

Patients and Methods

Out of a total of 31 patients with tumors we have reviewed the records of 15 children in the pineal region treated in our clinic from 1972 to 1982. There were 14 males and one female ranging in age from 3 to 18 years. All patients have a minimum follow-up period of three years.

Clinical and endocrinological disorders correlated closely with the location of the tumors in the pineal area or with extension into the hypothalamic region according to KAGIYAMA (Fig. 1). Most patients complained of headache and vomiting due to increased intracranial pressure. Ocular manifestations such as diplopia, paralysis of upward gaze and papilledema were frequent. Diabetes insipidus was observed in 3 of 15 patients.

Endocrinological disturbances were seen in 11 of 13 cases investigated before and after combined treatment (Table 1). In these patients the hypofunction of the somatotrophic axis was predominant. Total loss of hypothalamic-hypophyseal function existed in two cases and partial defects were found in two other children.

The endocrinological dysfunctions seem to be of hypothalamic origin, because in most cases a response of the pituitary gland could be observed after injection of releasing factors. Isolated defects of somatotrophic function were mainly found in cases of type 1 (KAGIYAMA), whereas pineal tumors of type 2 and 3 regularly show disturbances of more endocrine axes including diabetes insipidus. Only the latter need substitution of lacking hormones.

The clinical diagnosis was confirmed by roentgenographic contrast studies (including encephalotomograms, arteriograms and, more recently, computerized tomographic scans). The histological diagnosis was

Advances in Neurosurgery, Vol. 11
Edited by H.-P. Jensen, M. Brock, and M. Klinger

Table 1. Endocrinological findings in 13 patients with pineal tumors

Patient	Age	Tu-type	Somato-tropic	Cortico-tropic	Thyreo-tropic	Gonado-tropic	Diabetes insipid.	Substitution
K.M. ♂	4	2	↓	↓	NAD	↓	NAD	+
S.M. ♂	5	1	↓	NAD	NAD	?	NAD	
S.H. ♂	5	1	↓	NAD	NAD	NAD	NAD	
L.M. ♂	7	1	↓	NAD	NAD	?	NAD	
P.G. ♂	10	1	NAD	NAD	NAD	NAD	NAD	
L.T. ♂	11	3	↓	↓	↓	↓	+	+
K.V. ♂	14	1	NAD	NAD	NAD	NAD	NAD	
K.M. ♂	14	1	↓	NAD	NAD	?	NAD	
S.R. ♂	14	2	↓	↓	NAD	?	+	+
F.D. ♂	15	1	↓	↓	↓	↓	+	+
K.A. ♂	15	1	↓	NAD	NAD	NAD	NAD	
B.V. ♂	15	1	↓	NAD	NAD	NAD	NAD	
R.K. ♀	17	1	↓	NAD	NAD1	NAD	NAD	

verified in two patients who underwent partial tumor excision. Both pineal neoplasms were germinomas. The cerebrospinal fluid was examined in 12 patients and showed malignant cells on cytological study in four cases. Fourteen patients had a decompression by a ventriculo-atrial shunt before starting radiotherapy.

All 15 patients were irradiated with a ^{60}Co apparatus or a 6 MeV linear accelerator. Radiation doses ranged between 40 and 60 Gy and were given in four or five daily treatments per week in fractions of 1.6 - 2.0 Gy. The dose was delivered to the tumor-bearing area through bilateral opposing portals with a generous margin in nine patients. Five children received whole brain irradiation followed by a boost of the tumor region. Radiation therapy included the entire cerebrospinal axis in only one case.

Results

Eleven of 15 children or young adults are living at present and show no evidence of disease from 3 1/2 to 9 years after treatment. The 5-year survival rate calculated by life table method was 65% (Fig. 2, CUTLER et al. 1977). Eight of the surviving patients returned to a normal and useful life, although slight neurological or endocrinological disorders continue. Another two patients are in a fair condition, being capable of caring for themselves but suffering from severe neurological symptoms. One young boy attends a special school, because he is blind as a result of tumor bleeding after ventriculography.

Two of the first group with little or no disability developed frontal metastases outside the small radiation fields to the pineal tumor region. In both cases a second course of irradiation was given and the tumor mass disappeared completely. Two years later these boys are studying or attending high school, respectively, with no signs of active disease.

Four patients died at intervals ranging from four months to five years after treatment. One boy with diabetes insipidus and malignant cells in the cerebrospinal fluid (CSF) was given irradiation to the whole central nervous system. He improved but radiotherapy had to be interrupted because of myelosuppression. After an interval of four weeks the irradiation continued but was not completed, since septicemia led to death. In two of them progressive tumor growth and in the third patient a diffuse spread of metastatic tumor cells in the basal cisterns, meninges and ventricular ependyma were the cause of death. These three children had received irradiation only to the tumor area with doses less than 50 Gy.

Discussion

The risk of successful removal of tumors in the pineal region is high because of operative mortality and postoperative morbidity. If the patients survive the operative procedure, surgery alone has been reported to yield poor results for these neoplasms (5). Although surgical series support the importance of biopsy or partial excision for obtaining histologic diagnosis, a combined method of shunting for decompression and radiotherapy is the treatment of choice (4).

The beneficial effect of irradiation markedly depends on the histological type. Patients with germinomas have a much better prognosis with radiation therapy than those with teratoma or even glioma (6).

Many authors recommend tumor doses of 50 - 55 Gy, since significantly more recurrent primary tumors were seen in patients who received doses less than 50 Gy (4). The prognosis of pineal neoplasms seems also to be influenced by the size of the treatment field. Some radiotherapists irradiate only the tumor-bearing area with small portals, whereas others plead for larger radiation fields encompassing the entire ventricular system since local seeding of tumor cells may occur. Based on the subset of 67 patients from the literature and own series, SALAZAR et al. (7) found, that patients treated with whole brain irradiation survived better than those who received only partial brain irradiation. Two of our young patients developed metastases in the front of the third ventricle, which could have been avoided by radiation therapy of the entire brain volume.

The discussion concerning prophylactic spinal irradiation is not yet finished. In contrast to medulloblastomas the probability of spinal metastases is less than ten per cent of all patients in most series (6, 8). Because of the low incidence of spinal involvement and the associated morbidity of irradiation to the spinal axes, especially in children, elective radiation therapy of the spinal cord in only justified in patients with a positive CSF sample for malignant cells.

A fact not very well known seems to be the frequency of endocrinological disturbances in patients with neoplasms in the pineal region. More than one third of these patients needs hormonal substitution (2).

Conclusions

Based on our own experiences and results in the literature we recommend the diagnostic and therapeutic procedures which are shown in Table 2 and Fig. 3. After diagnosis by CT scan, supported if necessary by other roentgenographic contrast studies, a shunting system (ventriculo-atrial) for decompression is inserted. Repeated samples of CSF are taken (ventricular and spinal) to test for malignant cells. A complete endocrinological investigation follows and if necessary hormonal substitution.

Important tumor markers of pineal germ cell neoplasms are the beta chain of human chorionic gonadotropin (HCG) and alpha fetoprotein (AFP). Elevated levels of HCG and AFP in serum and CSF secreted by chorion carcinomas, embryonal carcinomas and endodermal sinus tumors have been reported to correlate with tumor growth or regression (Fig. 3). In these

Table 2. Tumors of pineal region (diagnostic and surgical procedures before irradiation)

Clinical and neurological status

↓

Computerized tomography (CT), (facultative: angiography, scintigraphy, encephalotomography)

↓

Shunting for decompression

↓

Cytology in CSF (ventricle, spinal cord)

Endocrine status

Tumor markers (HCG, AFP)

↓

Starting of irradiation

tumors a combined approach of irradiation and systemic chemotherapy should be chosen because of the absence of a blood brain barrier (4).

We recommend whole brain irradiation with doses of 36 Gy in five (!) weeks. This procedure takes note of the frequent failure rates in the brain outside irradiated areas and the sensitivity of brain tissue to irradiation in children. If the CT scan shows definite tumor regression after this dose has been given, radiotherapy should continue through smaller portals to the tumor region with an additional 14 to 20 Gy (Fig. 4). If there is no definite shrinking of the tumor after 36 Gy, the relative radioresistance of these neoplasms requires a higher dose up to a total of 60 Gy. The spinal axis should be irradiated only in patients with a positive CSF sample for malignant cells, to a dose of 36 Gy.

Operative removal can be attempted in patients who show clinical and radiographic evidence of growing pineal tumors despite irradiation or recurrent tumors.

After radiotherapy follow-up studies of the endocrine status are necessary in order to detect hormonal dysfunctions or radiogenic changes in the hypothalamic function.

References

1. Bamberg, M., Clar, H.E., Nahser, H.C., Benker, G.: Differenziertes Behandlungskonzept bei Tumoren der Pinealregion. Onkologie 1982, in press
2. Clar, H.E., Reinhardt, V., Gerhard, L., Hensell, V.: Clinical and morphological studies of pinealis tumors. Acta Neurochir. 46, 59 (1979)
3. Kagiyama, N., Belsky, R.: Ectopic pinealoma in the chiasma region. Neurology 11, 312 (1961)
4. Neuwelt, E.A., Glasberg, M., Frenkel, E., Clark, W.K.: Malignant pineal region tumors. J. Neurosurg. 51, 597 (1979)
5. Obrador, S., Soto, M., Gutierrez-Diaz, J.A.: Surgical management of the tumors of the pineal region. Acta Neurochir. 34, 150 (1976)
6. Onoyama, Y., Ono, K., Nakajima, T., Hiraoka, M., Abe, M.: Radiation therapy of pineal tumors. Radiology 130, 757 (1979)
7. Salazar, O.M., Castro-Vita, H., Bakos, R.S., Feldstein, M.L., Keller-B., Rubin, P.: Radiation therapy for tumors of the pineal region. Int. J. Radiation Oncology 5, 491 (1979)
8. Wara, W.M., Fellows, E.F., Sheline, G.E., Wilson, C.B., Townsend, J.J.: Radiation therapy of pineal tumors. Radiology 124, 221 (1977)

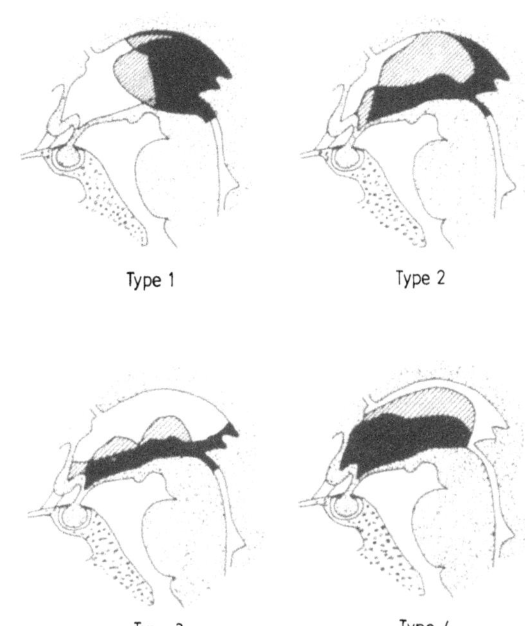

Fig. 1. Sites of pineal tumors according to KAGIYAMA

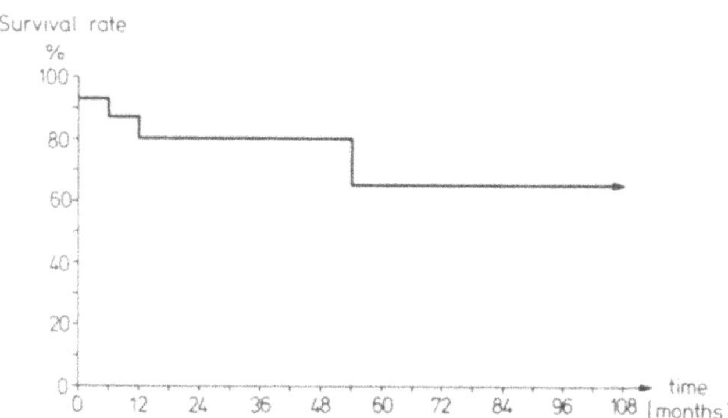

Fig. 2. Tumors of the pineal region. Five-year survival rate in 15 children calculated by the life table method

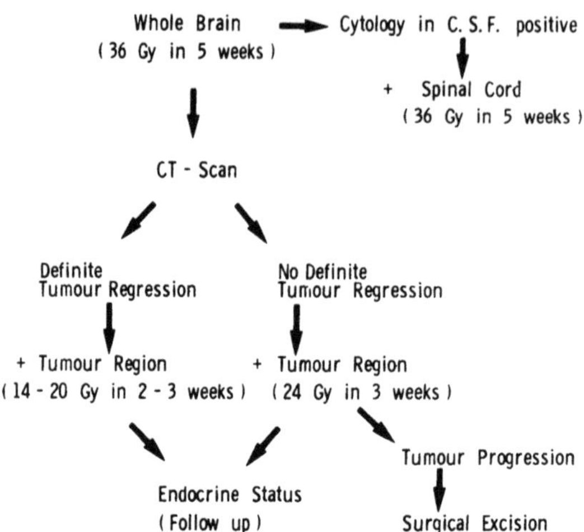

Fig. 3. Radiation treatment of tumors in the pineal region

Infratentorial – Supracerebellar Approach to the Pineal and Mesencephalic Region in Children

G. Pendl and W.Th. Koos
Neurochirurgische Universitätsklinik Wien, Alserstrasse 9, A-1090 Wien

Introduction

Since the introduction of microsurgery in the past ten years, space-occupying lesions in the oral and caudal brain stem are no longer considered inoperable; also, the more sophisticated instruments of diagnosis such as computer tomography are particularly suited to show lesions in the midline and permit much wider indications for surgical treatment than in the past (PENDL and KOOS). Nevertheless, the tumors of the pineal and mesencephalic region are still much under discussion, particularly because they are located within the range of extremely sensitive brain structures where there is always a problem of selecting the proper approach. The operability of these tumors will be discussed on the basis of the author's own successful experiences - using anatomical and topographical considerations; the proper approach will also be presented.

Anatomical Studies and Clinical Experience

Because of the inadequacy of existing anatomical studies regarding the regions of the pineal gland, the mesencephalon and the corpora quadrigemina, we have examined anatomical sections and cadaver brains in addition to clinical studies. It turned out that it is not so much the anatomical structure of the parenchyma of the brain as the vascular anatomy which makes the approach to the corpora quadrigemina so difficult. Figure 1 is a diagram showing the course of the arteries and veins in this region.

The following approaches are feasible for surgical intervention:

1. Parieto-occipital approach according to DANDY-FOERSTER.
2. Occipital lobe resection according to HORRAX.
3. Transventricular approach according to Van WAGENEN.
4. Suboccipital transtentorial approach according to HEPPNER-POPPEN.
5. Infratentorial-supracerebellar approach according to KRAUSE-BRUNNER.
6. Subtemporal approach.

On the basis of our anatomical studies and our experiences in microsurgery, we feel that the DANDY-FOERSTER approach is least suitable since the deep veins, in particular the great vein of GALEN, considerably impair the whole view and interfere with the dissection after splitting of the splenium of the corpus callosum. Resection of the

Advances in Neurosurgery, Vol. 11
Edited by H.-P. Jensen, M. Brock, and M. Klinger

occipital lobe according to HORRAX is only suited for very large lesions and we have used it only for adults with large meningiomas in the quadrigeminal region. The transventricular approach according to Van WAGENEN may be used especially for tumors growing on one side towards the lateral ventricle (area of the trigone) to which there is easy access through the lateral ventricles which are dilated due to hydrocephalus. The suboccipital transtentorial approach according to HEPPNER-POPPEN, which is done with the patient sitting up, affords a good view across the quadrigeminal region without any impairment from the deep veins which are situated above in a roof-like structure. In this case the tentorium has to be split generously. The subtemporal approach is suited only for lesions in the cerebral peduncle or in the region of the interpeduncular cistern. As regards tumors of the pineal and mesencephalic region in children we have confined ourselves almost wholly to the infratentorial supracerebellar approach according to KRAUSE-BRUNNER. This approach is particularly appropriate because of the few additional lesions caused by the surgical manipulations; only the bridging veins between tentorium and cerebellum have to be sacrificed which, according to the author's experiences, does not cause any additional morbidity. Through the tentorium the approach is maintained without the use of any additional spatula following the drop of the cerebellum caused by the drainage of the cisterna magna; the great vein of GALEN and Rosenthal's vein as well as the internal veins remain intact as a roof-like structure above the field of operation and may be well exposed and spared, the precentral cerebellar vein - if it still exists despite pathological changes caused by the tumor - may be sacrificed without any harm.

Figure 2 shows a diagram of the approach, Fig. 3 an autopsy specimen affording a view into the quadrigeminal region with the pineal gland - using this approach.

Table 1 shows ten cases of tumor of the pineo-mesencephalic region from the pediatric age group up to 16 years, taken out of a total number of 22 cases with pathological lesions in this region where operation was undertaken. It should be noted that among them there were four tumors in the mesencephalon itself (case 1, 2, 3, 5) which were extirpated by this approach. There has not been a single case of postoperative mortality in our cases in these past nine years. No increased morbidity possibly due to the approach was noted. The survival time as such depended on the histology of the tumor and, because of the malignant lesions - despite ensuing radiotherapy and chemotherapy in some was only 18 months until the tumor recurrence. Fig. 4 indicates the various tumor sites of these ten cases.

Discussion

Ever since HORSLEY in 1905, attempts have been made to expose tumors of the pineal gland but it was not until 1913 that KRAUSE (OPPENHEIM and KRAUSE) carried out a successful operation on a tumor of the pineal gland; already at that time the approach was infratentorial supracerebellar. Later on, however, DANDY's approach through the corpus callosum gained ground and was used until recently by most authors as the technique of choice. Most likely this very unsuitable approach with its macrotechnique, its considerable morbidity and mortality has also been responsible for the rather pessimistic attitude with regard to tumors of the quadrigeminal region. OBRADOR et al. as well as SCHÄFER et al. make reference to the good results published recently; there are more and more reports about the successful extirpation of tumors in the pineal region with the suboccipital transtentorial ap-

Table 1. Tumors of the pineo-mesencephalic region

Case No.	Age	Sex	Histology	Location	Surgical approach	Additional treatment	Survival time
1	11 years	f	Pilocytic astrocytoma	Left cerebral peduncle	Infratentorial-supracerebellar	Radiotherapy and v-a-shunt before operation	9 years, living
2	9 years	m	Ependymo-blastoma	Posterior 3rd ventricle and mesence-phalon	Infratentorial-supracerebellar	Radiotherapy and v-a-shunt before operation	Inoper-able after 1 1/2 years
3	3 years	f	Medullo-blastoma	Posterior 3rd ventr. and mesencephalon	Infratentorial-supracerebellar	Radiotherapy and v-a-shunt after operation	Spinal metasta-ses after 1 1/2 y.
4	16 years	m	Hamartoma	Interpeduncular	Subtemporal right	None	7 years, living
5	5 years	m	Astrocytoma II	Mesencephalon	Infratentorial-supracerebellar	Radio- and chemo-therapy after op.	2 years,
6	8 months	m	Medullo-blastoma	Pineal region posterior 3rd ventricle	Infratentorial-supracerebellar and right trans-ventricular	Radio- and chemo-therapy after operation	2nd re-currency after 2 1/2 y.
7	10 years	m	Germinoma	Pineal region	Infratentorial-supracerebellar	V-a-shunt before surgery, radio-therapy after op.	1 1/2 years, living
8	12 years	m	Teratoma	Pineal region	Infratentorial-supracerebellar	Radiotherapy after operation	7 months, living
9	12 years	f	Pineocytoma	Pineal region	Infratentorial-supracerebellar	None	6 months, living
10	16 months	f	Arachnoidal cyst	Pineal region	Infratentorial supracerebellar	V-a-shunt before operation	3 months, living

proach of HEPPNER and POPPEN, but also about the use of the infratentorial supracerebellar approach of KRAUSE-BRUNNER (PAGE, PENDL). STEIN and also SAND report in particular on the positive results following exposure by way of the infratentorial supracerebellar approach that had been developed by BRUNNER as early as 1913 after trials on the cadaver (RORSCHACH); TANDLER and RANZI also described this technique in their textbook in 1920, a technique that has been recently revived by ZAPLETAL and STEIN.

We agree with JAMIESON that the choice of a direct surgical approach to the pineal gland should depend on the following criteria:

1. Histological verification wherever possible since it might be a cystic lesion in which case radiation treatment would be ineffective; there are also other tumors which are absolutely resistent to radiation.
2. Open biopsy may have disastrous results - particularly bleeding into the tumor; this is far more dangerous than a total resection.
3. A suitable surgical approach - because of the possibilities afforded by the sophisticated microsurgical techniques - opens up many more possibilities.

Summary

Extensive micro-anatomical and microtopographic studies showed the infratentorial-supracerebellar approach to be the one of choice for microsurgery of tumors in the pineal region including the posterior third ventricle, the mesencephalon and the superior cerebellar vermis. In the past years tumors of different histological nature were successfully removed in 10 children and 12 adults without any postoperative mortality and, particularly, without increased postoperative morbidity. The approach which was reported by Fedor KRAUSE as early as 1912 offers an excellent view of the anatomy involved with its numerous vulnerable vessels and facilitates the exposure of both median and paramedian lesions.

References

1. Dandy, W.E., An operation for the removal of pineal tumors. Surg. Gyn. Obst. 33, 113 (1921)
2. Foerster, O.: Ein Fall von Vierhügeltumor, durch Operation entfernt. Arch. Psychiat. 84, 515 (1928)
3. Heppner, F.: Zur operativen Technik bei Pinealomen. Zbl. Neurochirurgie 19, 219 (1959)
4. Horrax, G.: Extirpation of a large pinealoma from a patient with pubertas praecox: a new operation approach. Arch. Neurol. Psychiat. 37, 385 (1937)
5. Horsley, V.: Discussion of paper by C.M.H. Howell on tumors of the pineal body. Proc. R. Soc. Med. 3: 65 (1910)
6. Jamieson, K.G.: Excision of pineal tumors. J. Neurosurg. 35, 550 (1971)
7. Obrador, S., Soto, M., Gutierrez Diaz, J.A.: Surgical management of tumors of the pineal region. Acta Neurochir. 34, 159 (1976)
8. Oppenheim, H., Krause, F.: Operativer Erfolg bei Geschwülsten der Sehhügel- und Vierhügelgegend. Berl. Klin. Wschr. 50, 2316 (1913)

9. Page, L.K.: The infratentorial-supracerebellar exposure of tumors in the pineal area. Neurosurgery 1, 36 (1977)

10. Pendl, G.: Infratentorial approach to mesencephalic tumors. In: Clinical microneurosurgery. Koos, W.Thl, Böck, F., Spetzler, R.F., (eds.), p. 143. Stuttgart: Thieme 1976

11. Pendl, G., Koos, W.Th.: Erfahrungen mit mikrochirurgischer Behandlung von Hirnstammtumoren im Kindes- und Jugendalter. Adv. Neurosurg. 8, 403 (1980)

12. Poppen, J.L.: An atlas of neurosurgical techniques. Philadelphia: Saunders 1960

13. Rorschach, H.: Zur Pathologie und Operabilität der Tumoren der Zirbeldrüse. Brun's Beitr. 83, 451 (1913)

14. Sano, K.: Diagnosis and treatment of tumors in the pineal region. Acta Neurochir. 34, 153 (1976)

15. Schäfer, C., Lapras, C., Ruf, H.: Experience with the direct surgical approach in 52 tumors of the pineal region. Adv. Neurosurg. 7, 97 (1979)

16. Stein, B.M.: The infratentorial supracerebellar approach to pineal lesions. J. Neurosurg. 35, 197 (1979)

17. Tandler, J., Ranzi, E.: Chirurgische Anatomie und Operationstechnik des Zentralnervensystems. In: Die Chirurgie (Kirschner, M., Normann, O., eds.). Berlin: Springer 1920

18. Van Wagenen, W.P.: A surgical approach for the removal of certain pineal tumors. Surg. Gynec. Obstet. 53, 216 (1931)

19. Zapletal, B.: Ein neuer operativer Zugang zum Gebiet der Incisura tentorii. Zbl. Neurochir. 16, 64 (1956)

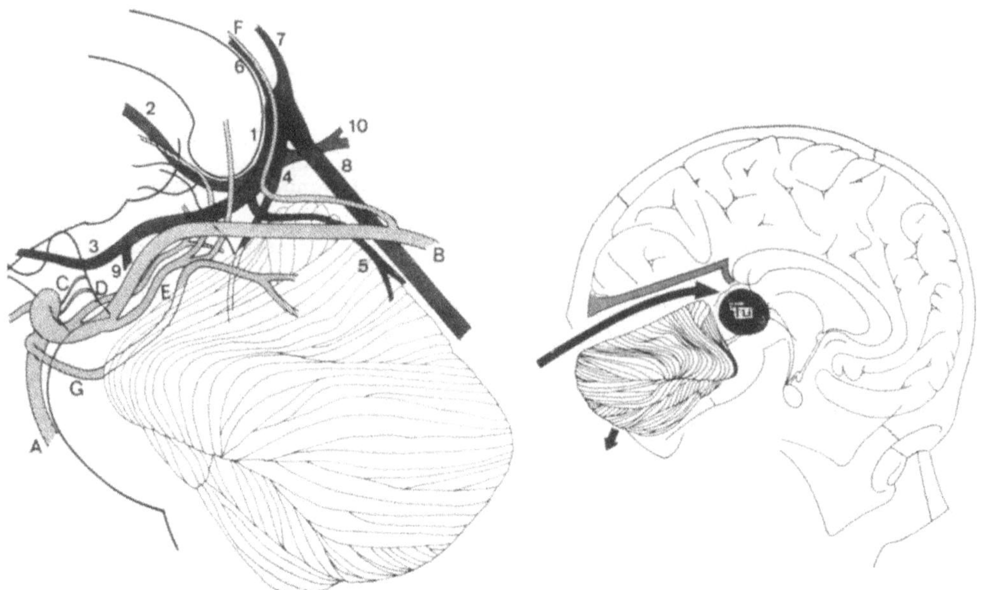

Fig. 1. *Left:* Diagram showing the course of the arteries and veins at and near the quadrigeminal cistern. A, basilar artery; B, posterior cerebral artery; C, posterior perforating thalamic arteries; D, quadrigeminal artery; E, posterior medial and lateral choroid arteries; F, posterior pericall. artery; 1, great vein of GALEN; 2, int. cerebral vein; 3, basal vein of ROSENTHAL; 4, precentral cerebellar vein; 5, superior cerebellar vein; 6, pericall. vein; 7, inferior sagittal sinus; 8, sinus rectus; 9, mesencephalic vein; 10, med. occipital vein

Fig. 2. *Right:* Diagram of the infratentorial-supracerebellar approach to the quadrigeminal cistern

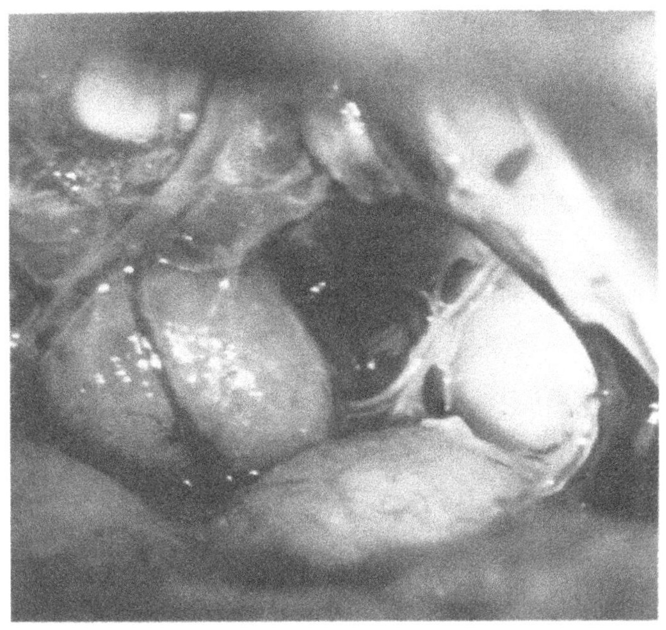

Fig. 3. Microsurgical view of the normal well-exposed anatomy of the quadrigeminal plate and pineal gland (situated to the right)

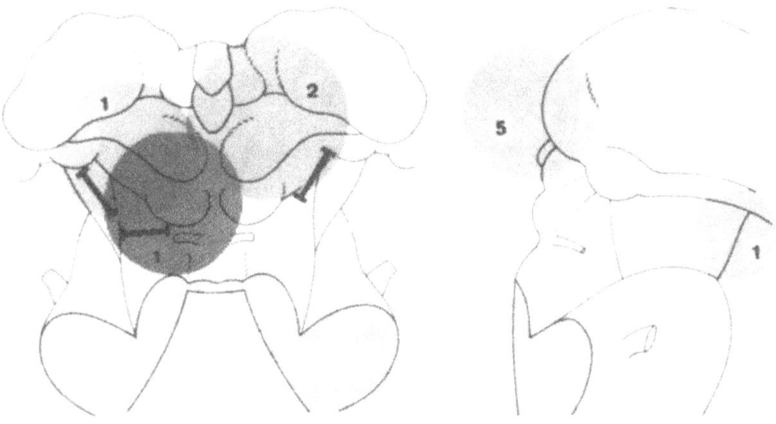

Fig. 4. Topography of various tumor sites in the pineal area (5 cases) and within the mesencephalon (5 cases)

Diagnosis and Treatment of Tumors of the Pineal Region, Posterior Third Ventricle and Quadrigeminale Plate. Present Value of the Stereotactic Exploration

J. R. Moringlane[1], C. Daumas-Duport[2], A. Musolino[3], and G. Szikla[3]

[1] Neurochirurgische Klinik im Nordstadt-Krankenhaus, Haltenhoffstrasse 41, D-3000 Hannover

[2] Laboratoire d'Anatomie pathologique Centre Hospitalier Sainte Anne – 1, rue Cabanis, F-75674 Paris Cedex 14

[3] Service de Neurochirurgie B, Centre Hospitalier Saint Anne – 1, rue Cabanis, F-75674 Paris Cedex 14

Introduction

A review of the literature concerning lesions of the pineal region, the quadrigeminal plate and of the posterior part of the third ventricle leads to the conclusion that the choice and the succession of diagnostic as well as of therapeutic procedures is very different from one neurosurgical center to another. Many reports deal only or essentially with the results of an operative approach, but little attention is paid to the biology of the lesions.

New operative techniques and their modifications have been described (1, 4, 10, 12). The development and propagation of microsurgery, the perfection of modern anesthesiology and progress in the field of postoperative monitoring and intensive care have encouraged a more aggressive attitude towards tumors of the pineal region (5-7, 13).

Some authors emphasize that 70% of the tumors of the pineal region are radiosensitive and recommend radiation therapy even without any histological diagnosis (9, 11). At present, however, most radiotherapists require a histological diagnosis before irradiation since a common histological nomenclature is now available thanks to the recent efforts at classification (12, 17).

Thus in our unit several patients who came from radiotherapy centers were admitted for stereotactic exploration including biopsies.

Material and Methods

From 1975 to 1982, 18 patients aged between 2 1/2 and 64 years (average: 27.8 years) presenting with this type of tumor underwent a stereotactic procedure at the Service de Neurochirurgie B of the Centre Hospitalier Sainte Anne in Paris. Among them were six children aged between 2 1/2 and 14 years. Eleven patients were male and seven were female.

There is little to say about the neurological aspect. The classical signs of raised intracranial pressure were initially present and in a few cases Parinaud's syndrom was observed.

Advances in Neurosurgery, Vol. 11
Edited by H.-P. Jensen, M. Brock, and M. Klinger

Present Technique of Stereotactic Exploration as used at Hospital Sainte Anne/Paris

1. The procedure is performed under general anesthesia. After surgical disinfection of the shaved head the stereotactic frame of TALAIRACH is fixed to the bony vault. The first X-ray *plain films* as well as all following radiographs are taken at a constant distance of 4.5 meters (teleradiography (15), so that there is practically no magnification. Thanks to the extremely careful control of the orthogonal incidence in the supine as well as in the sitting position, all images can be directly superimposed (16).

2. *Angiography* under stereotactic conditions is performed by puncture of the carotid and/or brachial arteries. All angiographies are made first with orthogonal incidence and second with an incidence of 6°. Stereoscopic three-dimensional effect is achieved by the cross-eyed technique or by the use of mirrors or especially constructed eyeglasses with adjustable mirrors. So it is possible to distinguish the superficial and the deep vessels and visualize the underlying surfaces of the brain (gyri, sulci, etc.) (17).

   One of the goals is to be able to reach any intracranial target knowing in advance that no vessel will be injured by the stereotactic probes.

3. First result of the angiography is the choice of a safe path to a frontal horn or better to the third ventricle. A CSF sample is taken for *cytological* and routine laboratory examinations. *Ventriculography* is made with air and Metrizamide. As for the angiography, pictures are made with orthogonal and 6° incidence in order to achieve stereoscopic viewing.

4. The next step is the *transfer of the CT-images* on X-ray film, allowing for a juxtaposition of the collected CT and other neuroradiological information, thus leading to a precise and comprehensive three-dimensional view of the lesion and of the surrounding brain structures (8).

5. Oriented by this exact localization of the tumor even quite small tumors (1 cm) can be reached safely with the probes. *Stepwise biopsies* furnish tissue samples from the different areas in and around the tumor. The importance of the stepwise biopsies must be emphasized because it is the unique method of determining the extent and the limits of a tumor as well as its nature. As a matter of fact CT-images cannot be interpreted in terms of histological diagnosis (3). Moreover they do not allow one to say whether a hypodense area at the periphery of a tumor is edema or infiltrated brain tissue (Fig. 1) (2). However, the volume of a tumor should be calculated as exactly as possible for radiotherapy and the limits should be clearly defined for surgical removal. Sampling of the tissue can - especially in the neighbourhood of important functional areas - be preceded by stimulation of motor fibers, *recording of depth EEG or of cellular activity* (macro- or microelectrodes). In cases where the smear technique does not allow an immediate histological classification of the tumor, there is enough tissue for paraffin sections which are systematically prepared. Every sample taken with the *SEDANVALLICIONI* trocar can reach the significant size of 10 x 1 mm.

Results

Table 1 shows the neuropathological findings obtained by the stereotactic procedure in 18 patients. Germinomas together with pineoblastomas and astrocytomas account for about 70%.

Treatment

Four of 18 patients had no previous CSF drainage. During the stereotactic procedure a ventriculocisternostomy was performed between the floor of the third ventricle and prepeduncular cistern (Fig. 2). No sign of recurrent hydrecephalus has been observed in these cases.

The colloid cyst of the posterior third ventricle was evacuated by *stereotactic aspiration*.

In the light of the information collected, out of the 18 patients one case with a meningioma was considered suitable for *open operation*. Subtotal removal was followed by external radiation therapy. In the case of a germinoma, open surgical exploration and biopsy performed subsequently in an other hospital was followed by external irradiation.

In the other 15 cases stereotactic exploration was followed by *radiation treatment* only: external radiation treatment (ERT), interstitial implantation of Ir 192 wires or I 115 seeds (Curie therapy) or a combination of both (ERT + Curie) (Table 2).

In cases of interstitial irradiation by implantation of radioisotopes the same stereotactic conditions and localisatory data were used as obtained by the previous stereotactic exploration. Actually the TALAIRACH frame allows repeated fixation of the skull in the same position in the frame, so that all visualized structures will have exactly the same coordinates as for the first time.

Mortality of the stereotactic procedures under the above-mentioned conditions is zero. Morbidity is low: one case of temporary hemiparesis after stereotactic 192 Iridium implantation.

Table 1. Tumors of the pineal region, quadrigeminal plate and of the posterior part of the third ventricle. 1975 - 1982; n = 18

Histological diagnosis	No. of patients 11 ♂ : 7 ♀ Average age: 27.8 years
Germinoma	7 (4 with mixed teratoma) 4 children)
Pineoblastoma	2 (2 children)
Malignant glioma	1
Astrocytoma	2
Ependymoma	1
Meningioma	3
Cyst	1
Cavernous angioma (3rd ventricle + frontal location)	1

Table 2. Results of radiotherapy of tumors of the pineal region, posterior part of the third ventricle and of the quadrigeminal plate in 17 patients 1975 - 1982

	Name	Sex, Age	Histological diagnosis	Radiotherapy Curie therapy	Status (May 1982)
1	WA	M 20 y	Germinoma	192 Ir - Juli 75 + ERT Nov. 1978 (2nd ectopic localization)	+ (late X-ray damage) moderate hemiparesis
2	CA	M 13 y	Germinoma	192 Ir - June 76 ERT June/July 76	+
3	MA	M 38 y	Meningioma	ERT July 1977 (after partial removal)	$\pm$ (Multiple meningioma)
4	POU	M 2 1/2 y	Pineoblastoma	192 Ir - May 77	Dead (tumor growth)
5	ROB	F 25 y	Ependymoma	192 Ir + ERT May 79	+
6	BEL	F 25 y	Astrocytoma	192 Ir + ERT May 79	+
7	GAU	F 47 y	Meningioma	192 Ir - October 79	+
8	SCHE	M 34 y	Germinoma	ERT - January 80 (after craniotomy)	+
9	BER	M 9 y	Germinoma	ERT - December 80	+
10	BOUR	M 23 y	Germinoma	ERT - March 81	+
11	PEL	F 14 y	Germinoma	ERT - May 81	+
12	LEC	M 49 y	Astrocytoma	125 I - July 81	+
13	BRU	M 51 y	Malignant glioma	ERT Jan./February 82	+

Table 2 (continued)

14	ALU	M 2 1/2 y	Pineoblastoma	ERT January 82	+
15	DASI 5 y	M	Germinoma	ERT January 82	+
16	TAS	F 39 y	Cavernous angioma post. 3rd ventricle + left frontal	ERT January 82 I 125 FEbruary 82	+
17	LAF	F 64 y	Meningioma	192 Ir - March 82	+

ERT, External radiotherapy; 192 Ir, 125 I, Interstitial Curie therapy by implantation of 192 Iridium (temporary) or 125 Iodine (permanent); +, normal life, no deficit; ±, Neurological deficit

The mortality of external radiotherapy alone or combined with interstitial Curie therapy is also zero. In one case a partially regressive hemiparesis appeared after external irradiation of a germinoma in an ectopic localization due to delayed X-ray damage.

From the six children there is one death to be mentioned between 1975 and 1982, six months after implantation of 192 iridium into a highly malignant pineoblastoma in a 2 1/2-year-old boy.

Figure 3 shows a return to normal of the internal cerebral vein and of CT-images after radiation therapy in two cases of germinoma, as well as of the posterior third ventricle in a case of ependymoma. The results of radiation treatment are shown in Table 2.

Comments and Conclusion

Considering that only 20% of tumors of the pineal region can be removed by open operation and that useless irradiation of the brain should be avoided, (children!) and considering that mortality and complications of craniotomy in radiosensitive tumors are not desirable, the authors feel that the following management is suitable for tumors of the pineal region, of the quadrigeminal plate and of the posterior third ventricle (Table 3):

1. In all cases choice of treatment has to be based on the results of a stereotactic diagnostic procedure. The latter includes stereotactic angiography and ventriculography with a reconstruction of the CT-images, together with stepwise sampling biopsies, which give precise and detailed information on the topography and histology of the tumor, the volume and the well or poorly circumscribed, infiltrating character of the growth; in our opinion the so-called simplified biopsy methods based exclusively on CT data yield only fragmentary information and are liable to complications.

2. In cases where surgical removal of the lesion seems possible, e.g. meningiomas, angiomas, dermoid cysts, etc.) precise anatomo-functional and in particular vascular topography of the tumor obtained by the stereotactic procedure contributes to optimal orientation at craniotomy.

3. External radiation therapy should be the best treatment for germinomas, pineoblastomas, poorly delimitated gliomas extending laterally and towards the floor of the third ventricle. In some cases the central tumor bulk can be destroyed by an additional isotope implant.

4. Interstitial radiotherapy (192 Iridium or 125 Iodine implantation) seems to be the best choice for small well-delimited inoperable lesions with low radiosensitivity (e.g. pilocytic astrocytomas). It can be a second choice in some cases of otherwise removable lesions where open operation is contraindicated by the age or poor general condition of the patient.

Table 3. Management of tumors of the pineal region, of the quadrigeminal plate and of the posterior part of the third ventricle

Stereotactic exploration (no mortality - low morbidity)

Avoids the disadvantages of usual methods of treatment	Provides thorough diagnostic data	Guides the choice of appropriate method of treatment
Mortality - Morbidity		
A. Primary open operation useless craniotomy for tumors not removable (Pinealoblastomas, Glioblastomas)	A. Detailed histological examination	A. Open operation (e.g. meningioma)
B. Primary irradiation without histological diagnosis useless irradiation of non-radiosensitive tumors (meningiomas, cysts, epidermoids)	B. Determination of zones of infiltration	B. Stereotactic aspiration (cyst)
	C. Evaluation of tumor volume	C. Stereotactic ventriculo-cisternotomy
	D. Exact topographical information	D. External radiotherapy (germinona)
		E. External + interstitial irradiation (infiltrating astrocytomas)

References

1. Capdeville, J.: Les tumeurs de la région pinéale. Thesis Lyon 1974
2. Daumas-Duport, C., Vedrenne, C., Szikla, G.: Contribution of stereotactic biopsies to the 3 D localization of brain tumors, gliomas in particular. In: Stereotactic cerebral irradiation. INSERM Symposium No. 12. Szikla, G. (ed.), pp. 33-42. Elsevier 1979
3. Futrell, N.N., Osborn, A.G., Cheson, B.D.: Pineal region tumors: Computed tomographic-pathologic spectrum. AJNR 2, 415-420 (1981)
4. Jamieson, K.G.: Excision of pineal tumors. J. Neurosurg. 35, 550-553 (1971)
5. Lazar, M.L., Clark, K.: Direct surgical management of masses in the region of the vein of Galen. Surg. Neurol. 2, 17-21 (1974)
6. McDonnell, D.E.: Pineal epidermoid cyst: its surgical therapy. Surg. Neurol. 7, 387-391 (1977)
7. Neuwelt, E.A., Glasberg, M., Frenkel, E., Clark, W.: Malignant pineal region tumors. A clinico-pathological study. J. Neurosurg. 51, 597-607 (1979)
8. Nguyen, J.P., Szikla, G., Missir, O.: Fiabilité d'une méthode simplifiée de transposition des images TDM: Corrélations du repérage stéréotaxique dans 30 cas. Neurochir. Masson. In press
9. Pertuiset, B.: Diagnosis of pinealoblastomas by positive response to cobalt therapy. Act. Neurochir. 34, 151-152 (1976)
10. Poppen, J.L.: The right occipital approach to a pinealoma. J. Neurosurg. 25, 706-710 (1966)
11. Rao, Y.T.R., Medini, E., Haselow, R.E., Jones, T.K., Levitt, S.H.: Pineal and ectopic pineal tumors: the role of radiation therapy. Cancer 48, 708-713 (1981)
12. Russell, D.S., Rubinstein, L.J.: Pathology of tumors of the nervous system. 4th ed. London: Edward Arnold 1977
13. Schäfer, M., Lapras, C., Ruf, H.: Experience with the direct surgical approach in 52 tumors of the pineal region. In: Advances in Neurosurgery 7. Marguth, F., Brock, M., Kazner, E., Klinger, M., Schmiedek, P. (eds.), pp. 97-103. Berlin, Heidelberg, New York: Springer 1978
14. Stein, B.M.: Surgical treatment of pineal tumors. Clin. Neurosurg. 28, 490-510 (1980)
15. Szikla, G., Bouvier, G., Hori, T., Petrov, V.: Angiography of the human brain cortex. Atlas of vascular patterns and stereotactic cortical localization. Berlin, Heidelberg, New York: Springer 1977
16. Talairach, J., David, M., Tournoux, P., Corredor, H., Kvasina, T.: Atlas d'anatomie stéréotaxique. Paris: Masson 1957
17. Talairach, J., Szikla, G., Tournoux, P., Prossalentis, M., Bordas-Ferrer, M., Covello, L., Iacob, M., Mempel, E.: Atlas d'anatomie stéréotaxique du télencéphale. Etudes anatomo-radiologiques. Paris: Masson 1967
18. Zülch, K.J.: Principles of the new World Health Organization (WHO) classification of brain tumors. Neuroradiology 19, 59-66 (1980)

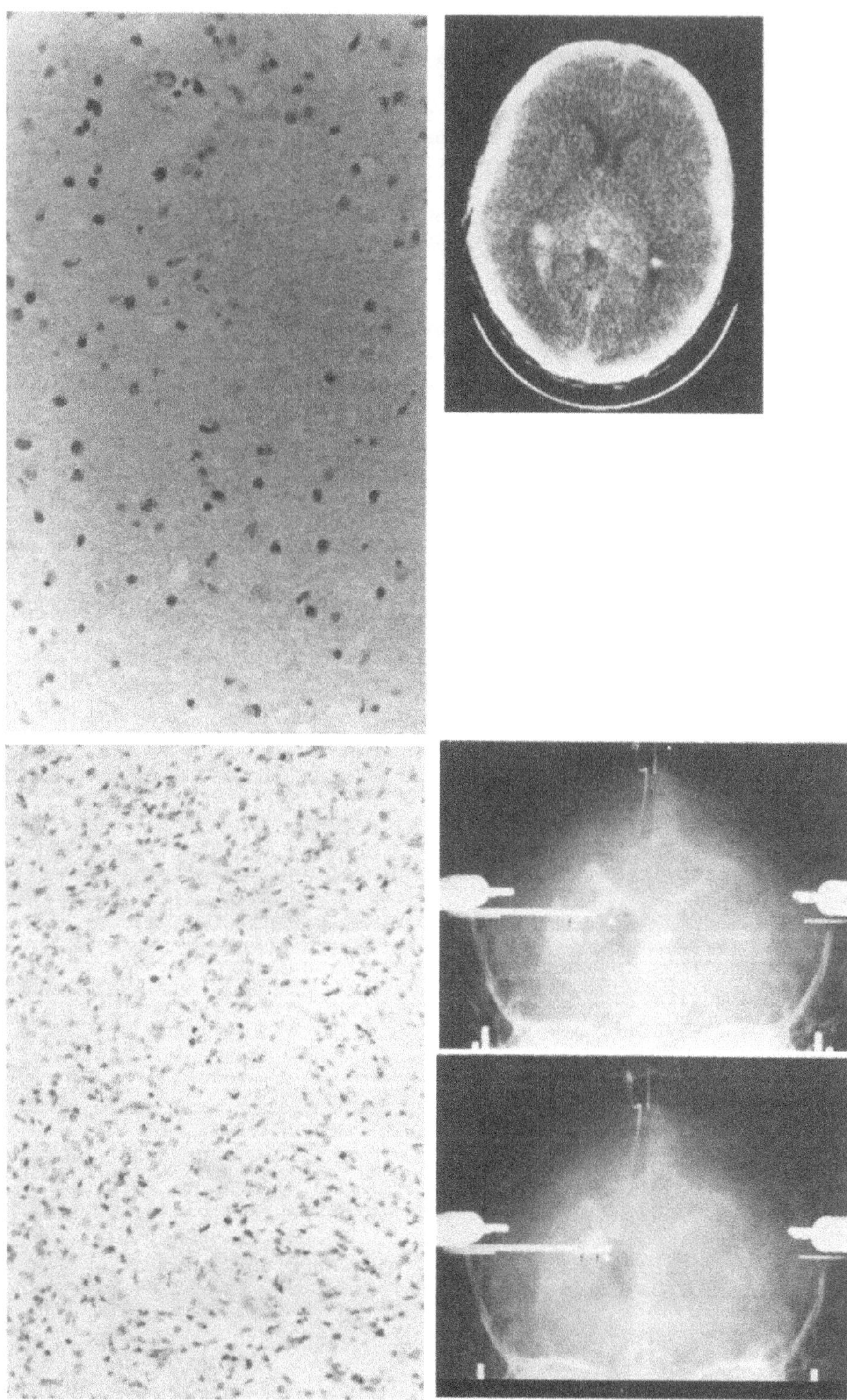

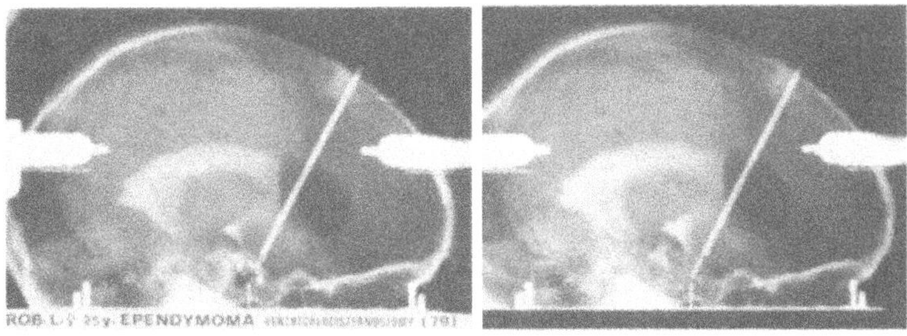

Fig. 2 a, b. Stereotactic ventriculocisternostomy for the treatment of hydrocephalus in a case of ependymoma of the third ventricle. a forceps in position for opening the floor of the third ventricle; b passage of contrast medium from the third ventricle into the prepeduncular cistern

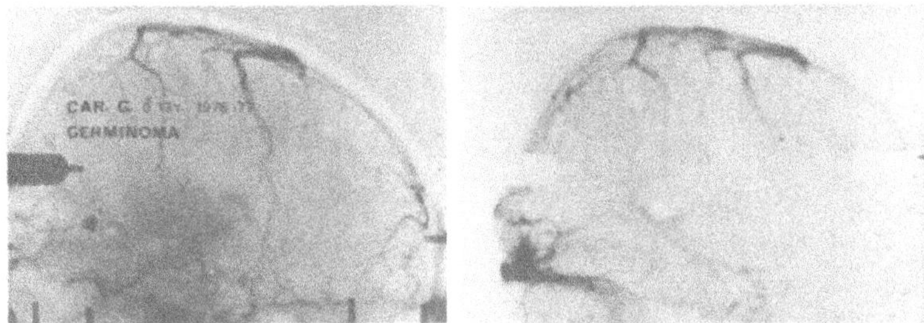

Fig. 3. a Internal cerebral vein returned to normal, one year after radiation therapy of a germinoma in a 13-year-old boy; *left:* before therapy, *right:* one year after irradiation

◄ Fig. 1. Glioblastoma of the pineal region: CT-image *(top, right):* no precise limit. *Upper pictures, left:* histological slide showing infiltration by tumor tissue; *right:* position of the biopsy trocar and its opening (arrow) at the level of infiltration; *lower picture, left:* histological slide of compact tumor mass; *right:* position of the biopsy trocar and its opening (double arrow) within the tumor 8 mm deeper than the infiltration zone

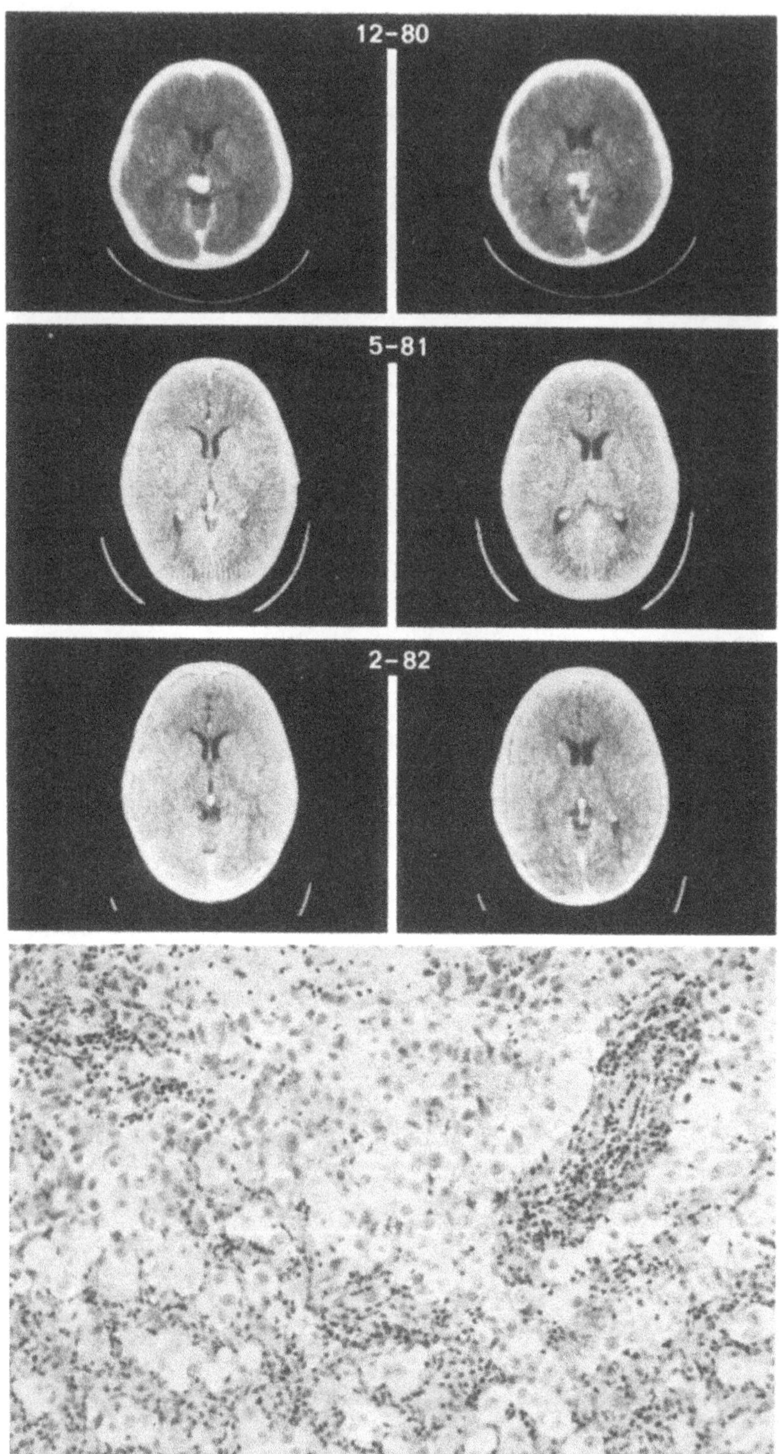
12-80
5-81
2-82

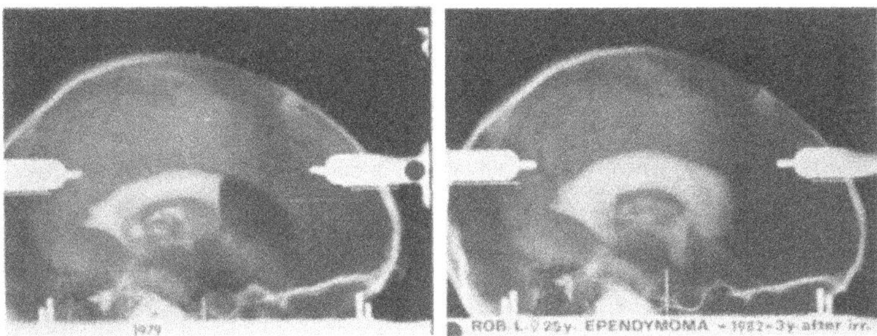

Fig. 3. c *Left:* Ependymona of the third ventricle; *right:* Normal shape of third ventricle three years after irradiation. Control stereotactic biopsies show no evidence of tumor tissue

◄
Fig. 3. b CT image returned to normal, one year after radiation therapy of a germinoma in a nine-year-old boy; *top:* CT image of the germinoma before therapy; *middle:* five months after irradiation; *lower:* CT-image 14 months after irradiation; *below:* histological slide of germinoma

CT Scanning for Long Term Follow-Ups After Microsurgery for Posterior Fossa Tumors in Children

P. Vorkapić, G. Pendl, W.Th. Koos, T. Reisner, and S. Engel
Neurochirurgische Universitätsklinik Wien, Alserstrasse 9, A-1090 Wien

Introduction

CT scanning has become especially valuable for the diagnosis of midline lesions and also of tumors in the posterior fossa. Great improvements in management have been achieved by follow-up CT studies of totally or partially excised tumors in the posterior fossa. A study of 33 cases of postoperative follow-ups by CT scanning substantiates this statement.

Case Material

Between the years of 1970 to 1980, 95 posterior fossa tumors in children up to the age of 16 years old were extirpated by microsurgical techniques at the Department of Neurosurgery, University of Vienna Medical School.

Table 1 shows the age and sex distribution of the cases, and indicates that there is an almost equal distribution of age as well as between

Table 1. Age and sex distribution of posterior fossa tumors (0 - 16 years)

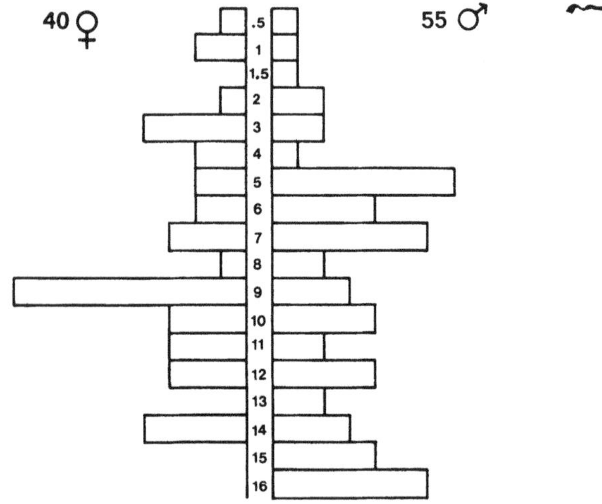

Advances in Neurosurgery, Vol. 11
Edited by H.-P. Jensen, M. Brock, and M. Klinger

Table 2. Histology of posterior fossa tumors (0 - 16 years)

Medulloblastoma	39
Astrocytoma	36
Ependymoma	8
Mixed gliomas	4
Sarcoma	2
Neurinoma/meningioma (von Recklinghausen disease)	1
Epidermoid	1
Ganglioma	1
Cystic brain stem glioma	1
Hemangioblastoma	2
Total	95

male and female. Histological studies (Table 2) revealed that the largest sub-group was the medulloblastoma - 39 cases, followed by astrocytomas with 36 cases, and ependymomas with 8 cases. The remainder of these subgroups appear in the table. During this same period 14 non-tumorous lesions in the posterior fossa were surgically explored, but were not included in this study. (Within this group the largest number of cases observed was of arachnoid cysts (five) This was followed in number by the Dandy-Walker malformation (three cases), hemorrhages, angiomas and inflammatory lesions).

One hundred and eight surgical procedures were performed on these 95 cases. Palliative measures such as shunting, or similar procedures were not included. The patients in this study had a postoperative mortality of 11%. This is in comparison with a mortality of 20% in 84 posterior fossa tumors in children, that were mainly macrosurgically excised at this institution during the years 1964 - 1969. Therefore, it can be seen that the use of microsurgical techniques produced strikingly better postoperative results. Forty of 95 children survived postoperatively, were followed up, and found to have a survival time of 2 to 12 years. Table 3 shows the topography of the tumors and their relationship to the cerebellar hemispheres and brain stem.

The following classification was used to describe the patients' postoperative quality of life:

Group 1: no handicap and with a normal quality of life
Group 2: socially deprived but integrated
Group 3: handicapped, but still integrated
Group 4: totally handicapped and in constant need of care.

Surprisingly only three of the 40 children were in Group 4, while the other 37 cases were in Group 1.

Table 3. Topography of cerebellar tumors in 40 cases of survival (0 - 16 years)

Cerebellar hemispheres and vermis	28
Brain stem	5
Hemispheres - vermis/brain stem	7
Total	40

Of the 40 surviving, 33 cases were followed up with CT scans. The histological distribution was as follows:

20 astrocytomas
6 medulloblastomas
3 ependymomas
1 hemangioblastoma
1 epidermoid tumor
1 isomorphic mixed glioma
1 malignant mixed glioma

The largest groups was composed of 20 cerebellar astrocytomas. CT scan results indicated that there was a distension of the CSF space in all patients at the site of the resected tumor. The pre-operative hydrocephalus could no longer be seen in those children treated with shunts. The ventricles were completely normal, and there was no recurrence of tumor (Fig. 1). In the majority of the cases, the surgically approached section of the cerebellum revealed that there was a very deep subarachnoid space - which generally proved to be patent.

The next largest group was made up of those patients with medulloblastomas. All six of these patients received up to 6000 rads of postoperative radiotherapy, and also chemotherapy. A survival time of 3 to 10 years was achieved. There was also a distension of CSF spaces at the site of the resected tumor in all cases, as in the group of the astrocytomas, and again hydrocephalus was no longer present (Fig. 2). However, in two cases, as a result of radiotherapy, there appeared to be spotted calcified enhancement of the parenchyma on the CT scan. Contrary to what might have been expected from the histology, there were no signs or symptoms of recurrence of tumors.

The last group consisted of seven other miscellaneous cases. CT scans revealed the same results as the other groups. However, in a single case a dormant remnant of tumor was discovered postoperatively, and found to be a mixed glioma (Fig. 3). This patient was treated with radiotherapy, subsequently achieved a normal quality of life, and had a survival time of ten years.

Discussion

The sophisticated microsurgical techniques used in these cases did not produce any greatly significant improvement in survival time. However, they did improve the postoperative morbidity and mortality.

Also the avoidance of metal clips during the microsurgical techniques has produced better quality CT scans. It had been previously reported by BOCK et al. that metal clips considerably decrease the quality of the CT scans, and make the analysis of the findings difficult and uncertain.

In accordance with the clinical findings on follow-up, the CT scans revealed that in the majority of patients no recurrence of tumor could be found. There was a good differentiation between the area of the excision of the tumor from the normal parenchyma of the cerebellum and brain stem. This could, of course, be correlated with the fact that there was no recurrence of tumor.

In only one case was there a residual tumor (mixed glioma). It remained clinically dormant and the patient had a survival time of ten years. This case is similar to one reported by HOFFMAN et al.

The number of surviving patients with medulloblastomas was quite remarkable (6/39). This can be attributed to either:

1. the microsurgical techniques of radical extirpation or
2. to the postoperative radiotherapy (HIRSCH et al., NORRIS et al.).

In comparing the results of the pre-microsurgical era to now, the authors believe that microsurgical technique has been mainly responsible for this success. The surviving patients with medulloblastomas all received postoperative radio- and chemotherapy as per the optimal therapy recommended by BONGARTZ et al. and by BLOOM.

On CT scan, two cases of spotted calcified enhancement of the parenchyma were noted after the radiotherapy. This has been previously reported by SCHIFFER et al.

It is particularly worth mentioning that in all follow-up CT scans there was no longer any hydrocephalus, and above all there was a restoration of the subarachnoid CSF spaces of the posterior fossa (particularly in those hemispheres that were operated on). It is the authors' belief that the lack of hydrocephalus was a result of the routine water tight dural closure usually made by means of the lyophilized dural grafts.

Summary

Of 89 children undergoing microsurgery for medulloblastoma, ependymoma and astrocytoma of the cerebellum and fourth ventricle 33 were followed up post-operatively for two to 11 years.

The most significant findings included:

1. the absence of tumor tissue at the operative site,
2. calcification was sometimes observed post-radiotherapy in the region of the basal ganglia,
3. most interestingly, there was a restoration of the subarachnoid CSF spaces in the posterior fossa in all cases. This, the authors' believe, was the result of the routine water-tight dural closure by means of the lyophilized dural graft, and furthermore,
4. in the majority of cases, the post-operative quality of life was found to be satisfactory.

References

1. Bloom, H.J.G.: Recent concepts in the conservative treatment of intracranial tumors in children. Acta Neurochir. 50, 103 (1979)

2. Bock, W.J., Clar, H.-E., Weichert, C., Gerhard, L.: Follow-up studies in the posterior fossa in children, using computerized tomography. Neurochirurgia 24, 2 (1981)

3. Bongartz, E.B., Bamberg, M., Nau, H.E., Schmitt, G., Bayindir, C.: Optimal therapy in medulloblastoma. Acta Neurochir. 50, 117 (1979)

4. Hirsch, J.F., Renier, D., Czernichow, P., Benveniste, L., Pierre-Kahn, A.: Medulloblastoma in childhood - survival and functional results. Acta Neurochir. 48, 1 (1979)

5. Hoffman, H.J., Becker, L., Craven, M.A.: A clinically and pathologically distinct group of benign brain stem gliomas. Neurosurgery 7, 243 (1980)

6. Norris, D.G., Bruce, D.A., Byrd, R.I., Schut, I.L., Littman, P., Bilaniuk, L.I., Zimmerman, R.A., Capp, R.: Improved relapse-free survival in medulloblastoma utilizing modern techniques. Neurosurgery 9, 661 (1981)

7. Schiffer, D., Giodana, M.T., Soffietti, R., Tarenzi, L., Milani, R., Vasario, E., Paoletti, P.: Radio- and chemotherapy of malignant gliomas. Pathological changes in the normal nervous tissue. Acta Neurochir. 58, 37 (1981)

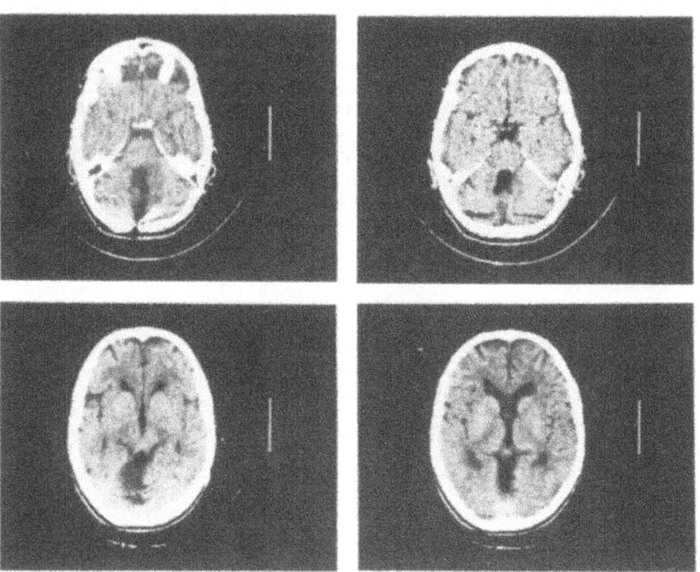

Fig. 1. CT scan of the posterior fossa six years after resection of an astrocytoma. There is an almost normal subarachnoid space, but there is distension of the CSF spaces at the site of the resected tumor. No hydrocephalic enlargement of the ventricular system is seen

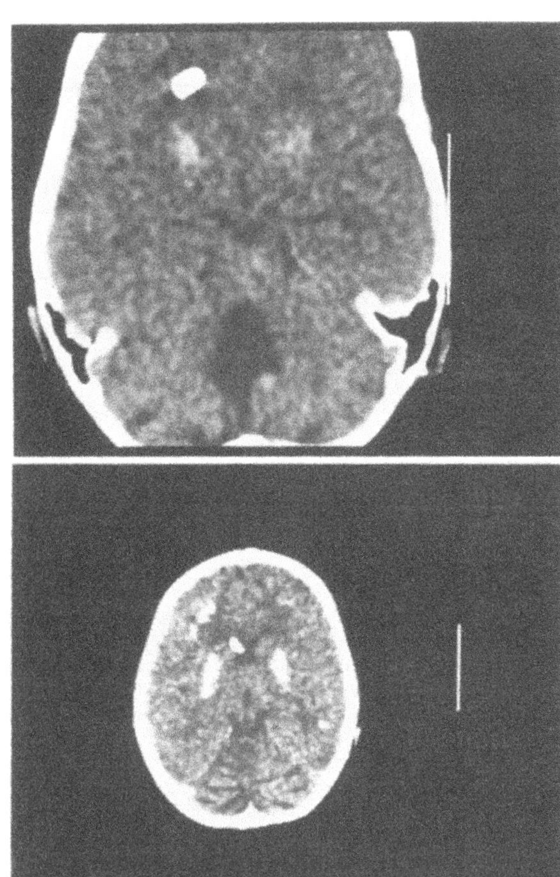

Fig. 2. A CT scan seven years after resection of a medulloblastoma. The posterior fossa reveals no recurrence of tumor, but there are areas of calcification in the region of the basal ganglia, a possible result of radiotherapy

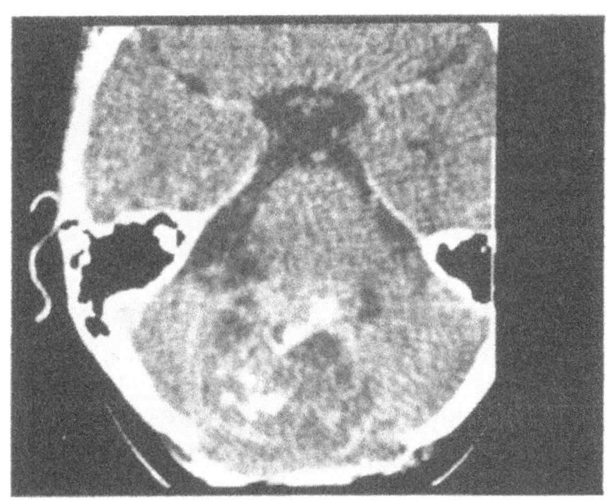

Fig. 3. CT scan ten years after operation on a mixed glioma, reveals a dormant tumor at the site of operation despite the patient's good clinical condition

Long-Term Survival of Medulloblastomas

G. C. Andrioli, I. Iob, M. Zuccarello, and G. Trincia
Instituto di Neurochirurgia dell Universitá di Padova, Via Giustiniani 5, I-35100 Padova

Medulloblastoma was once generally regarded as a rapidly and uniformly fatal brain tumor in children.

More recently it has been increasingly recognised that a significant number treated by a complete cerebrospinal axis radiotherapy and chemotherapy survive for years without evidence of recurrence. Survivals up to 50% at three years, 40 - 41% at five years, 22 - 30% after ten years have been recently described (MEALEY and HALE 1977; BLOOM 1975, 1977; QUEST et al. 1978, SCUCCIMARRA et al. 1978; CARTERI et al. 1979; GEROSA et al. 1980). Some authors (GULLOTTA 1979, GULLOTTA and NEUMANN 1980) are very critical about this long term survival of an increasing number of medulloblastomas. They report that the peculiar behaviour of some of these malignant tumors has to be ascribed to their histological structure and to the age of the patient more than to modern radio- and chemotherapy.

Furthermore, some of these long survival cases have been misdiagnosed; the variety of histological appearances of medulloblastomas as well as the small amount of tissue often available for microscopical examination may cause an incorrect diagnosis (GULLOTTA 1979). According to GULLOTTA's guidline about histogenesis and histology of this tumor we have identified 122 cases of medulloblastoma who underwent operation at the Institute of Neurosurgery in Padua from 1951 to 1978.

A complete follow-up investigation has been only possible in 71 cases (Table 1). One quarter of the patients died within one month of operation and 15% had unreliable postoperative or histological data.

The survival rate of 71 cases is illustrated by Table 2.

If we consider that our material has been collected over a long period in which different postoperative treatments have been employed and that every case has been histologically reviewed according to a unique principle, it seems reasonable to assess the various factors which have in-

Table 1.

Surgical mortality		25%
Not reliable data		15%
Follow-up investigations	71 cases	60%

Advances in Neurosurgery, Vol. 11
Edited by H.-P. Jensen, M. Brock, and M. Klinger

Table 2. Survival rate

53%	> 3 years
23%	> 5 years
8%	> 10 years

fluenced the prognosis. We will compare the survival rate with age, histological picture, site of tumor, type of operation and treatment after operation.

Age (Fig. 1)

The age and sex distribution is nearly typical. The oldest patient was a 57-year-old man and the youngest was a five-month-old female. The mean age is 11 years; 22% of the cases were more than 16 years old at time of admission.

Comparing the age with survival (Table 3) we may conclude that this factor has no importance in survivals of more than five years; up to this period the older group seems to have a better prognosis.

Surprisingly, all cases surviving more than ten years are less than 16 years old.

Histology

According to GULLOTTA's histogenetic theory (1967), the medulloblastoma has to be regarded as an embryonic mixed tumor with a neuroectodermal and mesenchymal tissue component. The pure medulloblastoma, with only a neuroectodermal component is extremely rare. Often, the mesenchymal component overgrows the neuroectodermal one and this phenomenon is mainly related to the age of the patient (Table 4). This variant of tumor (desmoplastic or arachnoidal sarcoma) is characterized by a very rich, tumoral reticulin network surrounding islands of cells (Fig. 2). Seventeen per cent of our cases were of the "desmoplastic" type.

Table 3. Age

Survival rate (%)	Age (%)	
52 > 3 years	>16 years	55
	<16 years	45
	mean age 11 years	
23 > 5 years	>16 years	25
	<16 years	23
	mean age 10 years	
8 >10 years	Case No. 1	5 years
	Case No. 2	4 years
	Case No. 3	7 years
	Case No. 4	13 years
	Case No. 5	2 years
	Case No. 6	2 years

Table 4. Medulloblastoma

Classic		Desmoplastic
No.	101 (83%)	21 (17%)
Mean age	10,5 years	19 years
>16 years	16%	61%

Table 5. Histology

Survival rate (%)		Histological structure (%)	
> 3 years	52	Classic	45
		Desmoplastic	90
> 5 years	23	Classic	24
		Desmoplastic	30
>10 years	8	Classic	5/6
		Desmoplastic	1/6

As regards (Table 5) the survival rates, our data suggest that this variant has a really better prognosis up to five years but no longer. After this period the survival does not seem to be related to the histological structure.

Site of Tumor

Medulloblastoma is a typical tumor of the midline (Table 6). The relatively small number of laterally situated tumors are more frequently found in young adults than in children. The shift of the tumor site from the midline to the cerebellar hemispheres is connected with overgrowth of the mesenchymal component which is similarly age-dependent.

The tumor site really seems to influence the survival mainly up to five years.

Type of Operation

We have considered only two kinds of operation: total removal (more than 90% of the tumor excised) and partial removal (a pure biopsy has been performed in only three cases).

From our data (Table 7) the total removal of this tumor is an important factor determining the duration of survival of the patients.

Table 6. Site of tumor

		Vermis or paramedian	Hemisphere
		84%	16%
Survival rate	> 3 years	47%	84%
	> 5 years	25%	50%
	>10 years	6%	16%

Table 7. Operation

	Total		Partial
Mean survival	> 3 years		< 2 years
Survival rate	> 3 years	57%	40%
	> 5 years	30%	5%
	>10 years	12%	-

Post-Operative Treatment

During the first 15 years of activity of our Service the patients received post-operatively only conventional radiotherapy to the posterior cranial fossa with various doses ranging from 2500 to 5000 rads.

After this period the majority of cases received Co80 radiotherapy to the entire neuraxis and chemotherapeutic treatment according to various trials (GEROSA et al. 1980). A comparison between the two groups of patients is therefore possible. From our data we may conclude that cobalt therapy and chemotherapy are important factors in the increase in the survival rate up to five years but no more (Table 8).

Conclusions

We may summarize that all five factors considered are significantly important in determining the survival rate; all factors but one (total removal) are of diminishing significance in the true long-term survival cases (Table 9).

Table 8. Postoperative treatment

Survival rate			
		Rx-th	Co^{80} + Chemotherapy
> 3 years	52%	38%	84%
> 5 years	23%	22%	23%
>10 years	8%	100% (?)	-

Table 9. Survivors >10 years

Case No.	Age y.	Sex	Site	Operation	Histology	Treatment	Survival
1	5	m	Vermis	Total	Classic	Rx-th	16 years alive
2	4	m	L.hem.	Total	Classic	Rx-th	10 years alive
3	5	m	Vermis	Total	Classic	Rx-th	11 y. later: metastasis in frontal midline
4	13	f	L.hem.	Total	Classic	Rx-th	17 years alive
5	2	f	Vermis	Total	Classic	Rx-th	12,5 years alive
6	2	f	Vermis	Total	Desmopl.	Rx-th	11 y. later: spinal intradural condroma 1 y. later: alive

Most of these patients, according to Collin's law, have to be considered definitely cured. This unexpected behaviour can be only explained by the coincidence of different factors such as an abnormal immunological response and/or a true, total removal of the tumor.

References

1. Bloom, H.J.: Combined modality therapy for intracranial tumors. Cancer 35, 111-120 (1975)
2. Bloom, H.J.: Medulloblastomas: prognosis and prospects. Int. J. Radiot. Oncol. Biol. Phys. 2, 1031-1033 (1977)
3. Carteri, A., Scuccimarra, A., Giordano, R., Giovannelli, M., Gaini, S.M., Villani, R.: Histological aspects of medulloblastomas in infancy and their correlations with survival. J. Neurol. Sci. 22, 29-33 (1978)
4. Gerosa, M., Di Stefano, E., Carli, M., Iraci, G.: Combined treatment of pediatric medulloblastoma. Child's Brain 6, 262-273 (1980)
5. Gullotta, F.: Das sogenannte Medulloblastom. Berlin, Heidelberg, New York: Springer 1967
6. Gullotta, F.: Morphological structure and biological behaviour of medulloblastoma in relation to age. In: Multidisciplinary aspects of brain tumor therapy, pp. 69-76. North Holland, Biomedical Press: Elsevier 1979
7. Gullotta, F., Neumann, J.: Medulloblastome und zerebelläre Sarkome. Eine histologisch katamnestische Untersuchung. Neurochirurgia 23, 35-40 (1980)
8. Mealey, J., Hall, P.: Medulloblastoma in children. J. Neurosurg. 46, 56-64 (1977)
9. Quest, D., Brisman, R., Antunes, J., Housepian, E.: Period of risk for recurrence in medulloblastoma. J. Neurosurg. 48, 159-163 (1978)
10. Scuccimarra, A., Carteri, A., Giordano, R., Crotti, F.M., Tomei, G., Zavanone, M.L.: Medulloblastoma in children; review on 119 cases. J. Neuro. Sci. 23, 37-45 (1979)

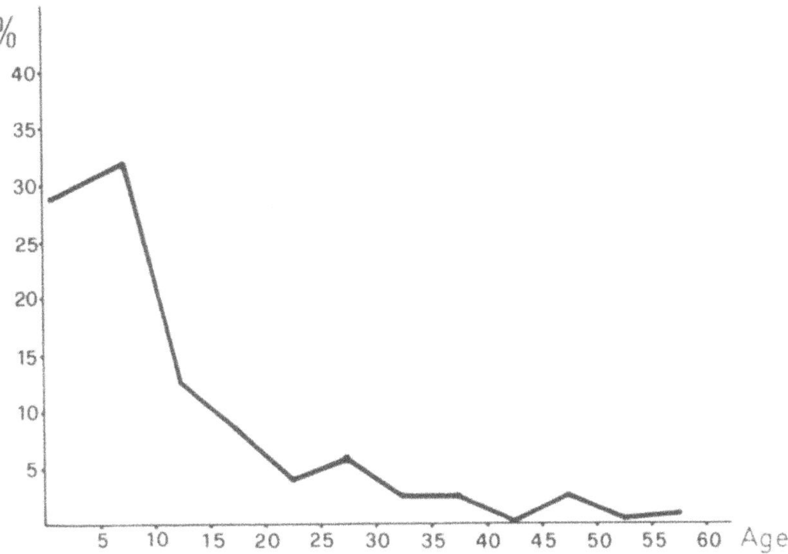

Fig. 1. Age. Cases No. 122 (1951 - 1978); male 60%, female 40%
>16 years 22%

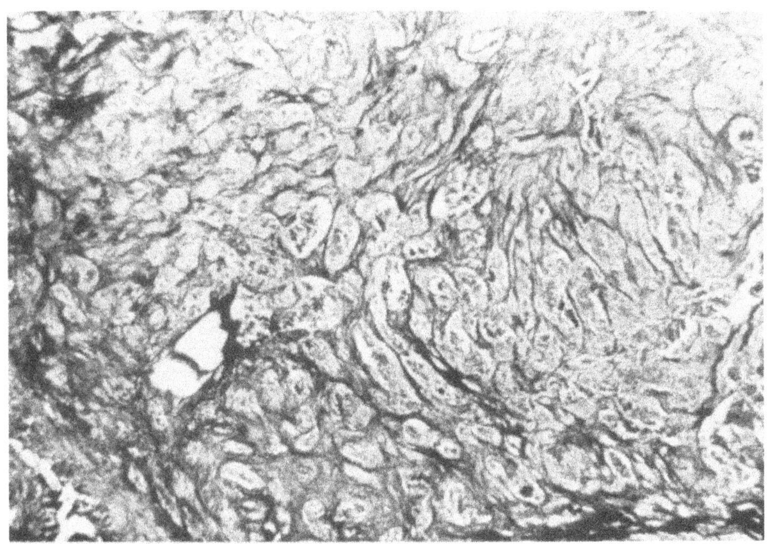

Fig. 2. Histological picture of the "desmoplastic" type of medulloblastoma. Note reticulin network surrounding islands of cells

Comparative Studies of Clinical Findings in Children and Adults with Tumors of the Posterior Fossa

A. Spring, H. Dietz, and I. Mund

Neurochirurgische Klinik der Medizinischen Hochschule Hannover, Karl-Wiechert-Allee 9, D-3000 Hannover 61

From a consecutive series of 146 patients, who were operated on for tumors of the posterior fossa, we analysed the clinical findings in 45 children (up to 18 years) with medulloblastoma or spongioblastoma and 48 adults with acoustic neurinoma. These tumors represent the typical diseases of the two groups. The other tumors are not analysed in this study, because they are relatively rare (11 angioblastomas, five angiomas, five ependymomas, nine meningiomas, three metastases and nine other tumors). The question arises, if there are typical neurological signs or symptoms, which correlate with the site and the size of the tumor.

Results

In the group of children (group I) there were 17 females (38%) and 28 males (62%). Of these patients 60% were less than nine years old, the youngest one year. The patients with acoustic neurinoma (group II) had an average age of 51 years. The youngest was 26, the oldest 65 years old. The sex ratio was 30 female (62.5%) and 18 male (37.5%). The average length of history was nine months for children and 36 months for adults. Sex ratio and length of history are typical for the two groups. Of the children 37 (82%) complained of headache and vomiting while 25 adults (52%) suffered from headache and 14 (29%) from vomiting. In 36 (80%) of the children and 16 (30%) of the adults papilledema was seen. The classical triad of papilledema, headache and vomiting was observed in 28 (62%) children and 7 (15%) adults. Corresponding to the signs of increased intracranial pressure 17 children (38%) and 9 (19%) adults were drowsy. Of the patients with acoustic neurinoma 26 (54%) felt dizzy, but only 9 (20%) of the children.

Naturally the nerves of the cerebellopontine angle are affected predominantly in patients with acoustic neurinomas. All patients had disturbances of the vestibulocochlear nerve. Additionally the trigeminal and facial nerve were affected on the same side in 28 (58%) and 23 (48%) patients respectively. On the other hand none of the children had disturbances of the eighth nerve. Symptoms of the trigeminal and facial nerves were found in two, and five cases respectively. Abducens paralysis was seen in 12 patients of group I and in eight of group II. Hypoglossal paresis in two children and four adults, dysphagia in five adults and accessory nerve paresis in two children demonstrate the rare involvement of the lower cranial nerve group (IX - XII). Pyramidal signs with positive Babinski reflex or Trömner's sign were recorded

Advances in Neurosurgery, Vol. 11
Edited by H.-P. Jensen, M. Brock, and M. Klinger

in seven (16%) children and four (8%) adults. Muscle tone was decreased in four cases of group I.

Cerebellar symptoms with loss of balance, ataxia and dysdiadochokinesia was very common in all patients. In group I 17 (38%) children had loss of balance, 33 (73%) ataxia and 18 (40%) dysdiadochokinesia. In group II 24 (50%) patients showed loss of balance, 21 (44%) ataxia and 12 (25%) dysdiadochokinesia. Horizontal or rotatory nystagmus was observed in 18 children and 30 adults.

The question is, if there are strong correlations between the neurological findings and the site or the size of the tumor (Tables 1 and 2). Disturbances of the cranial nerves V, VII, XI and XII are relative rare in the group of children and are nearly always explained by involvement of the brain stem. In addition the brain stem is shifted laterally by the tumor. In those cases with abducens paralysis, there were only 4 without brain stem involvement. This can be explained by the unfavorable course of this nerve and the increased intracranial pressure.

It is self-evident that all patients with acoustic neurinomas have disturbances of the eighth nerve. Abducens paralysis was noted in eight adults. Disturbance of the trigeminal and facial nerves indicates the existence of a larger tumor with involvement of the brain stem. Nevertheless there are some tumors with a diameter of less than 3 cm, which cause disturbances of the fifth, seventh, ninth and twelfth nerves. The peripheral part of the nerves close to the acoustic nerve is impaired by the tumor.

Motor disturbances on one side always correlate with tumors of the same cerebellar hemisphere in children. Nevertheless midline tumors of the vermis and the neighbouring cerebellum can cause the same symptoms. Some tumors, which are lateralised, are associated with bilateral motor disturbances.

In children bilateral ataxia is very often associated with tumors of the vermis (Table 3). On the other hand, in 11 children with bilateral ataxia the tumor was situated in only one cerebellar hemisphere. The tumors affected or infiltrated the brain stem in 13 cases with ataxia. The brain stem was not involved in nine children with this symptom. These findings are in contrast to the adults (Table 4). In all patients with bilateral ataxia the brain stem was affected by the tumor. Homolateral ataxia was seen in nine patients with tumors, affecting the brain stem. In four patients with hemiataxia the brain stem was not involved. Ataxia is associated with large acoustic neurinomas.

Summary

These investigations show that there are many similar neurological disturbances in the two groups, although the location and nature of the tumors are very different. Abducens paralysis, disturbances of the caudal cranial nerves and cerebellar motor disturbances are seen in both groups. The existence of these signs is of different significance. Paralysis of cranial nerves indicates an infiltration and shift of the brain stem in patients with medulloblastoma and spongioblastoma. Small acoustic neurinomas can cause disturbances of the nerves in the cerebello-pontine angle. Above all, ataxia in an acoustic neurinoma is an unfavorable sign, indicating a large tumor with involvement of the brain stem.

Table 1. Correlation between neurological symptoms, site and size of medulloblastomas and spongioblastomas in 45 children

	No.	Vermis	Cerebellar hemisphere lat.	bilat.	Brain stem Not affected	Affected	Infiltrated	Size <3 cm	3 - 5 cm	>5 cm
		24	20	9	18	7	20	2	26	17
Disturbance of cranial nerve V	2	1	1	-	-	-	2	-	2	-
VI	12	7	4	4	3	2	7	1	7	4
VII	5	1	4	-	1	1	3	-	3	2
XI	2	1	2	-	-	-	2	-	1	1
XII	2	2	2	-	-	-	3	-	1	1
Loss of balance	17	9	8	3	6	3	8	1	11	5
Ataxia	32	18	16	5	11	6	14	1	20	11
Dysdiadochokinesia	18	10	11	2	10	2	6	1	11	6
Nystagmus	18	8	11	2	8	3	7	1	12	5

Table 2. Correlation between neurological symptoms, brain stem involvement and size of acoustic neurinomas in 48 adults

		Brain stem			Size		
		Not involved	Involved	Shifted	<3 cm	3 - 5 cm	>5 cm
	No.	18	19	11	17	25	6
Disturbance of cranial nerve V	28	8	10	9	6	18	4
VI	8	3	1	4	2	4	2
VII	23	7	9	7	7	13	3
IX	5	2	1	2	2	2	1
XII	4	2	1	1	2	1	1
Loss of balance	24	6	11	7	5	14	5
Ataxia	21	4	10	7	2	13	6
Dysdiadochokinesia	12	2	6	4	1	7	4
Nystagmus	30	11	11	8	10	16	4

Table 3. Correlation between ataxia, site and size of medulloblastomas and spongioblastomas in 32 children

	Hemiataxia (n = 10)	Bilateral ataxia (n = 22)
Vermis	2	15
Cerebellar hemisphere		
unilateral	5	11
bilateral	1	4
Brain stem		
affected	3	3
infiltrated	3	10
Size		
<3 cm	-	1
3 - 5 cm	6	14
>5 cm	4	7

Table 4. Correlation between ataxia, brain stem involvement and size of acoustic neurinomas in 21 adults

	Hemiataxia (n = 13)	Bilateral ataxia (n = 8)
Brain stem		
affected	6	4
shifted	3	4
Size		
<3 cm	1	1
3 - 5 cm	8	5
>5 cm	4	2

Diagnosis and Treatment of Cystic Non-Tumorous Lesions of the Posterior Fossa in Children

D. Dorsic[1], H. Altenburg[2], and Th. Herter[1]

[1] Neurochirurgische Universitätsklinik, Jungeblodtplatz 1, D-4400 Münster

[2] Neurochirurgische Abteilung des Ev. Stift St. Martin, Johannes-Müller-Strasse 7, D-5400 Koblenz

Cystic lesions of the posterior fossa cannot easily be classified according to etiology, location, symptoms and treatment. We are proposing a better classification mainly of the benign infratentorial lesions.

We tried to combine the site and type of the cyst in a common classification. According to this classification, four groups of cysts can be distinguished (12):

The first group is constituted by the cystically dilated fourth ventricle as seen in the Dandy-Walker syndrome or other membranous occlusions of the fourth ventricle (14). The Arnold-Chiari malformation (3), showing a dilation of the fourth ventricle due to interference with the flow of CSF caused by the low position of the cerebellar tonsils, is also considered one of this group. Obstructive hydrocephalus caused by a high tumor of the cervical cord at the cranio-spinal junction is reckoned among this group, as also the so-called "trapped ventricle" (15) characterized by obstruction not only of the outlets but also of the Sylvian aqueduct.

The second group is made up by the cystically dilated cistern occurring with a paradoxical upward herniation of the contents of the posterior fossa.

Occasionally, dilatation of the cisterns is also met with in atrophy or low pressure hydrocephalus (12).

The third group is composed of the so-called true infratentorial cysts not showing any connection with the ventricular system in ventriculography with positive contrast radium. They can be located in the midline behind the cerebellum, in the region of the clivus above the cerebellar hemispheres, in the cerebello-pontine angle and also infratentorially above the cerebellum. These almost invariably benign cysts belong to the arachnoidal (5) or leptomeningeal cysts (2); trauma, inflammation and bleeding are considered as etiologic factors, as well as congenital conditions (9).

The fourth group consists the so-called pseudo-cysts, practically empty cavities filled with fluid and possibly lined with a thin membrane, caused by cerebellar hypoplasia or aplasia according to the *e vacuo* principle (8, 12).

The symptoms of the benign cystic lesions of the posterior fossa are characterized by an almost inevitable disturbance of fluid circulation (1) and the resulting signs of increased intracranial pressure.

Advances in Neurosurgery, Vol. 11
Edited by H.-P. Jensen, M. Brock, and M. Klinger

The extraventricular, extracisternal cysts also lead to a certainly slow but steadily progressive space occupation and impede the flow of CSF after having reached a certain size. Cerebellar symptoms are relatively infrequent. More common are acute crises of increased intracranial pressure, which necessitate diagnosis and treatment on an emergency basis if there is evidence of commencing herniation. On the other hand, pseudocysts resulting from underdevelopment of the cerebellum and the medulla oblongata are most likely to manifest themselves with the symptoms of the retardation complex (8).

The diagnosis of cystic infratentorial lesions became very exact and relatively easy with the introduction of computerized tomography, CT (11). By using contrast medium it is occasionally possible to differentiate between purely cystic and tumorous-cystic lesions (e.g. lack of enhancement in arachnoid cysts as compared with cystic hemangioblastomas).

The exact position and extent of infratentorial cysts can also be diagnosed relatively well from the CT scan.

Frequently, CT cannot provide proof of any connection between the cystic lesion and the ventricular system.

Occasionally it is impossible on the CT scan (1) to differentiate retrocerebellar cysts from a cystically dilated fourth ventricle. In this respect control ventriculography is of great help. It clearly demonstrates or excludes a possible communication between cyst and ventricular system, as well as any obstruction to the passage of CSF.

Ventriculography is also of advantage, because a possibly urgent operation for the drainage of CSF can be carried out under the same anesthesia. Vertebral angiography is of less importance in diagnosing non-tumorous lesions of the posterior fossa.

The operability of the cystic lesion can also be judged from ventriculography rather than from CT. The relations of the cyst to aqueduct or fourth ventricle and the exact location of these connections is decisive for the further treatment which will consist of CSF drainage or/and a direct operative approach to the cystic lesion of the posterior fossa.

References

1. Archer, C.R., Darwish, H., Smith, K.: Enlarged cisternae magnae and posterior fossa cysts simulating Dandy-Walker Syndrome on Computed Tomography. Radiology 127, 681-686 (1978)
2. Dunsker, S.B., McCreary, H.S.: Leptomeningeal cyst of the posterior fossa. J. Neurosurg. 34, 687-692 (1971)
3. Huber, G., Emde, H., Piepgras, U.: Röntgenologische und szintigraphische Befunde beim Arnold-Chiari-Dandy-Walker- und Syringohydromyeliesyndrom. Nervenarzt 48, 156-164 (1977)
4. Ito, J., Yamazaki, Y., Honda, H., Ishikawa, H., Ueki, K.: Angiographic appearance of a huge retrocerebellar arachnoid cyst in an infant. Neuroradiology 13, 115-119 (1977)
5. Little, J.R., Gomez, M.R., McCarty, C.S.: Infratentorial arachnoid cysts. J. Neurosurg. 39, 380-386 (1973)
6. Lipton, H.L., Preziosi, T.J., Moses, H.: Adult onset of the Dandy-Walker syndrome. Arch. Neurol. 35, 672-675 (1978)

7. Lourie, H., Shende, M.C., Krawchenki, J., Stewart, D.H.: Trapped fourth ventricle: A report of two unusual cases. Neurosurgery 7, 279-282 (1980)

8. Mercuri, S., Curatolo, P., Giuffre, R., Di Lorenzo, N.: Agenesis of the vermis cerebelli and malformations of the posterior fossa in childhood and adolescence. Neurochirurgia 22, 180-188 (1979)

9. Mori, K., Hayashi, T., Handa, H.: Infratentorial retrocerebellar cysts. Surg. Neurol. 7, 135-142 (1977)

10. Mori, K., Hayashi, T., Handa, H.: Radiological manifestations of infratentorial retrocerebellar cysts. Neuroradiology 13, 201-207 (1977)

11. Naidich, T.P., Lin, J.P., Leeds, N.E., Pudlowski, R.M., Naidich, J.B.: Primary tumors and other masses of the cerebellum and fourth ventricle: Differential diagnosis by computed tomography. Neuroradiology 14, 153-174 (1977)

12. Peeters, F.L.M., Thijssen, H.O.M.: Die Differentialdiagnose von Zysten in der hinteren Schädelgrube. Fortschr. Röntenstr. 121, 146-153 (1974)

13. Sharpe, J.A., Deck, J.H.N.: Neuroepithelial cyst of the fourth ventricle. J. Neurosurg. 46, 820-824 (1977)

14. Touitou, D., Rakapopoulos, D.S., Wackenheim, A.: Subarachnoid cyst of the posterior fossa and Dandy-Walker syndrome. J. belge Radiol. 61, 335-340 (1978)

15. Zimmerman, R.A., Bilaniuk, L.T., Gallo, E.: Computed tomography of the trapped fourth ventricle. Am. J. Roentgenol. 130, 503-506 (1978)

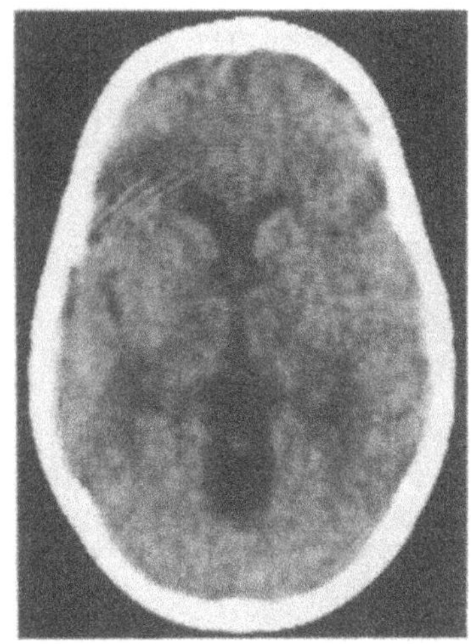

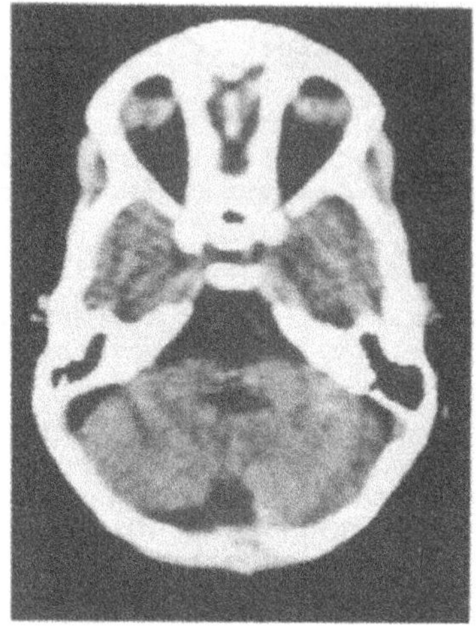

Fig. 1. *Left:* Combination of group I, II and IV. During the operation a small vermis and small cerebellar tonsils were found

Fig. 2. *Right:* Legend see Fig. 1

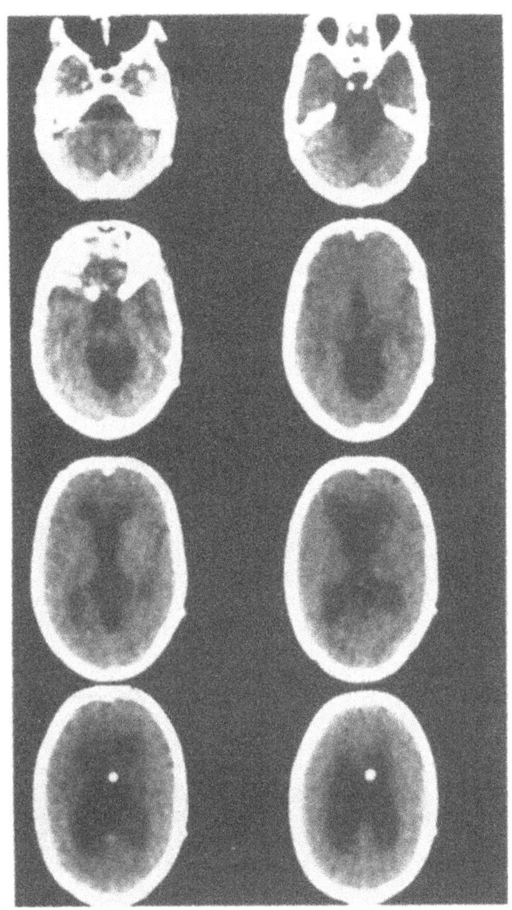

Fig. 3. Group I Cystically dilated fourth ventricle

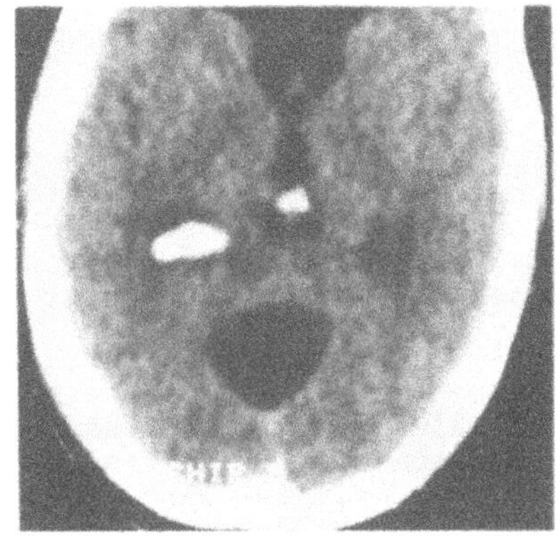

Fig. 4. Not group I but a retrocerebeliar cyst at a higher level

Effects of Dexamethasone on Physical and Biochemical Properties of the Normal Non-Edematous Brain

H.-W. Bothe[1], Th. Wallenfang[2], and K. Schürmann[2]

[1] Max-Planck-Institut für Hirnforschung, Ostmerheimer Strasse 200, D-5000 Köln 91
[2] Abteilung für Neurochirurgie, Medizinische Universitätsklinik, Langenbeckstrasse 1, D-6500 Mainz

Introduction

Vasogenic brain edema associated with brain tumor or brain abscess responds well to steroid treatment. However, in the management of edema due to ischemia or trauma, the efficiency of glucosteroid therapy declines. Therefore, the mechanisms involved in the steroid therapy of tumor or abscess must differ from those involved in the treatment of ischemia or trauma. To improve the treatment of ischemic or traumatic brain edema, e.g. by supplementary measures supporting the glucocorticoid effect, it is necessary to know the pathophysiological mechanisms and pathways concerned in the steroid management of brain edema. Yet it is a common experience that the evaluation of pathophysiological pathways is often helped by the investigation of the normal physiological background. Surprisingly, the only investigations of which we are aware involve steroid effects under pathological conditions. The task, therefore, is to estimate the effect of glucocorticoids in normal brain tissue by taking into consideration physical and biochemical factors.

Material and Approach

Material. Twelve balanced and randomized adult male cats (weighing $\bar{x}$ = 3.5 kg, s_x = 0.3 kg) were divided into two groups, each with six animals.

Treatment. One group was treated by daily intramuscular injection of 0.5 mg/kg dexamethasone (Decadron) given as two single doses of 0.25 mg/kg. Treatment was continued for 13 days. The other group served as normal untreated controls.

Operation. After this period the animals were anesthetised (15 mg/kg ketamine, Ketanest), tracheotomized, ventilated with room air and immobilized (0.2 mg/kg hexcarbacholinbromide, Imbretil). The intraventricular pressure (IVP) was determined by means of a needle (diameter 0.5 mm) inserted stereotactically into the right ventricle. Subsequently, the brain was frozen *in situ*, by pouring liquid nitrogen over it, and then removed.

Estimations carried out: The determination of water content, sodium and potassium was performed in the grey and white matter of the left hemisphere according to the method described by REULEN (10), whereas glucose, lactate, creatine phosphate and ATP content of brain tissue

 Advances in Neurosurgery, Vol. 11
Edited by H.-P. Jensen, M. Brock, and M. Klinger

were estimated according to the methods described by LOWRY (7). The animals of the control group underwent the same procedures. Statistical evaluation was based upon the non-parametrical u-test of Mann-Whitney-Wilcoxon for independent random sampling (6).

Results

Physical Findings

After treatment of normal animals with dexamethasone, IVP shows no significant changes in comparison to that of the untreated group (Fig. 1). The values obtained amount to $\bar{x}$ = 2.4 mm Hg for treated cats and $\bar{x}$ = 2.2 mm Hg for untreated cats. However, the water content of the white matter (Fig. 1) declines after treatment from $\bar{x}$ = 70.7 g/100 g w.w. to $\bar{x}$ = 67.2 g/100 g w.w. ($p<0.05$), whereas the water content of grey matter shows no changes. The change of water content in the white matter corresponds to a mass-volume decrease of 11.9% (for calculation see 2).

Biomechanical Findings

The animals treated with dexamethasone show slight though not significant changes of sodium content (Fig. 1), whereas an obvious alteration of potassium content is detectable in the white matter of the treated group. Sodium content of white matter amounts to $\bar{x}$ = 193.7 mmol/kg d.w. in the untreated group and $\bar{x}$ = 169.4 mmol/kg d.w. in the treated group. Potassium content of white matter increases from $\bar{x}$ = 197.1 mmol/kg d.w. in the untreated group, to $\bar{x}$ = 249.3 mmol/kg d.w. in the dexamethasone-treated group ($p<0.05$).

Concerning the energy metabolism (Fig. 2) no differences between the treated and the control group are observable. Glucose content as well as lactate, creatine phosphate and ATP content are unchanged after administration of dexamethasone both in the grey and white matter.

Discussion

In considering the results presented above, it seems to be advantageous to summarize the metabolic pathways of glucocorticoids, especially dexamethasone, previously described by numerous authors (Fig. 3).

The current concept of dexamethasone's action in nervous tissue involves specific gene activation after administration of the steroid. Nowadays, the degree of steroid gene activation is nearly understood and the substances synthesized after activation are partly estimated: The first step of the specific gene activation by dexamethasone is the binding of the steroid to cytoplasmic receptor proteins (4, 11). The binding of dexamethasone is followed by a cytoplasmic nuclear translocation of the steroid-receptor complex with the transformed receptor protein (8, 5). After nuclear gene expression a changed set of proteins is released to the steroid affected nervous tissue (1). Some classes of these proteins are consistently increased, other classes of proteins are decreased. However, the metabolic objectives of the single protein released after steroid administration are not yet known, although many results are available concerning the gross physical and biochemical changes of neuronal tissue after treatment with dexamethasone (Fig. 3).

One gross alteration of normal non-edematous brain is presented in this paper; supporting evidence of the chain of causality, which possibly produces that alteration, may be provided by the above described intermediary metabolism of dexamethasone. After long term administration of dexamethasone the water content of white matter decreases. This loss of water being 3.5 g/100 g w.w. corresponds to a mass-volume decrease of 11.9% in brain tissue. Since the diminution in tissue volume is not to be explained by a reduced vascular volume, which extends maximally to 5% of the white matter of the entire brain (3), a decrease of neuropil itself must be required. Considering the sodium and potassium content after dexamethasone treatment it can be concluded, that the mass-volume of brain declines mainly in the extracellular space (ECS) without any change in the intracellular compartment. Therefore, a mass-volume shifting from extracellular to the fluid spaces is to be gathered from the "Monro-Kellie doctrine". Since disturbances of energy metabolism are not apparent, the effect of dexamethasone on normal (presumably as well as on altered) brain tissue seems at least to result from biomechanical changes in the brain. Because of the reduction of the ECS the hydraulic condictivity, which is a function of the diameter of ECS (9), deteriorates, i.e., the resistance of ECS to bulk flow of brain water increases. Consequently, in pathological conditions causing vasogenic brain edema the spread of edema fluid into unaffected areas of the brain may be prevented. However, the present experiment suggests not only a prevention of water accumulation in brain tissue, but also an active reduction of extracellular water. Although the mechanisms involved in that extracellular fluid reduction by dexamethasone treatment remains rather unclear, there is properly only one possible hypothesis to explain the active reduction of extracellular fluid by dexamethasone: the volume capacity must decline, that means, the elasticity of brain tissue is raised after glucocorticoid therapy. To evaluate this hypothesis measurement of tissue compliance and cerebro-vascular resistance after dexamethasone therapy will be undertaken in a subsequent investigation.

Conclusion

Administration of dexamethasone over a period of 13 days reduces water content of normal non-edematous brain tissue. Since the decrease of water content involves mainly the extracellular compartment of the brain, it is concluded, that dexamethasone causes an alteration of biomechanical cerebral properties, i.e., hydraulic resistance and brain tissue compliance.

References

1. Arenander, A.T., De Vellis, J.: Glial-released proteins: Two-dimensional identification of proteins regulated by hydrocortisone. Brain. Res. 224, 105-116 (1981)
2. Baethmann, A., Van Harreveld, A.: Water and electrolyte distribution in gray matter rendered edematous with a metabolic inhibitor. J. Neuropathol. Exp. Neurol. 32, 408-423 (1973)
3. Bothe, H.W., Wallenfang, Th., Schürmann, K.: The relationship between brain edema, energy metabolism, glucose content and rCBF investigated by artificial brain abscess in cats. In: Advances in Neurosurgery. Vol. 10, Driesen, Brock, Klinger (eds.), pp. 214-221. Berlin, Heidelberg, New York: Springer

4. Coutard, M., Osborne-Pellegrin, M.J., Funder, J.W.: Tissue distribution and specific binding of tritiated dexamethasone *in vivo:* autoradiographic and cell fractionation studies in the mouse. Endocrinology 103, 1144-1152 (1978)

5. Izawa, M., Ichii, S.: Interaction of ^{3}H-dexamethasone-receptor complex with nuclei: Inhibition by ethidium bromide. Endocrinol. Japon. 27, 101-106 (1980)

6. Lienert, G.A.: Verteilungsfreie Methoden in der Biostatik, Vol. I, pp. 212-234. Meisenheim: Anton Hain KG ;§)=

7. Lowry, O.H., Passonneau, J.V., Hasselberger, F.X., Schulz, D.W.: Effect of ischemia on known substrates and cofactors of the glycolytic pathway in brain. J. Biol. Chem. 239, 18-29 (1964)

8. Papamichail, M., Tsokos, G., Tsawda-Roglouw, N., Sekeris, C.E.: Immunocytochemical demonstration of glycocorticoid receptors in different cell types and their translocation from the cytoplasm to the cell nucleus in the presence of dexamethasone. Exp. Cell. Res. 125, 490-493

9. Rapaport, S.I.: A mathematical model for vasogenic brain edema. J. Theor. Biol. 74, 439-467

10. Reulen, H.J., Medzuhradsky, F., Enzenbach, R., Marguth, F.: Electrolytes, fluids and energy metabolism in human cerebral edema. Arch. Neurol. 21, 517-525 (1969)

11. Warembourg, M., Otten, U., Schwab, M.E.: Labelling of Schwann and satellite cells by ^{3}H dexamethasone in a rat sympathetic ganglion and sciatic nerve. Neuroscience 6, 1139-1143 (1981)

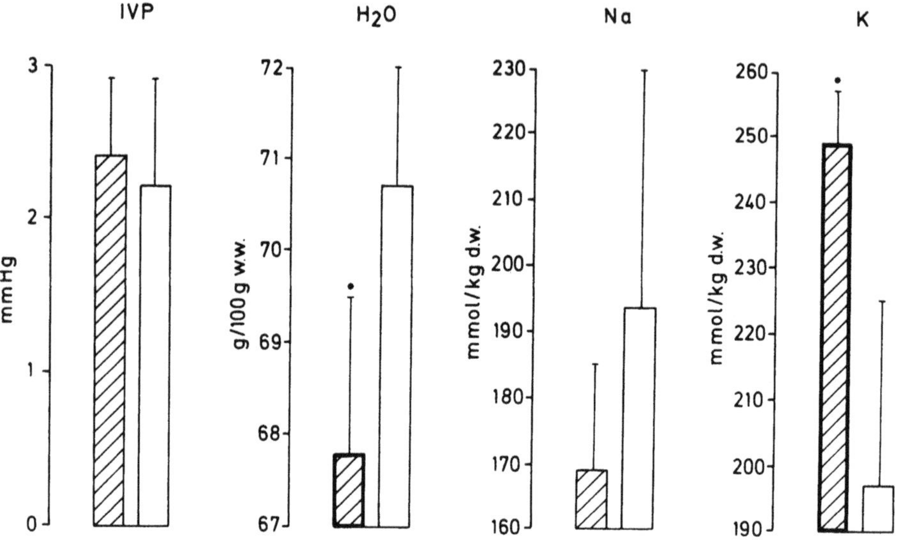

Fig. 1. Intraventricular pressure (IVP), water content (H_2O), sodium content (na) and potassium content (K) of white matter in the treated (hatched bars) and control (open bars) group. Values are given as mean and SD; *, $p<0.05$

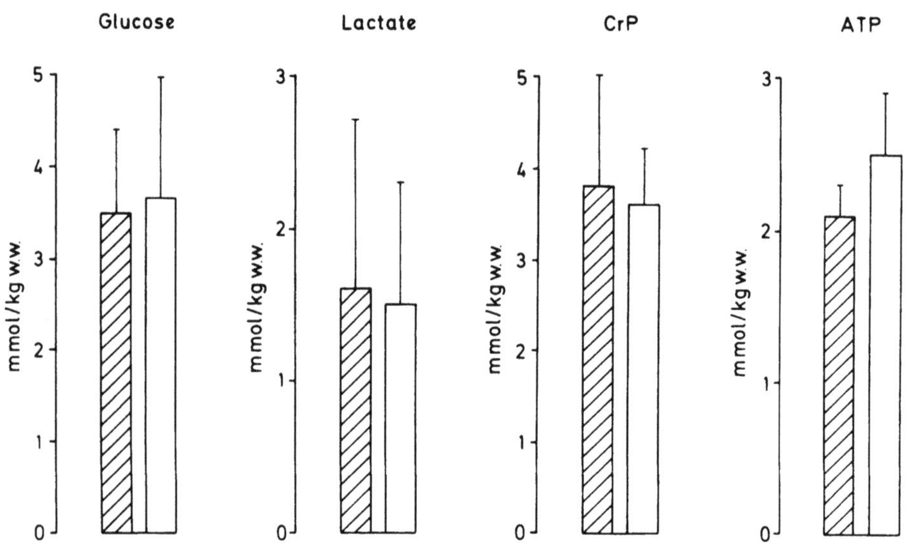

Fig. 2. Glucose content, lactate content, creatine phosphate content (CrP) and ATP content of white matter in the treated (hatched bars) and control (open bars) group. Values are given as mean and SD; *, $p<0.05$

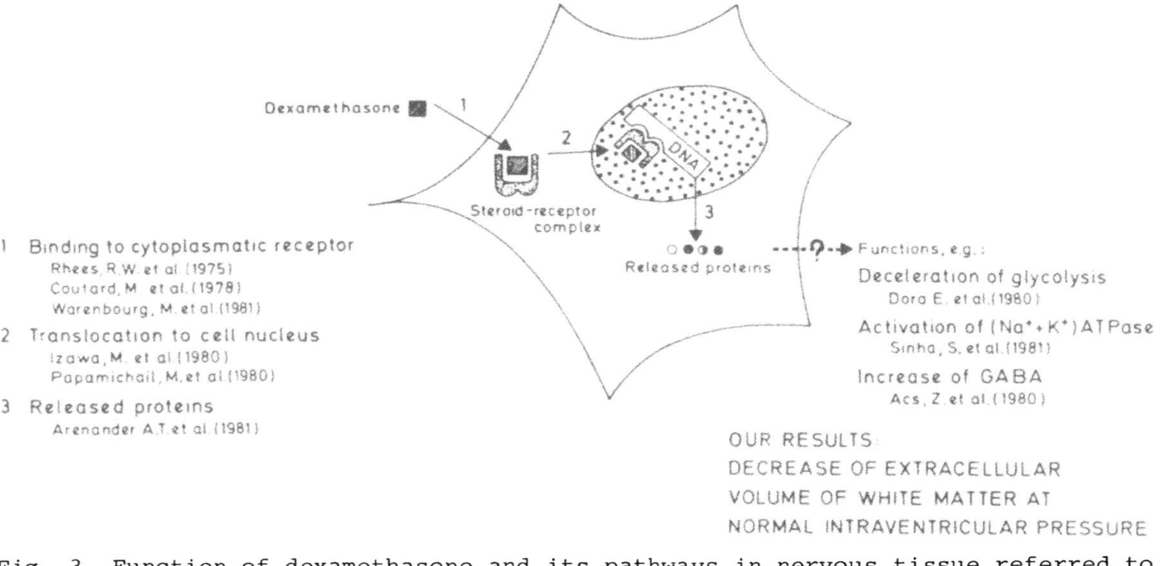

Fig. 3. Function of dexamethasone and its pathways in nervous tissue referred to the underlying literature

Effect of Dexamethasone on Regional Cerebral Blood Flow in Patients with Cerebral Tumors and Brain Edema

A. Hartmann[1], R. v. Kummer[1], J. Menzel[1], and C. Buttinger[2]

[1] Neurologische Klinik der Universität Bonn, Sigmund-Freud-Strasse 25, D-5300 Bonn-Venusberg
[2] Neurochirurgische Klinik der Universität Heidelberg, Vosstrasse 2, D-6900 Heidelberg 1

Steroids such as dexamethasone are used routinely against brain edema in patients with cerebral tumors. With sufficiently high doses the clinical effect starts to become recognizable within 24 hours. However, intracranial pressure recordings indicate that the raised cerebrospinal fluid pressure falls later (1, 2, 4), whereas so-called plateau waves diminish in number and intensity within 12 - 18 hours (Fig. 1). Therefore mechanisms other than ICP-reduction alone are probably involved in the beneficial action of steroids in patients with brain edema. Since regional cerebral blood flow (rCBF) is an important factor in brain function and since it is influenced not only by intracranial pressure but also by other factors such as metabolism and nerve function, the study of rCBF during dexamethasone therapy might give some clue to its action.

Method

The nontraumatic method of measuring rCBF over both hemispheres was used for the study. After brief inhalation of 25 - 30 mCi Xenon 133 from an airbag 32 desaturation curves were recorded from both sides of the skull. All curves were corrected for recirculation by simultaneous measurement of Xenon 133 activity in the end expiratory air. After correction for extracerebral contamination the computer printed out the data about flow in the so-called fast perfused tissue (which is almost identical with the grey matter).

The blood-tissue partition coefficient for Xenon 133 is altered in pathological tissue such as that of neoplasms and in tissue altered by edema. Therefore a flow-factor was used, which does not take this coefficient into account. This factor - the initial flow index ISI - is mainly influenced by the grey matter and only to a minor extent by the white matter. All data were corrected by a factor of 4% per $PaCO_2$-difference between two studies, if necessary.

Further technical details can be read elsewhere (3).

Seventeen patients were included into the study and treated according to the following protocol (Fig. 2):

All patients received dexamethasone 1 mg/kg bodyweight, on day 1 divided into four doses. Initially one third of the total daily doses was given intravenously and the rest divided into three doses for

Advances in Neurosurgery, Vol. 11
Edited by H.-P. Jensen, M. Brock, and M. Klinger

intramuscular administration. From day 2 on, the total doses of dexamethasone was reduced to 0.5 mg/kg/day according to Fig. 2.

rCBF was measured in group I (n = 10) before the start of treatment. 24 hours after the start (day 1) and on day 7.

In group II (n = 7) rCBF was measured in the same manner but on day 2 instead of day 1.

Results

Steady State Data

Compared to a group of patients with no organic central nervous system disease mean rCBF of the diseased hemisphere was normal in five out of ten patients (group I) and in three out of seven (group II). On the non-tumor side mean rCBF was normal in four patients of group I and in four of group II.

From this it could be concluded that mean rCBF - which reflects the hemispheric flow - is unaltered in about half of the patients despite clinical signs of increased intracranial pressure and transhemispheric tissue shift. However, looking at the regional data it became obvious that all patients had a pathological regional pattern with increased *and* decreased intratumoral flow and perifocal reduced flow. In more than three-quarters of all 17 patients the rCBF in the contralateral hemisphere showed signs of disturbed regional tissue perfusion.

Alteration of Flow in Group I

After 24 hours of dexamethasone therapy rCBF did not show any significant change: -0.4% (± 10.3) in the tumor-hemisphere and 0.5% (± 10.9) on the contralateral side. On day 7 rCBF increased significantly by +14.6% (± 12.5) in the diseased side and by +13.1% (± 7.2) in the healthy side.

Despite the unaltered flow 24 hours after start of treatment some of the patients showed clinical improvement such as reduced complaints about headache or tinnitus and improvement in the neurological status.

Alteration of Flow in Group II

On day 2 (48 hours after start of treatment) rCBF on the tumor side increased by +8.4% (± 9.6) and by +9.2% (± 8.0) in the non-tumor side. On day 7 rCBF had increased in a similar manner than in group I: +16.2% (± 9.7) on the diseased side and +17.6% (± 11.5) on the healthy side. In the patients where a second CT was performed at this time we were unable to detect any change of the amount of edema (Fig. 3).

Regional Patterns

Figure 4 indicates the percentage of all areas with a significant change of rCBF (more than 15%) during the period involved. In group I on day 1 about 20% of all areas in both hemispheres showed a significant rCBF-increase and about 20% a significant rCBF reduction. On day 7 only 8% showed a significant rCBF-reduction but more than 40% of all areas presented with a significant rCBF-increase.

In group II 10% of all areas showed on day 2 a decrease in rCBF, but 27% showed rCBF increase. On day 7 only 2% of all areas presented with a reduction of rCBF and 37% with increased rCBF.

Discussion

The results indicate that rCBF is decreased in patients with cerebral tumors, accompanied by edema, which leads to tissue shift and mass displacement. rCBF reduction occurs not only on the tumor side but also on the contralateral side. There might be several reasons for this:

1. transneural depression (which has not so far been described in brain tumors)
2. generalized increase of ICP, as shown by different authors (1, 4)
3. development of brain edema on the contralateral side. This was almost ruled out in this series by computer tomography.

Since all but one patient only suffered from a tumor on one side of their brain and since both hemispheres presented with decreased flow the study clearly indicates the importance of treatment of the basic disease and its consequences, viz. brain edema and tissue displacements.

rCBF started to increase from the second day on and was even higher on the seventh day of treatment. REULEN et al. (5) observed an increase of rCBF between 25.3 and 36.3 ml/100 g/min after five days of treatment using a daily dose of 24 mg dexamethasone. Since they performed their study with the intra-arterial Xenon 133 injection technique (as was usual during that period of rCBF-research) a third study between steady state and day 5 was largely impossible.

Our results show that the increase of rCBF - probably due to removal of edema - does not appear before the second day of high dose therapy. This delayed increase of rCBF parallels the reduction of ICP (1) but is preceded by some clinical improvement and reduction of plateau waves (1). Therefore in patients with tumors and edema, blood flow in the brain seems to be mostly triggered by increased intracranial pressure. The rather early action of dexamethasone may well be due to mechanisms other than ICP and rCBF.

References

1. Alberti, E., Hartmann, A., Schütz, H.-J., Schreckenberger, F.: The effect of large doses of dexamethasone on the cerebrospinal fluid pressure in patients with supratentorial tumors. J. Neurol. 217, 173-181 (1978)

2. Brock, M., Zillig, C., Wiegand, H., Zywietz, C., Mock, P.: The effect of dexamethasone on ICI in cases of posterior fossa tumors. In: ICP III. Beks, J.W.F., Bosch, D.A., Brock, M. (eds.), pp. 236-243. Berlin, Heidelberg, New York: Springer 1976

3. Hartmann, A., v. Kummer, R.: Die atraumatische Messung der regionalen Gehirndurchblutung: Methodik und Zuverlässigkeitsprüfung. Fortschr. Neurol. Psychiat. (in press, 1982)

4. Kullberg, G.: Clinical studies on the effect of corticosteroids on the ventricular pressure. In: Steroids and Brain Edema. Reulen, H., Schürmann, K. (eds.), pp. 253-258. Berlin, Heidelberg, New York: Springer 1972

5. Reulen, H.J., Hadjidimos, A., Schürmann, K.: The effect of dexamethasone on water and electrolyte content and on rCBF in perifocal brain edema in man. In: Steroids and Brain Edema. Reulen, H.J., Schürmann, K. (eds.), pp. 239-252. Berlin, Heidelberg, New York: Springer 1972

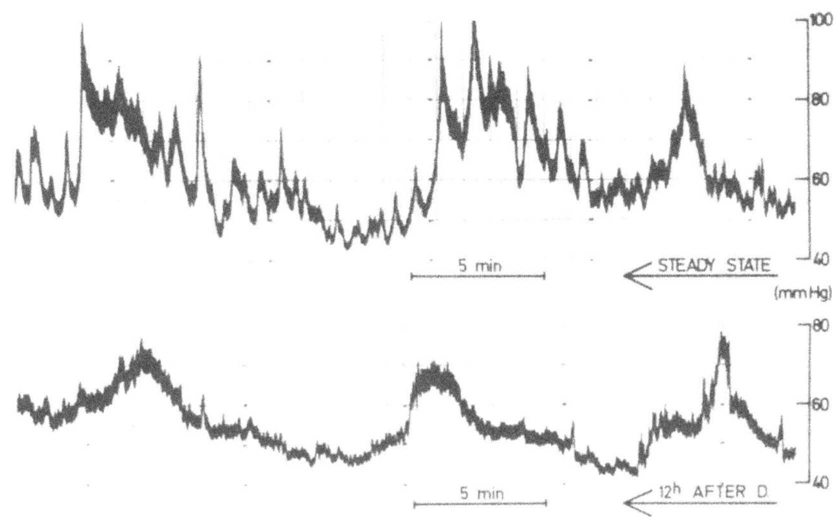

Fig. 1. Effect of dexamethasone (1 mg/kg/day) on cerebrospinal fluid pressure (CSFP) in a patient with glioblastoma multiforme. The curve reads from right to left. Twelve hours after start of treatment there is no significant reduction in the basic pressure, but there is reduction in height and frequency of the pressure waves

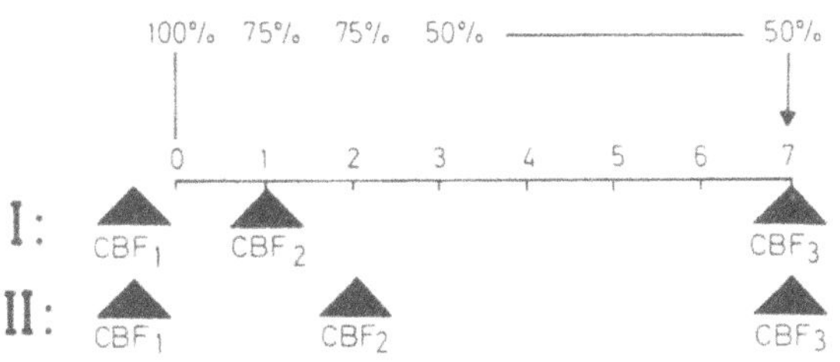

Fig. 2. Protocol of the study. In group I rCBF was studied before start of dexamethasone treatment, on day 1 and 7. In group II CBF was studied before the start of treatment, on day 2 and 7

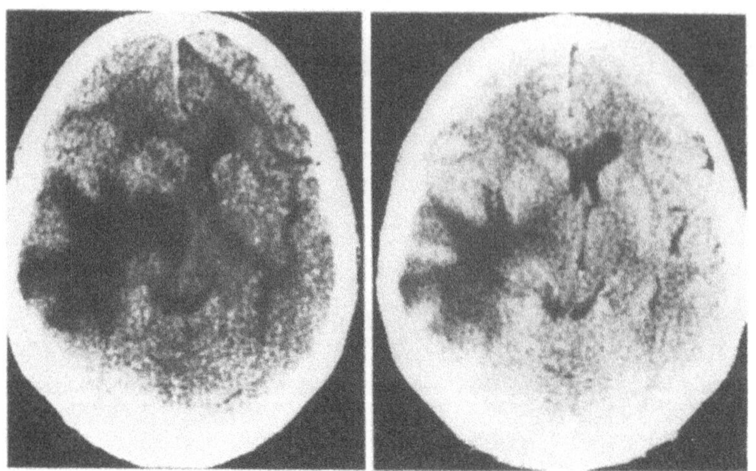

Fig. 3. Computer tomography before *(left)* and 24 hours after treatment *(right)* with 1 mg/kg/day dexamethasone. There is no significant change in the appearance of the edema. However, at the time of the second CT scan the patient was more awake and complained less about his headache

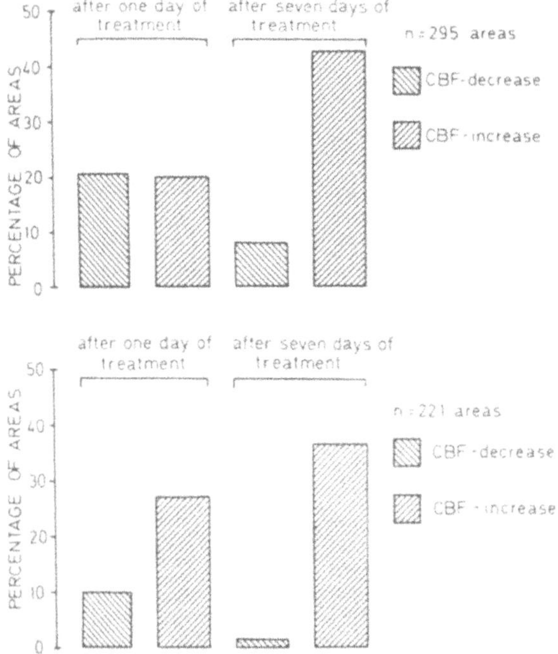

Fig. 4. Percentage of areas with significant rCBF-chage during dexameth-asone treatment. Each column represents the percentage amount of areas with rCBF-increase or rCBF-decrease of more than 15%. The rCBF of each area was compared to its own steady state value. *Upper:* group I (second CBF-measurement on day 1); *lower:* group II (second CBF-measurement on day 2)

Determination of the Effect of Dexamethasone on Peritumoral Brain Edema

D.-P. Lim, W. J. Bock, and D. Reinhardt
Neurochirurgische Universitätsklinik, Moorenstrasse 5, D-4000 Düsseldorf 1

Introduction

Dexamethasone (Decadron) in various dosages is recommended nowadays all over the world for the basic treatment of zones of perifocal edema surrounding a tumor, cerebral infarction and traumatic cerebral oedema.

Although this form of treatment is generally accepted, many points relating to the mechanism of action are still unclear. Investigations have been carried out in recent years into the binding of glucocorticoids to specific receptors in the target cells. Detailed knowledge on the structure and function of the glucocorticoid receptors is fundamental to the understanding of the mechanism of action of steroids on cerebral edema. A number of authors have already reported results in animal experiments dealing with these glucocorticoid receptors and on their significance for patients.

Using radioimmunoassay, we have determined the tissue and serum levels of dexamethasone in 21 patients with brain tumors under treatment and carried out a correlation with the morphological changes in the computed tomogram. In addition, the catecholamines in the serum of six patients were measured. The object of the measurement was to construct dosage schedules and to obtain pharmacokinetic data.

Materials and Method

Twenty-one patients aged from 8 to 72 years (6 women and 15 men) with various types of cerebral tumors were examined and treated. There were seven with astrocytoma, of which two were in the posterior fossa, five with glioblastoma, two with oligodendroglioma, five with metastases, one of which was in the posterior fossa, one patient with a meningioma and one with a pineal tumor (only CSF examined). The procedure was always in accordance with the following protocol: 3 x 8 mg were administered intravenously before operation (3 x 4 mg for children under 12 years of age). During operation, a bolus intravenous injection of 40 mg was given (20 mg intravenously for children), and after the operation, 4 x 8 mg intravenously for three days (children 4 mg), the dose being reduced to 3 x 8 mg per day on the fourth to the sixth day. In addition, infusion therapy was given which varied from patient to patient. This treatment was carried out in each case irrespective of the type of tumor and irrespective of the finding of manifest cerebral edema, (Tables 1 and 2).

Advances in Neurosurgery, Vol. 11
Edited by H.-P. Jensen, M. Brock, and M. Klinger

Table 1

Classification	No.	Sex	Age
Astrocytoma	7	6 ♀	8 to 72
Glioblastoma	5	15 ♂	
Oligodendroglioma	2		
Metastasis	5		
Meningioma	1		
Pinealoma	1		
Total	21		

Table 2

Dosage (intravenous) of dexamethasone[a] in brain tumor

	Pre-operative	Intra-operative	Postoperative day 1.-3.	4.-7.
Adult	3x8 mg/die	40 mg (Bolus)	4x8 mg	3x8 mg
Children under 12 years	3x4 mg/die	20 mg (Bolus)	4x4 mg	3x4 mg

a Decadron

The morphological changes in the computed tomogram served as the criterion for the duration of treatment with dexamethasone. Tissue samples for measuring the concentrations were collected during operation in the direct neighbourhood of the tumor. In one patient only a CSF investigation was carried out. The serum samples were collected after 30 minutes, and after one, two, four and eight hours and on the first and second days after operation (Fig. 1).

Stripped plasma treated with various concentrations of dexamethasone, and tissue and plasma of patients were treated with ^{3}H-labelled dexamethasone and then with the antibody. After an incubation time of more than 12 hours at a temperature of $4^{o}C$, charcoal separation was carried out, followed by measurement in a liquid scintillation counter.

All patients were investigated by computed tomography before and after treatment. For the cerebral tumors, the extent of the edema zone related to the diameter of the tumor was reported. Blood samples for determination of catecholamines were taken from a cubital vein before operation and two and 24 hours afterwards. The catecholamines were determined radioenzymically.

Results

The highest tissue concentrations found on determination of the perifocal zones were in cases of metastases (39.5 ng/100 mg of tissue and 27.59 ng/100 mg tissue). These patients already showed before operation a marked improvement in hemiparesis and the state of consciousness under dexamethasone treatment (3 x 8 mg/day). Under dexamethasone treatment there was already a regression of the perifocal cerebral edema in the computed tomogram before operation. The level in the perifocal edema zone of oligodendroglioma was 27 ng/100 mg of tissue. This figure was between 17 and 25 ng/100 mg of tissue for astrocytomas. The values

Table 3. Dexamethasone concentration levels in various brain tumors

Classification	Patient	Age	Sex	Surrounding brain tissue concentration in ng/100 mg tissue
Astrocytoma[a]	TR	13	F	16.75
Astrocytoma	AC	40	M	19.30
Astrocytoma	KH	41	M	25.80
Astrocytoma	EW	34	M	20.00
Astrocytoma	WH	30	M	21.70
Astrocytoma[a]	PA	8	F	18.50
Astrocytoma	HG	41	F	19.00
Glioblastoma	EB	59	M	36.10
Glioblastoma	PT	45	M	6.40
Glioblastoma	UM	40	M	35.00
Glioblastoma	AB	70	M	32.00
Glioblastoma	DK	60	F	24.90
Oligodendroglioma	EA	46	M	27.00
Oligodendroglioma	HR	69	M	26.50
Metastases	MS	42	F	37.30
Metastases[a]	MP	45	F	38.43
Metastases	NH	27	M	27.59
Metastases	KB	51	M	39.50
Metastases	LH	47	M	10.48
Meningioma	DH	72	M	17.70
Pineal tumor[b]	NJ	56	M	22.00 ng/dl[o]

a, posterior fossa; b, CSF

for glioblastoma showed no difference from the astrocytomas on average. For meningioma, there was a relatively low value of 17.5 ng/100 mg of tissue. After operation, there were clinical signs of a frontal psychosyndrome, and the investigation by computed tomography showed morphological signs of an increase in local swelling of the brain, for which reason a bolus intravenous administration of 40 mg of dexamethasone was again necessary. The patient recovered on the fourth day. On CT follow-up, the swelling of the brain was seen to have regressed in the region of the operation. Table 3 shows the levels of dexamethasone concentration in the tissue for various types of tumors.

On bolus injection, the serum level is between 80 and 130 μg/dl, and is between 40 and 20 μg/dl after four to eight hours. As regards the pharmacokinetics and the therapeutic range of level in the serum, the results are shown in Table 4 and Figs. 2, 3 and 4.

Table 4. Results

No.	α (1/H)	β (1/H)	T_α (H)	T_β (H)
2	0.35	0.33	1.98	2.10
3	13.69	0.092	0.05	7.53
5	3.52	0.265	0.196	2.62
6	0.98	0.105	0.71	6.6
13	19.0	0.16	0.036	4.33

n = 5; $\bar{x}$ = 4.69 h; s = 2.39; Phase: Max. = 7.53 h; Min. = 2.10 h

The most important causes of cerebral edema are endogenous and exogenous effects which may be recognised morphologically in the computed tomogram of the cranium.

Twenty-one tumor patients showed varying extents of perifocal edema in the CT. After treatment with dexamethasone and operation, the rate of regression of the swelling varied. The regression of the zones of edema after giving dexamethasone was most marked with the metastases, less for glial tumors and the least for meningioma. In our patient, a morphological improvement was seen in the CT after a repeat bolus injection (Fig. 5).

Figure 6 shows the edema volume of brain tumor plotted against brain tissue concentration.

The basal levels of catecholamines in the plasma of the tumor patients before operation were within the normal range: noradrenaline 198 ± 24 pg/ml of plasma ($\bar{x}$ ± SEM), adrenaline 44 ± 13 pg/ml, n = 6. The plasma catecholamines were raised after operation, noradrenaline 385 ± 165 pg/ml, adrenaline 91 ± 25 pg/ml, n = 3. This is an observation which has also been made by BERGER et al., and HALTER and PFLUG. No decrease in the noradrenaline level was detected in our patients 24 hours after operation, but there was a drop in the adrenaline level into the range of the concentration before operation, noradrenaline 363 ± 106, adrenaline 46 ± 18 pg/ml, n = 3 (Fig. 7).

Discussion

According to the correlation of the computed tomography findings and the determination of the level of dexamethasone in tissue, the best results on edema treatment are achieved with metastases, less good in glial tumors and least good for meningiomas. On determination of the basal catecholamine level in the plasma, there was no substantial difference between patients with and those without dexamethasone treatment in our small number of patients. According to our pharmakokinetic investigation, the half-life of dexamethasone is 4.69 hours. After a bolus injection, a maintenance dose of 8 mg 4 to 6x per day in the first six days is recommended, in order to keep the dexamethasone concentration in the tissue at the same level for this period. The variable response of various types of tumor to dexamethasone may be explained by specific factors which are probably responsible for the opening of the membrane of the blood-brain barrier and intercellular electrolyte exchange.

It is possible with tumors to differentiate between the characteristics of an early and a late cerebral edema. The early edema develops as a result of the sudden occurrence of hypoxia with an increase in the permeability of the plasma membrane. The consequence is an initial accumulation of intracellular water associated with a decrease in the intercellular space, but without there being any clear signs of damage to the blood-brain barrier. The tight endothelial junctions of the vessels are still retained.

It is possible in this type of situation that dexamethasone is more effective, for example, for metastases with rapidly increasing cerebral edema than for more slowly growing meningioma (late edema). Our results on the determination of the levels of dexamethasone in tissue are consistent with this.

References

1. Berger, H.G., Kraas, E., Bittner, R., Lohmann, F.W.: Plasma catecholamine, Insulin und Glucose in der postoperativen Phase. Chirurg. 52, 225-230 (1981)
2. Drayer, B.P., Rosenbaum, A.E.: Brain edema defined by cranial computed tomography. J. Comput. Assist. Tomogr. 3, 317-323 (1979)
3. Fiscman, R.A.: Brain edema. N. Eng. J. Med. 293, 706-711 (1975)
4. Gerber, A.M., Salvolaine, E.R.: Modification of tumor enhancement and brain edema in computerized tomography by corticosteroid. Neurosurgery 6, 282-284 (1980)
5. Halter, J.B., Pflug, A.E.: Relationship of impaired insulin secretion during surgical stress to anesthesia and catecholamine release. J. Clin. Endocrinol. Metab. 51, 1093-1098 (1980)
6. Ignelzi, R.J.: Cerebral edema: Present perspectives. Neurosurgery 4, 338-342 (1979)
7. Klatzo, I.: Neuropathological aspects of brain edema. J. Neuropath. Exp. Neurol. 26, 1-14 (1967)
8. Meinig, G., Aulich, A., Wende, S., Reulen, H.J.: The effect of Dexamethasone and diuretics on peritumoral brain edema: Comparative study of tissue water content and CT. In: Dynamics of brain edema. Pappius, H.M., Feindel, W. (eds.), pp. 301-305. Berlin, Heidelberg, New York: Springer 1976
9. Nagel, M., Schümann, H.J.: A sensitive method for determination of conjugated catecholamines in blood plasma. J. clin. chem. clin. biochem. 18, 431-432 (1980)

10. Pappius, H.M.: Pathophysiology of cerebral edema: The search for the ideal therapeutic agent. In: Current controversies in neurosurgery. Morley, T.P. (ed.), pp, 587-594. Philadelphia: Saunders Co. 1976

11. Reulen, H.J., Tsuyumu, M., Tack, A., Fenske, A.R., Prioleau, G.R.: Clearance of edema fluid into cerebrospinal fluid. A mechanism for resolution of vasogenic brain edema. J. Neurosurg. 48, 754-764 (1978)

12. Vecsei, P., Akangbou, C., Joumaah, A., Sallum, N.I.: Studies on antibodies against corticoid hormones. Acta Endocr. (Kbh) 69 (Suppl. 159) 33 (1972)

13. Yamaguchi, M., Shirakata, S., Taomoto, K., Matsumoto, S.: Steroid treatment of brain edema. Surg. Neurol. 4, 5-8 (1975)

14. Yamada, K., Bremer, A.M., West, C.R.: Effects of dexamethasone on tumor-induced brain edema and its distribution in the brain of monkeys. J. Neurosurg. 50, 361-367 (1979)

15. Yu, Z.-Y., Wrange, Ö., Boëthius, J., Hatam, A., Granholm, L., Gustafsson, J.-A.: A study of glucocorticoid receptors in intracranial tumors. J. Neurosurg. 55, 757-760 (1981)

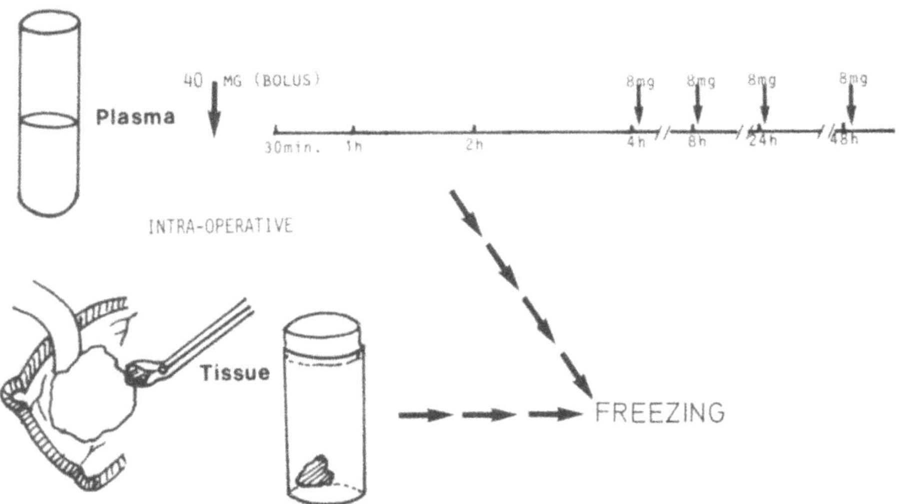

Fig. 1

dexa 5
Plasma dexamethasone concentration ng/100 ml
100
80
60
40
20
0
0
5
10
15
20
25
Hours

Fig. 2

E2
Plasma dexamethasone concentration ng/100 ml
3
2.5
2
1.5
1
0.5
0
0
2
4
6
8
10
Hours

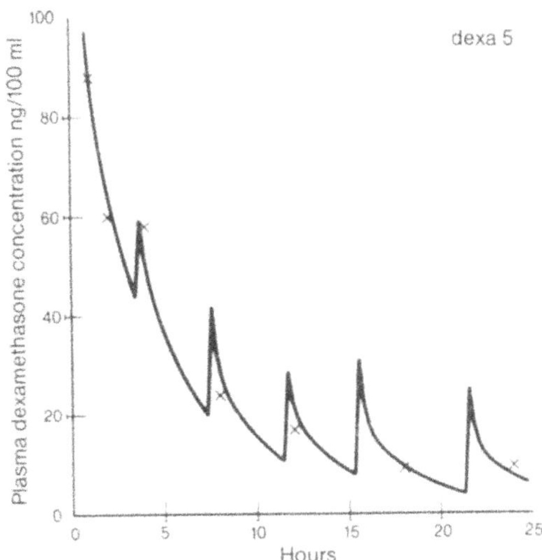

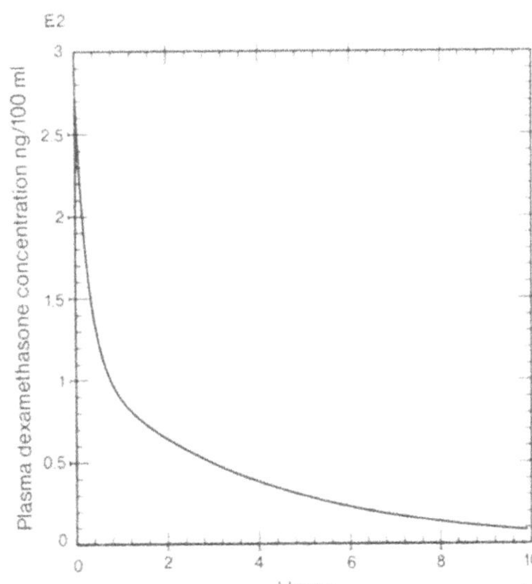

Fig. 3

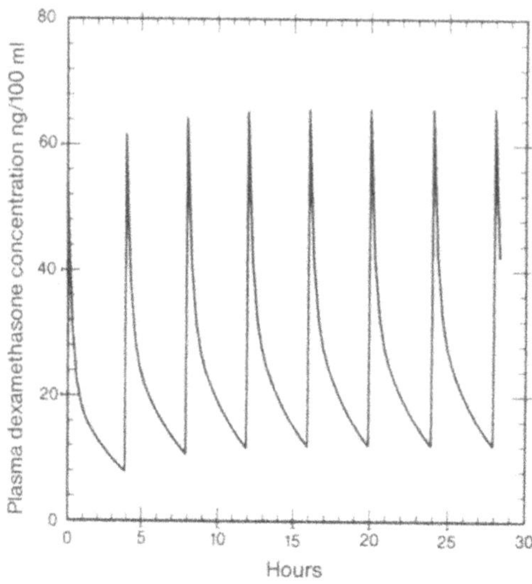

Fig. 4. Pharmacokinetics. Four-hourly dexamethasone

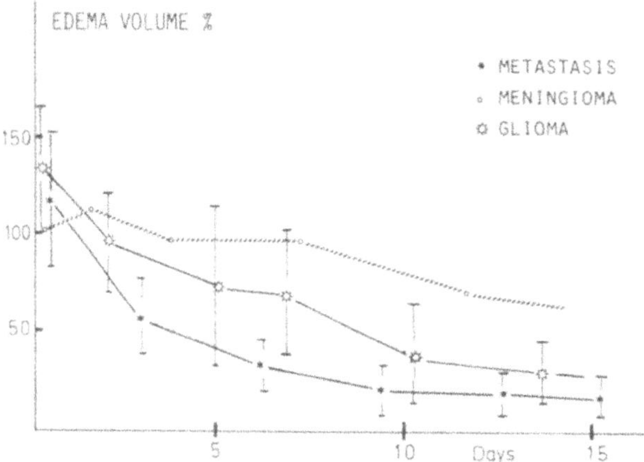

Fig. 5. Edema volume of brain tumor plotted against time after starting treatment with dexamethasone

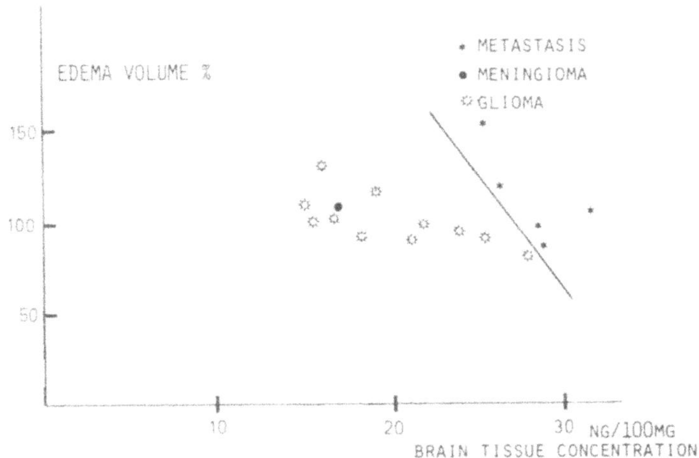

Fig. 6. Edema volume of brain tumor plotted against brain tissue concentration

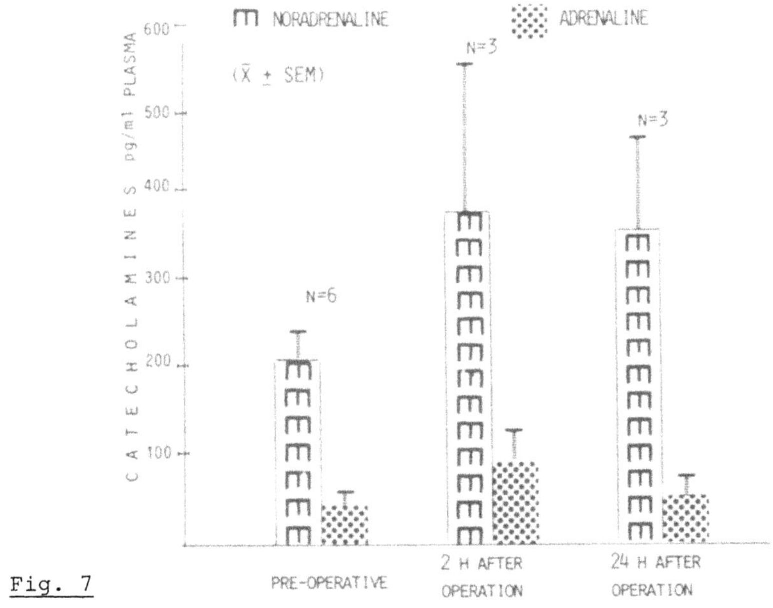

Fig. 7

Cerebral Oxygen Tension and Microcirculation in Barbiturate Treatment. An Experimental Investigation

B. Poch[1], V. Heller[1], M. R. Gaab[1], M. Sold[2], and H. E. Heissler[1]

[1] Neurochirurgische Klinik, Josef-Schneider-Strasse 11, D-8700 Würzburg
[2] Institut für Anästhesiologie, Josef-Schneider-Strasse 2, D-8700 Würzburg

Introduction

The results of barbiturate treatment after acute brain injury are encouraging. To elucidate the mechanism of barbiturate action, direct measurement of oxygen pressure in tissue (tpO_2) offers new possibilities. Neither the venous pO_2 nor the arterio-venous pO_2 difference are reliable values for the determination of brain oxygenation (6, 7). Only the measurement of tpO_2 *in situ* and the calculation of pO_2-histograms provides a good estimate of tissue oxygen supply. Therefore we used an eight channel surface-microelectrode (produced by Eschweiler, Kiel) (6). Looking at the rates of O_2 disappearance, after an externally induced total arrest of cerebral blood flow, allows direct conclusions to be drawn regarding the O_2 metabolism (9). Moreover, with the help of a H_2 clearance technique a simultaneous determination of local cerebral blood flow becomes possible (2, 10).

Methods

O_2/H_2 Electrode

The O_2 electrode consists of a reference electrode, made of Ag/AgCl, about 4 mm in diameter. In its centre Pt-wires, each 15 µm thick, are encased. A 12 µm teflon membrane separates the electrode surface from the tissue. A solution of KCl fills the space underneath the membrane. With the Pt-wires set to a voltage of 750 mV oxygen is transformed to OH^-. The current generated, measured by specially developed electronics, is proportional to the actual pO_2. By changing the voltage, it should be possible to measure H_2 with the same electrode, but as the resulting current of 0.5 pA/kPa is 100 times lower than in the case of O_2, the measurement is affected by many disturbances. For that reason we use another electrode with four 15 µm Pt-wires for measurement of pO_2, four 100 µm thick Pt-wires for H_2 measurements and a third 200 µm thick Pt-wire, that can be used as fifth H_2 electrode or, without a membrane, as H_2 generator for the combined O_2/H_2 determination (2, 10).

Animals and Experiments

We use male SPD rats weighing 250 to 400 g. They are anesthetised with 320 mg/kg Ketamine-HCl or 65 mg/kg Na-Pentobarbital, and breathe spontaneously. A standardized brain injury is produced by focal freezing, as described previously (1, 3). Twenty-four hours later, the rats are

Advances in Neurosurgery, Vol. 11
Edited by H.-P. Jensen, M. Brock, and M. Klinger

trepanned between the coronal, sagittal and lambdoid sutures. PO_2 histograms are then determined under normoxic (21% O_2) and hyperoxic (100% O_2) conditions, *outside* the necrosis. For subsequent recording of the O_2 disappearance rate, the blood flow to the brain is suddenly stopped by a cuff, fixed around the neck of the animal, that is blown up to 55 kPa within one second. This experiment is carried out under hyperoxic conditions.

EEG and ECG were recorded in some animals. From the acquired data, including atmospheric pressure, temperature of rat and gauging fluid and a calibration program before and after each measurement, the computer (IN 110) calculates the pO_2-histograms, O_2 disappearance rates and their first derivations, as well as all statistical values and displays the results as a graphic and a numerical print-out (8).

The H_2 clearance measurements start with the rats breathing a mixture of 8% H_2, 20% O_2 and 72% N_2. After saturation of the tissue with hydrogen the gas mixture is changed to air and the half-life ($t_{1/2}$) of H_2 is determined (Fig. 4a, b).

Results

Regarding normal brain and normoxia the pO_2-histograms are normal and show only minor differences in narcosis (Fig. 1a, b). Breathing of 100% O_2 shifts the pO_2-histogram to the right (5). During Ketanest anesthesia, this shift is significantly ($p < 0.01$) less pronounced than with Na-Pentobarbital (Fig. 1c, d). The fall in O_2 pressure during complete ischemia (a direct measure of the momentary, regional O_2 consumption (9)) is significantly steeper ($p < 0.01$) under Ketanest anesthesia than under Na-Pentobarbital (Fig. 3a, b). The CBF measurement with the H_2 clearance technique shows a 30% shorter $t_{1/2}$ under Ketanest than under Na-Pentobarbital (Fig. 4a, b).

Compared to normal rats the traumatized animals exhibit a significant ($p < 0.01$) *right shift* of their pO_2-histograms. It means that areas with a *low* tpO_2 *decrease* and areas with a *high* tpO_2 *increase*. This right shift is less pronounced under Ketanest (Fig. 2a, b). The O_2 disappearance rates after complete ischemia indicate a significantly ($p < 0.01$) lower O_2 utilization in the traumatized brains. Whereas in traumatized animal it is impossible to demonstrate regions with normal O_2 utilization under Ketanest anesthesia, this is at least partially possible with Na-Pentobarbital anesthetised rats (Fig. 3c,d).

Discussion

Our results demonstrate, that Ketanest increases O_2 metabolism of the brain. Under normal conditions the O_2 supply to the brain is not compromised, as the increased metabolism is compensated by a higher blood flow. After complete cerebral ischemia the amount of O_2 is more rapidly diminished under Ketanest, because of the higher metabolism even in the healthy brain. This is demonstrated by the steeper course of the O_2 disappearance curves in our cuff experiments (Fig. 3a, b).

After cold brain injury, a well-known technique for producing traumatic, vasogenic edema (1, 3), a shift of the pO_2-histograms *to the right* and low oxygen disappearance rates after total ischemia indicate diminished O_2 utilization. This could be the result of a primary disturbance of aerobic glycolysis. Thus O_2 availability in traumatized brain tissue *is not diminished*, at least as long as intracranial pressure is normal. The pO_2-histograms together with the cuff experiments show that barbiturate treatment after acute brain injury protects the metabolism in the injured tissue at least partially, while the metabolism of the normal areas is diminished.

From these results *one may argue against the therapeutic use of hyperbaric oxygen* (4) for treating brain injury, as the O_2 availability is not diminished. We suppose that the right shift of the pO_2-histogram demonstrates an *essential* difference of this primary vasogenic edema compared with an ischemic edema. According to our first H_2 measurements, a simultaneous record of local cerebral blood flow will improve the capabilities of this dynamic method of investigation.

Summary

The aim of our experiments is to obtain information about O_2-pressure-gradients, O_2 utilization and local cerebral blood flow in normal and injured brains. Therefore the experiments are carried out using polarographic micromethods. Compared to Na-Pentobarbital, O_2 metabolism in the healthy brain is significantly decreased by barbiturate. In vasogenic edema a *right shift* of the pO_2-histogram is seen and O_2 disappearance after acute total cerebral circulatory arrest is retarded. This means that O_2 utilization is impaired in vasogenic brain edema, perhaps by a primary disturbance of the aerobic O_2 metabolism. Contrary to the use of Ketanest the O_2 metabolism is partly preserved with Na-Pentobarbital. The simultaneous measurements of local cerebral blood flow using H_2 clearance technique considerably increases the capabilities of this dynamic technique for further investigation of pathophysiological and therapeutical concepts in brain edema.

References

1. Gaab, M., Pflughaupt, K.W.: Experimentelle und klinische Untersuchungen zur intravenösen Glycerintherapie beim Hirnödem. Acta neurochir. 37, 17-31 (1977)
2. Haining, J.L., Don Turner, M., Pnatall, R.: Measurement of local cerebral blood flow in the unanaesthetized rat using a hydrogen clearance method. Circ. Res. 13, 313-324 (1968)
3. Herrmann, F., Gaab, M., Pflughaupt, K.W., Gruss, P.: Medikamentöse Therapie beim experimentellen Hirnödem. Neurochirurgia 24, 39-46 (1981)
4. Holbach, K.H., Gött, U.: Effect of hyperbaric oxygenation therapy (HOT) on neurosurgical patients. In: Proceeding of the 4th Intern. Congr. on hyperbaric medicine. WADA, J., Iva, T. (eds.), pp. 442-447. Tokyo: Igaku Shoin 1970
5. Leniger-Follert, E., Lübbers, D.W., Wrabetz, W.: Regulation of local tissue pO_2 of the brain cortex at different arterial O_2 pressures. Pflügers Arch. 359, 81-95 (1975)
6. Lübbers, D.W.: Local tissue PO_2: Its measurement and meaning. In: Oxygen supply. Kessler, M., Bruley, D.F., Lübbers, D.W., Silver, I.A., Strauss, J. (eds.), pp. 151-155. München, Berlin, Wien: Urban und Schwarzenberg 1973
7. Lübbers, D.W.: Die Bedeutung des lokalen Gewebesauerstoffdruckes und des PO_2-Histogrammes für die Beurteilung der Sauerstoffversorgung eines Organes. Prakt. Anästh. 12, 184-193 (1977)

8. Ödman, S., Lund, N.: Data acquisition and information processing in MDO oxygen electrode measurement of tissue oxygen pressure. Acta anaesth. Scand. 24, 161-165 (1980)

9. Reneau, D.D., Halsey, J.H.: Interpretation of oxygen disappearance rates in brain cortex following total ischaemia. In: Oxygen transport to tissue, Vol. 3. Silver, I.A., Erecinska, M., Bicher, H.I. (eds.), pp. 189-198. New York: Plenum Press 1978

10. Wise Young, M.D.: H_2 Clearance measurement of blood flow: a review of technique and polarographic principles. Stroke 11, 552-564 (1980)

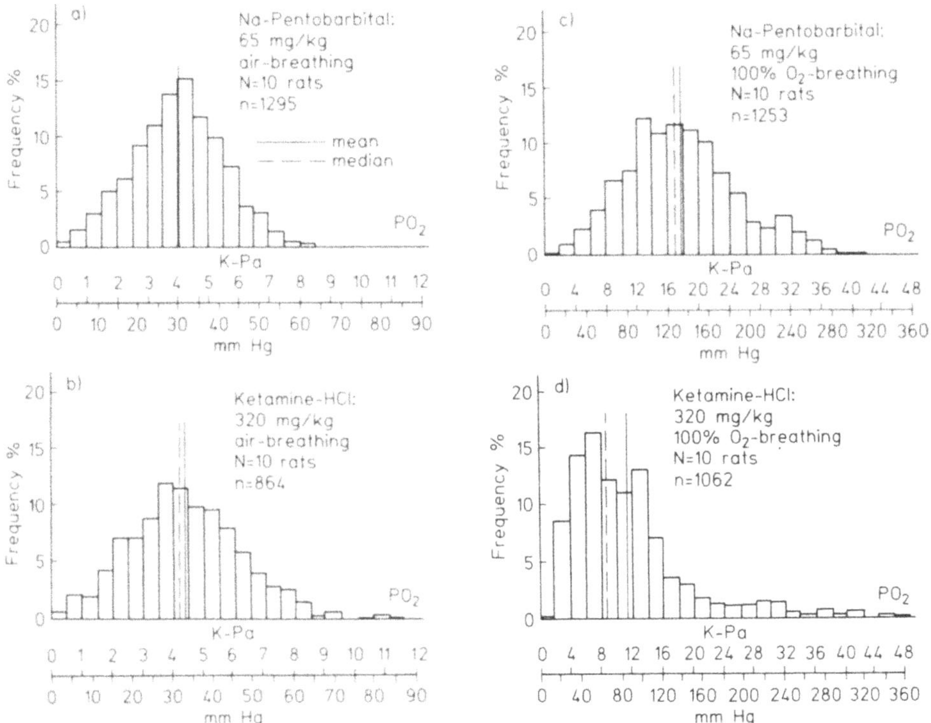

Fig. 1 a-d. PO_2-histograms of normal rat brains anesthetised with Na-Pentobarbital (a, d), compared to those narcotised with Ketamine-HCl (b, d). a and b are measured under normoxic, c and d under hyperoxic conditions. Significantly higher O_2 tension under Barbiturate anesthesia in hyperoxia due to increased brain metabolism with Ketamine

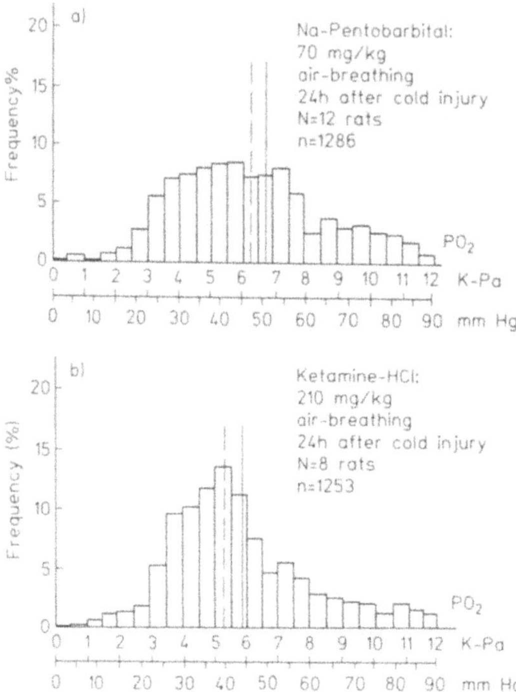

Fig. 2 a-b. PO_2-histograms after cold brain injury. In this edema the pO_2 tension is *increased* (right shift, comp. Fig. 1); this shifting of the histogram to the right is more pronounced with Pentobarbital (a) than with Ketamine (b)

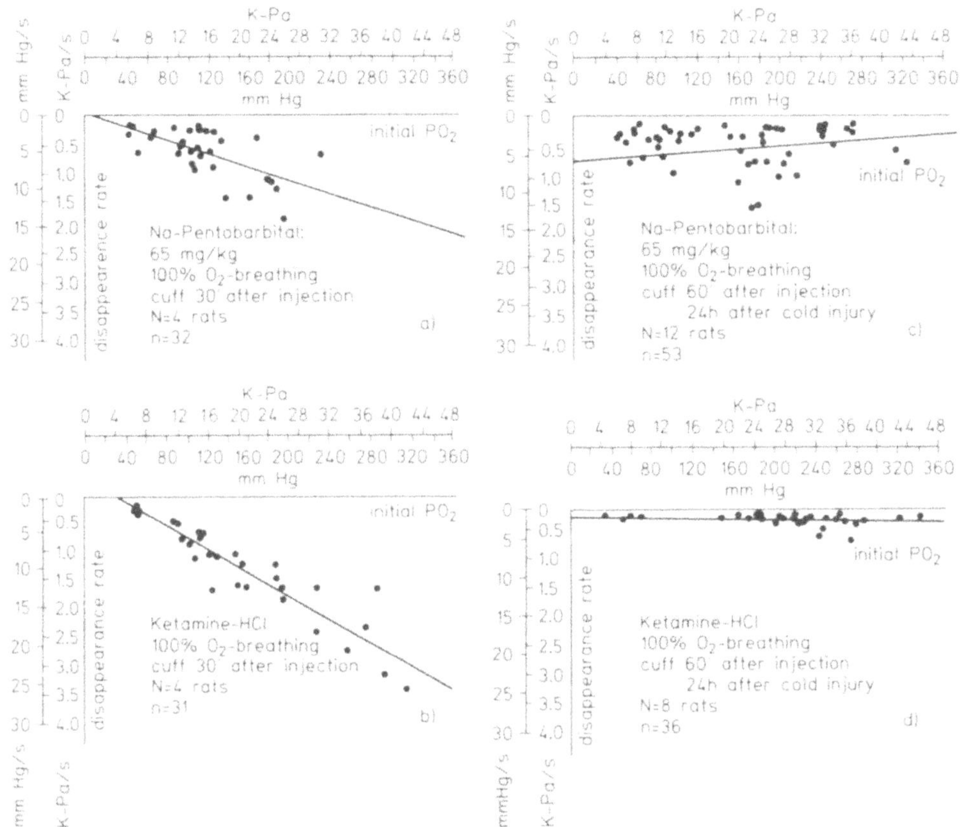

Fig. 3 a-d. The O_2 disappearance rate after sudden complete brain ischemia (cuff inflation) is smaller under Na-Pentobarbital (a) than under Ketamine-HCl (b). After cold injury the O_2 disappearance rate under Pentobarbital (c) and Ketanest (d) is diminished. Na-Pentobarbital shows, however, some areas of normal O_2 utilization

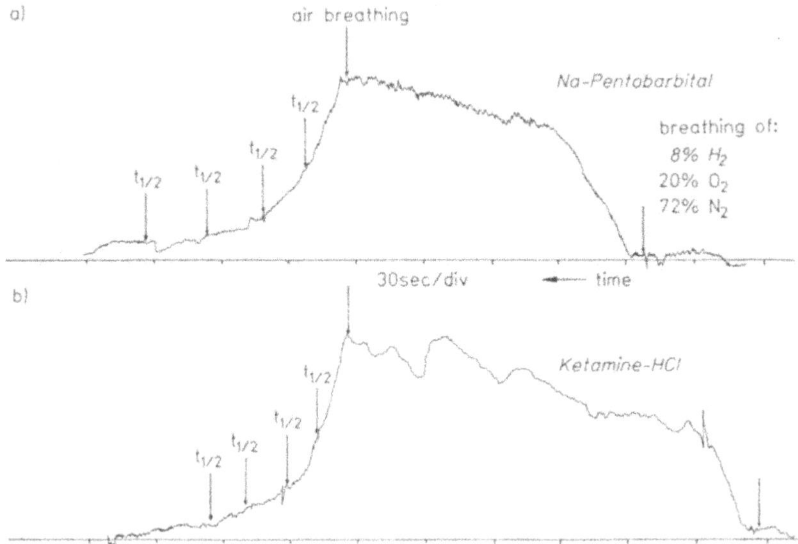

Fig. 4 a-b. H_2 clearance curves under Na-Pentobarbital (a) and Ketamine-HCl (b) in the healthy brain. Changing the breathing to normal air, after H_2 saturation, the half-life of H_2 is shorter under Ketanest than under Na-Pentobarbital, due to a higher blood flow under Ketanest anesthesia

Barbiturate Effect on Intracranial Pressure After Experimental Cold Lesions

D. Stolke, U. Kunz, and H. Dietz
Neurochirurgische Klinik der Medizinischen Hochschule Hannover, Karl-Wiechert-Allee 9, D-3000 Hannover 61

Within recent years many reports have been published dealing with the beneficial influence of barbiturates in experimental and clinical ischemic cerebral injuries when administered before, during or shortly after the onset of the ischemic or traumatic event (1, 2, 6, 7, 12, 14-16, 21).

These protective properties were explained as being primarily a consequence of reduction in cerebral metabolic rate, of a decrease of cerebral blood flow and oxygen consumption (3, 5, 12-14, 15). Our own studies produced results which demonstrated that barbiturates are able to stabilize lysosomal membranes in experimental cold lesions (18-20). MARSHALL et al. (9-11) reported a decrease of ICP in severely brain injured patients when given barbiturate therapy. We added, to explain this effect, the stabilizing effect of the barbiturates on subcellular membranes. This effect was regarded as a barrier against a spreading brain edema and therefore an increasing ICP.

As we had always seen a decrease in systemic blood pressure (SBP) in severely brain injured patients using barbiturates clinically we did the following experimental study.

Material and Methods

We had two experimental groups of ten cats each of 2 1/2 - 4 kg body weight. The first group was anesthetised by 30 mg/kg body weight, the second by thiopental (60 mg/kg body weight).

Half of this dose per hour was given intravenously over the total experimental period of four hours. The cats were relaxed and under controlled ventilation.

A cryogenic lesion was made, through a burr hole onto the intact dura mater, according to the KLATZO method (8). The right parietal lobe was cooled down to -70° C for 45 seconds. A Statham element was placed directly onto the frozen area. The bone defect around the element was closed by dental cement.

Intracranial pressure (ICP) and systemic blood pressure (SBP) were observed and recorded over a four hour period.

Advances in Neurosurgery, Vol. 11
Edited by H.-P. Jensen, M. Brock, and M. Klinger

Results

From the beginning until the end of the experimental period the SBP of the animals under ketamine anesthesia is significantly higher (about 100 - 125 mm Hg) than that of the cats under thiopental anesthesia (about 70 - 90 mm Hg).

In both groups SBP is higher at the beginning and decreases slightly towards the end of the period (Fig. 1).

During the whole period ICP shows a more pronounced increase and remains higher in the ketamine group compared with the animals under thiopental anesthesia (Fig. 2).

The statistical analysis reveals a very significant close correlation between ICP and SBP in both groups but on different levels (Figs. 3 and 4).

Conclusions

Regarding the close correlation between ICP and SBP we have to return to KLATZO's results which say that developing brain edema spreads in close correlation to SBP, i.e. the higher the SBP the more extensive the brain edema. ICP is expected to increase with the spread of the brain edema.

But ICP is not to be regarded as purely dependent on blood pressure. CLUBB and co-workers (4) studied ICP under hypotension with sodium nitroprusside. They recorded no decrease of ICP - but on the contrary a definite increase. Nitroprusside was regarded as a cerebral vasodilator and was therefore contra-indicated in cases of brain edema and space-occupying lesions.

The beneficial effect of barbiturates on ICP observed in our experimental studies is explained by the reduction in cerebral metabolic rate, in the decrease of oxygen consumption, in the decrease of cerebral blood flow as reported by other authors and by the membrane stabilizing effect on lysosomal organelles as reported in our own studies. Only these additional effects make possible the beneficial correlation of SBP and ICP in severely brain injured patients under the influence of barbiturates.

References

1. Bleyaert, A.L., Nemoto, E.M., Safar, P., Stezoski, S.W., Mossy, J., Rao, G.R., Micekll, J.: Thiopental amelioration of postischemic encephalopathy in monkeys. Acta Neurol. Scand. 56, Suppl. 64, 144-145 (1977)
2. Breivik, H., Fabritius, R., Lind, B., Lust, P., Mullie, A., Orr, M., Renck, H., Safar, P., Snyder, J.: Brain resuscitation; clinical feasibility trials with barbiturates. Crit. Care Med. 6, 93 (1978)
3. Cucchiara, R.F., Michenfelder, J.D.: The effect of interruption of the reticular activations system on metabolism in canine cerebral hemisphere before and after thiopental. Anesthesiol. 39, 3-12 (1973)
4. Clubb, R.J., Maxwell, R.E., Chou, S.N.: Experimental brain injury in the dog. The pharmacological effects of pentobarbital and sodium nitroprusside. J. Neurosurg. 52, 189-196 (1980)

5. Fink, B.R., Haschke, R.H.: Anesthetic effects on cerebral metabolism. Anesthesiology 39, 199-215 (1972)

6. Hoff, J.-T., Pitts, L.H., Spetzler, R., Wislon, C.B.: Barbiturates for protection from cerebral ischemia in aneurysm surgery. Acta Neurol. Scand. 56, Suppl. 64, 158-159 (1977)

7. Hoff, J.-T., Smith, A.L., Hankinson, H.L., Nielson, S.L.: Barbiturate protection from cerebral infarction in primates. Stroke 6, 28-33 (1975)

8. Klatzo, J., Wisniewsky, H., Steinwall, O., Streicher, E.: Dynamics of cold injury. In: Brain edema. J. Klatzo (ed.), pp. 554-563. Berlin, Heidelberg, New York: Springer 1967

9. Marshall, L.F., Bruce, D.A., Bruno, L., Schut, L.: Role of intracranial pressure monitoring and barbiturate therapy in malignant intracranial hypertension. J. Neurosurg. 47, 481-484 (1977)

10. Marshall, L.F., Shapiro, H.M.: Barbiturate control of intracranial hypertension in head injury and other conditions: iatrogenic coma. Acta Neurol. Scand. 56, 157-157 (1977)

11. Marshall, L.F., Shapiro, H.M., Smith, R.W.: Barbiturate treatment of intracranial hypertensive states. In: Neural trauma. Popp, A.J., Bourke, R.S., Nelson, L.A., Kimmelberg, H.K. (eds.), pp. 347-351. New York: Raven Press 1979

12. Michenfelder, J.D., Theye, R.A.: Cerebral protection by thiopental during hypoxia. Anesthesiology 39, 510-517 (1973)

13. Nemoto, E.M., Kofke, W.A., Wessler, P., Hossmann, K.A., Stezoski, W.S., Safar, P.: Studies on the pathogenesis of ischemic brain damage and the mechanism of its amelioration by thiopental. Acta Neurol. Scand. 56, 142-143 (1977)

14. Nilson, L.: The influence of barbiturate anesthesia upon energy state and upon acid-base parameters of the brain in arterial hypotension and in asphyxia. Acta Neurol. Scand. 47, 233-253 (1971)

15. Nordström, C.-H., Calderini, G., Rehncrona, S., Siesjö, B.K.: Effects of pentobarbital anesthesia on post-ischemic cerebral blood flow and oxygen consumption in the rat. Acta Neurol. Scand. 56, Suppl. 64, 146-147 (1977)

16. Simeone, F.A., Frazer, G., Lawner, B.S., Lawner, P.: Ischemic brain edema, comparative effects of barbiturates and hypothermia. Stroke 10, 8-12 (1979)

17. Smith, A.L., Hoff, J.T., Nielson, S.L., Larson, P.: Barbiturate protection in acute focal ischemia. Stroke 5, 1-7 (1974)

18. Stolke, D., Dietz, H.: Membrane stability. Changes of subcellular organelles after barbiturate treatment. In: Advances in Neurosurgery, Vol. 9. Schiefer, W., Klinger, M., Brock, M. (eds.), pp. 331-337. Berlin, Heidelberg, New York: Springer 1981

19. Stolke, D., Dietz, H.: Barbiturate influence on subcellular organelles in experimental cold lesion. Advances in Neurosurgery, Vol. 10. Driesen, W., Brock, M., Klinger, M. (eds.), pp. 259-265. Berlin, Heidelberg, New York: Springer 1982

20. Stolke, D., Seidel, B.U., Hartmann, N.: Barbiturate treatment and membrane stability of subcellular organelles. Anaesthesist 29, 539-541 (1980)

21. Yatsu, F.U., Diamond, J., Graziano, C., Lindquist, P.: Experimental brain ischemia; protecting from irreversible damage with a rapid-acting barbiturate (Methohexital). Stroke 3, 726-732 (1972)

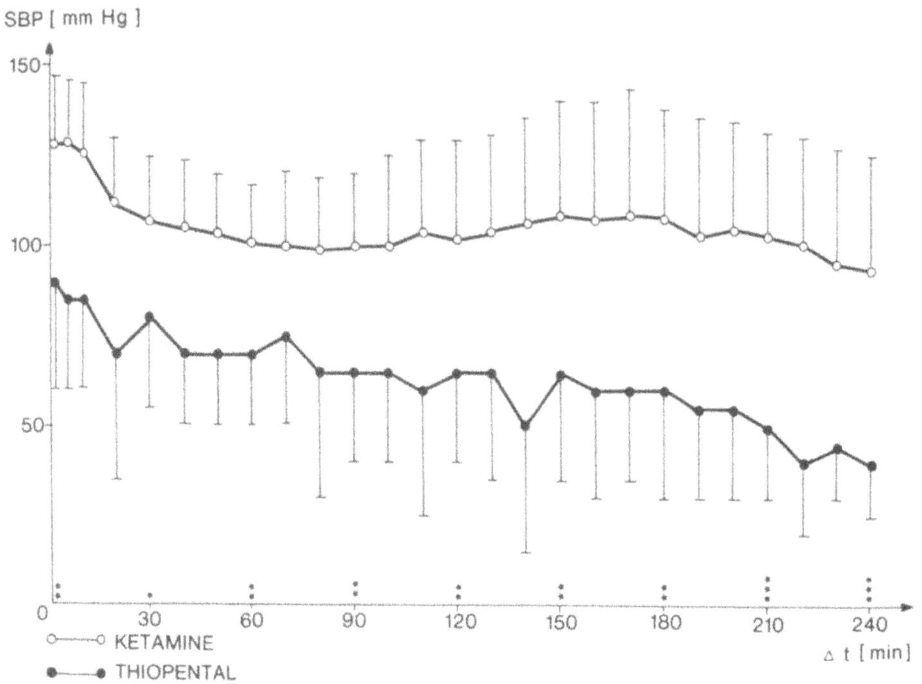

Fig. 1. Course of systemic blood pressure (SBP) over the experimental period. Comparison of animals with ketamine and thiopental anesthesia

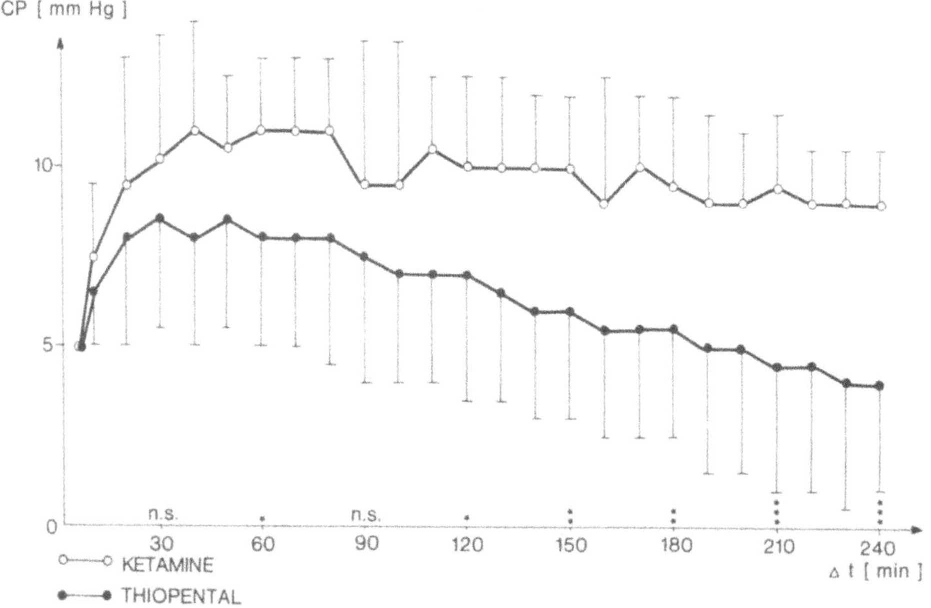

Fig. 2. Course of intracranial pressure (ICP) over the experimental period. Comparison of the animals with ketamine and thiopental anesthesia

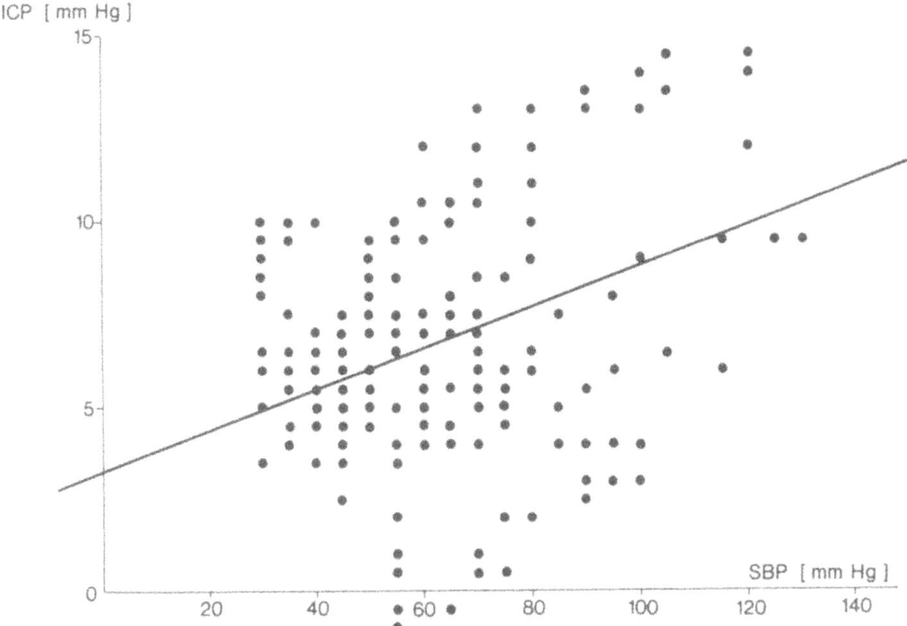

Fig. 3. Correlation of ICP and SBP under thiopental anesthesia

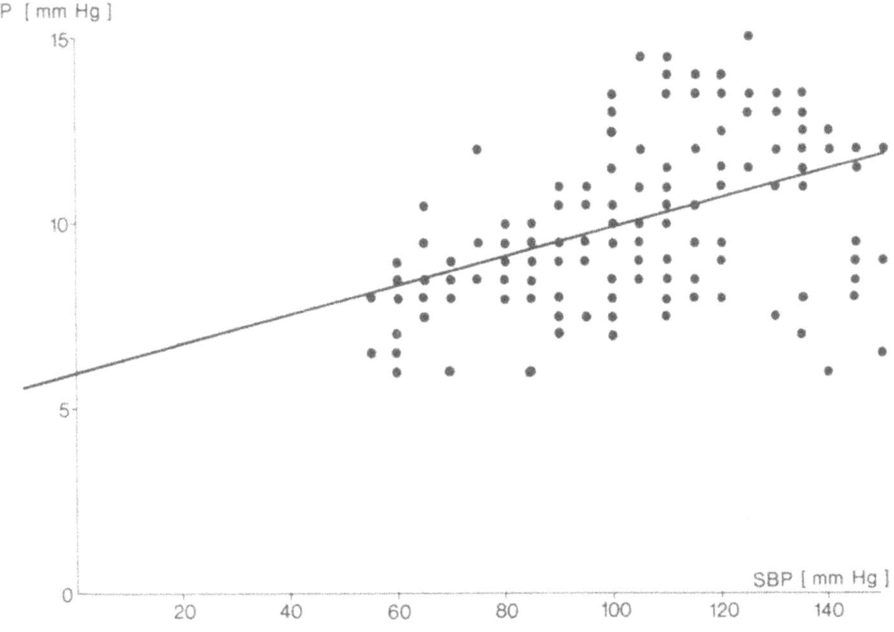

Fig. 4. Correlation of ICP and SBP under ketamine anesthesia. Comparison of Fig. 3 and Fig. 4 reveals the shifting of the values to a higher (Fig. 4) and lower level (Fig. 3)

Does Barbiturate "Mono-Therapy" Favorably Influence Brain Survival After Traumatic Coma?

P. Sehested, O. Fedders, M. Tange, and J. Overgaard*
Department of Neurological Surgery, Odense University Hospital, DK-5000 Odense C

Introduction

Increasing evidence suggesting possible brain protecting properties of barbiturates has accumulated over the past 10 to 15 years (5, 9, 10, 12, 14, 15). In 1970 OVERGAARD et al. (12) recommended the use of phenobarbitone together with hyperventilation in the early management of patients with severe head injuries. The object of treatment was to prevent or reduce brain edema. Based on the observation, that the short acting thiopental could abruptly reduce elevated intracranial pressure (ICP) during anesthesia (14), SHAPIRO et al. employed barbiturates and hypothermia to reduce increased ICP in five critically ill patients in whom unexpected neurological recoveries occurred (15). Other investigators have since confirmed, that thiopental is useful in reducing ICP in intensive care patients (5). Further results from the San Diego group concerning the ICP-reducing effect of short acting barbiturates have recently been reported (9, 10). Treatment resulted in a higher than expected rate of good quality survival, however, in addition to barbiturates, the patients were also aggressively treated with high doses of steroids and mannitol.

The approach of the present study is quite different from recent reports and deals with treatment by barbiturates alone. The object was to examine the effect upon outcome of phenobarbital, when sedative doses were given to patients in traumatic coma.

Material and Methods

This study was carried out on a total of 143 patients with severe head injury admitted during the years 1977 and 1978. All patients were alive on admission and fulfilled the criteria of traumatic coma (16). None of the patients have been excluded from the study. Male to female ratio was 2.3 to 1, and the median age was 18 years (range: 1 to 85 years; interquartile range: 10 to 48 years). Fifty-four per cent were one to 20 years old and 65% were younger than 30 years of age. There were 82% who were transferred from 15 different local hospitals. The majority of the patients (83%) arrived at the Neurosurgical Department within six hours of the injury. Approximately one-third (35%) arrived less than two hours after the injury.

*The authors acknowledge the valuable discussions with SHARON A. BOWERS, M. Sc. and Professor LAWRENCE F. MARSHALL, M.D. of The San Diego Head Injury Study Group

Advances in Neurosurgery, Vol. 11
Edited by H.-P. Jensen, M. Brock, and M. Klinger

Forty per cent had focal, surgical lesions, that were promptly treated, while the remainder had diffuse brain lesions or contusions. Apart from surgery, routine general therapeutic measures were:

1. Endotracheal intubation plus controlled ventilation to moderately hypocapnic levels;
2. Sedative doses of slowly eliminated barbiturate (phenobarbitone = phenemal);
3. External drainage of ventricular cerebrospinal fluid; and
4. No steroids or osmotic diuretics.

Phenobarbitone was given for sedation, primarily for elimination of abnormal motor activity, especially extensor and spastic flexion responses. Individual doses were 100 to 200 mg. Rapid bolus injections were never given, and phenobarbitone administration was not related to raised ICP. The members of staff were free to make their own choice whether or not to give barbiturates. We are therefore not reporting a randomized study of barbiturate in head injury. Different dosage levels of phenobarbitone were decided in two periods, i.e. within 24 hours and within seven days of the injury. Serum levels of phenobarbitone were estimated in the same periods.

Conditions Tested for Barbiturate Effect

The effect of barbiturate treatment was tested against the early deaths related to primary brain injury which occurred within the first 30 days after injury. Mortality rates were determined in the two periods mentioned above and according to the treatment given, i.e. with or without phenobarbitone. Thus, testing for differences in mortality rates between *treatments* and *periods* was the fundamental principle of the statistical analysis. The approach was supported by the fact, that the population, according to survival rates (Fig. 1), was almost stable after the first week of injury, with 100 of the patients still alive.

Recognizing that several factors influence the outcome, patients have been distributed according to: coma score on arrival, age, and type of cranio-cerebral injury. Analysis has shown, that appropriate subgroups were:

1. Coma score 3 to 4, 5 to 7, and 8;
2. Age $\leq$ 20 years, and age >20 years; and
3. Diffuse brain lesions or contusions, and focal, surgical lesions.

Therefore, patients could be arranged hierarchically in different prognostic groups according to the treatment given.

Statistics

Differences between subgroups of the three prognostic factors, distributed according to treatments, were tested by the X^2-test. Early, i.e. within 30-days, mortality rates were calculated in each subgroup, and for each of the three factors influencing outcome, a comprehensive statistic was used to test for differences in mortality rates between treatments. For these statistics, Cochran's test was chosen. In this test, differences in mortality rates between subgroups were pooled in form of a weighted mean with the property, that more weight was given to subgroups with larger numbers. The calculated critical ratio: $\bar{d}/SE(\bar{d})$, is approximately a standardized normal deviate (or its square a X^2-deviate with 1 df) (1).

In addition, differences in subgroup-mortality between treatments have been tested by a X^2-test, allowing a direct comparison of mortality rates in patients with the same prognostic features. The strength of relationship was measured as the differences of percentages. The interpretation is straightforward. Thus, the differences of differences of percentages between the two 'periods' indicates whether the relationship increases or decreases in strength, or, whether the direction of the differences obtained in this way is uniform or not (3).

The same procedure was adopted in testing the influence of treatment upon outcome. It was found appropriate to condense the five outcome groups (13) into two broad categories:

1. Recovery = good recovery + slight deficits; and
2. Non-recovery = severe deficits + dead.

All tests used are two-tailed with a probability level less than 5% ($p < 0.05$) taken as significant.

Results

Deaths related to the primary brain injury, which occurred within the first month after injury, were 30%. In June 1981, approximately 3 1/2 years after injury, the patients were re-examined and classified into five outcome groups according to the method of OVERGAARD et al., (13). This approach is not fully comparable to JENNETT and BOND's outcome scale (7). At follow-up examination 27% had made a good recovery, 13% had slight deficits, 23% had severe deficits, and the mortality rate had increased to 37%. None were vegetative (Fig. 2).

In addition to general principles for intensive care of comatose head injured patients, 114 (80%) were treated with various amounts of phenobarbitone within 24 hours of the injury, whereas 29 (20%) were given no barbiturates. Only four patients exceeded a total of 1000 mg phenobarbitone. Table 1 shows the mortality rates at different levels of barbiturate medication. There were no differences in mortality between different dosage levels in the treated group ($X^2 = 1.37$, df = 2, $p>0.5$). When medication was tested against no medication, a significantly higher mortality rate was found among the latter ($X^2 = 23.90$, df = 1, $p < 0.001$).

The number of patients who were being treated increased during the first week to 120 (84%). Among these patients, 88 (73%) were given more than 1000 mg phenobarbitone. Mortality rates varied considerably among the different dosage levels, being 8 - 14% in patients receiving 1000 to 3000 mg phenobarbitone against 34 - 35% in patients outside this dosage level. These percentages are significantly different ($X^2 = 10.04$, df = 3, $p < 0.02$). By contrast, the mortality rate in the non-treated group was 78%, which was significantly higher than in the treated group ($X^2 = 27.60$, df = 1, $p < 0.001$).

In 90 (75%) of the 120 patients who were treated, serum-phenobarbitone levels were measured on day 7. Serum-levels higher than 25 mg/l were found in 63 (70%) of these patients. There were no differences in mortality rates between groups with measurable serum-levels ($X^2 = 2.73$, df = 2, $p > 0.1$).

Table 2 lists subgroup-distributions of the three prognostic factors as well as the mortality rates within the subgroups. Additional distinctions have been made between treatments and 'periods'. Within

Table 1. Patients and mortality rates distributed according to different dosage-levels and serum-levels of phenobarbitone. Distributions of dosage-levels and their associated mortality rates are listed in 2 'periods', i.e. within 24 hours and within 7 days of the injury, whereas different serum-levels plus mortality rates were determined approximately at day 7. Figures in brackets are 95% confidence limits

Dosage-levels, Serum-levels, and periods	No. of cases	Dead No.	%
Phenobarbital 0 - 24 hours			
Phenobarbital 0 mg	29	20	69 (49 - 85)
Phenobarbital 1 - 500 mg	76	17	22 (14 - 33)
Phenobarbital 501 - 1000 mg	34	6	18 (7 - 35)
Phenobarbital $\geq$ 1001 mg	4	0	0 (0 - 60)
Phenobarbital 0 - 7 days			
Phenobarbital 0 mg	23	18	78 (56 - 93)
Phenobarbital 1 - 1000 mg	32	11	34 (19 - 53)
Phenobarbital 1001 - 2000 mg	35	5	14 (5 - 30)
Phenobarbital 2001 - 3000 mg	36	3	8 (2 - 22)
Phenobarbital $\geq$ 3001 mg	17	6	35 (14 - 62)
Serum-phenobarbital 0 - 7 days			
No phenobarbital	23	18	78 (56 - 93)
Serum-phenobarbital not analyzed	30	12	40 (23 - 59)
Serum-phenobarbital 0 - 25 mg/l	27	5	19 (6 - 38)
Serum-phenobarbital 25 - 50 mg/l	46	4	9 (2 - 21)
Serum-phenobarbital > 50 mg/l	17	4	24 (7 - 50)

24 hours of the injury, a significantly higher proportion of adverse prognostic features was found among coma scores ($X^2 = 17.36$, df = 2, $p < 0.001$) and age-groups ($X^2 = 5.45$, df = 1, $p < 0.02$). Only traumatic pathology was equally distributed between treatments ($X^2 = 2.14$, df = 1, $p > 0.1$). Taking this skewness of subgroup-distributions into consideration, mortality rates were significant lower among treated patients (coma-scores: $\bar{d}/SE(\bar{d}) = 3.17$, $p < 0.002$; age-groups: $\bar{d}/SE(\bar{d}) = 4.62$, $p < 0.001$; and traumatic pathology: $\bar{d}/SE(\bar{d}) = 4.91$, $p < 0.001$).

At day 7 the only subgroup-distribution showing a significant difference between treatments, was the distribution of coma scores ($X^2 = 21.22$, df = 2, $p < 0.001$). Mortality rates, however, showed differences between treatments at the same levels as for the first 24 hours (coma-scores: $\bar{d}/SE(\bar{d}) = 3.24$, $p < 0.002$; age-groups: $\bar{d}/SE(\bar{d}) = 5.13$, $p < 0.001$; and traumatic pathology: $\bar{d}/SE(\bar{d}) = 5.50$, $p < 0.001$).

Table 3 shows the results of subgroup analysis, patients with the same prognostic features being tested for differences in mortality rates according to the treatment given. Only significant relationships, which all are in favour of phenobarbitone treatment, are shown. Furthermore, a comparison of 'periods' reveals, that the differences of differences of percentages ranged from -4% to +16%. With the exception of coma score 3 to 4, they were all positive, indicating an increase, during the first week, in strength of the relationships between survival and phenobarbitone administration. A positive relationship between phenobarbitone treatment and recovery of brain function was also found, although a slight decrease in the strength at day 7 was observed.

Table 2. The three factors influencing outcome - coma score on arrival, age, and traumatic pathology - distributed into subgroups and mortality rates within subgroups. Additional distinctions have been made between *treatments* and *periods*. The contracted outcome groups, i.e. *recovery* = good recovery + slight deficits; and *non-recovery* = severe deficits + dead, have likewise been distributed between treatments and 'periods'

Subgroups, outcomes, and groups (post-injury)	Treatment No. of cases	Dead No.	%	No Treatment No. of cases	Dead No.	%
Phenobarbital 0-24 hours						
Coma scores 3 - 4	28	14	50	19	18	95
5 - 7	72	9	13	9	2	22
8	14	0	0	1	0	0
Age groups ≦ 20 years	67	8	12	10	7	70
> 20 years	47	15	32	19	13	68
Traumatic pathology						
Contusions	72	10	14	14	10	71
Focal lesions	42	13	31	15	10	67
Outcome groups						
Recovery	54	-	-	3	-	-
Non-recovery	60	-	-	26	-	-
Phenobarbital 0-7 days						
Coma scores 3 - 4	30	16	53	17	16	94
5 - 7	75	9	12	6	2	33
8	15	0	0	0	0	0
Age groups ≦ 20 years	67	8	12	10	7	70
> 20 years	53	17	32	13	11	85
Traumatic pathology						
Contusions	73	10	14	13	10	77
Focal lesions	47	15	32	10	8	80
Outcome groups						
Recovery	54	-	-	3	-	-
Non-recovery	66	-	-	20	-	-

Discussion

The object of this study was to examine whether or not pharmacological monotherapy with sedative doses of the slowly eliminated phenobarbitone influenced outcome after traumatic coma. Evidence suggests a favorable effect of treatment upon both survival and recovery of brain function. Not only had treated patients a higher survival rate than non-treated patients with similar prognostic features, but the strength of relationship between treatment and survival increased almost uniformly during the first week after injury, where most of the deaths related to the primary brain injury had occurred. The increase however is slight suggesting that the additional advantage of starting barbiturate treatment beyond the first 24 hours after injury is relatively small. The stability of the population at day 7 indicates, that phenobarbitone

Table 3. Results of bivariate analysis between treatments and outcome. Patients with subgroups - and therefore similar as regards prognostic features - were analysed in 2 x 2 Tables according to the treatment given and to outcome. Thus, the calculated X^2 value has 1 df. The differences (in percentages) of survival rates (or recovery rates) between *treated* patients and *non-treated* patients, measures the strength of the relationship between the prognostic variable and the condition tested. The differences of %-differences between the two 'periods' measures whether the relationship increases or decreases in strength, or whether the direction of these differences is uniform or not. The slight variations among %-differences listed in Table 3, and those, which can be obtained directly from Table 2, are due to rounding errors

Condition tested	Prognostic variable	Subgroup	X^2	P	% Difference
Survival	Phenobarbital treatment 0-24 h		23.90	< 0.001	49
Survival	Phenobarbital treatment 0-24 h	Coma score 3-4	8.47	< 0.01	45
Survival	Phenobarbital treatment 0-24 h	Age $\leqq$ 20 years	15.18	< 0.001	58
Survival	Phenobarbital treatment 0-24 h	Age > 20 years	5.96	< 0.02	37
Survival	Phenobarbital treatment 0-24 h	Contusions	18.64	< 0.001	58
Survival	Phenobarbital treatment 0-24 h	Focal lesions	4.47	< 0.05	36
Recovery	Phenobarbital treatment 0-24 h		11.72	< 0.001	37
Survival	Phenobarbital treatment 0-7 days		27.60	< 0.001	57
Survival	Phenobarbital treatment 0-7 days	Coma score 3-4	6.54	< 0.02	41
Survival	Phenobarbital treatment 0-7 days	Age $\leqq$ 20 years	15.18	< 0.001	58
Survival	Phenobarbital treatment 0-7 days	Age > 20 years	9.75	< 0.01	53
Survival	Phenobarbital treatment 0-7 days	Contusions	21.30	< 0.001	63
Survival	Phenobarbital treatment 0-7 days	Focal lesions	6.05	< 0.02	48
Recovery	Phenobarbital treatment 0-7 days		6.94	< 0.01	32

had also a positive effect upon recovery of brain function. It is in fact possible at this point to predict final outcome groupings with a high degree of confidence (17).

It was not possible to ascertain, whether the lowered mortality among treated patients was dose dependent, although significantly fewer deaths occurred among patients receiving 1000 to 3000 mg phenobarbitone during the first week after injury.

At present, no randomized clinical trial concerning barbiturate therapy in severely brain injured patients has been published. A recent report from San Diego (10) suggest that short-acting barbiturates administered, if the ICP was uncontrollably raised using conventional treatment, resulted in an unexpectedly high rate of good quality survival. Their approach was quite different from that in the present study; high doses of the short-acting pentobarbital were employed after aggressive therapy with hyperventilation, steroids and mannitol with the aim of lowering a critically high ICP had been unsuccessful. Our use of phenobarbitone in moderate doses was not related to raised ICP, and neither steroids nor mannitol were given. Thus, apart from general therapeutic measures for intensive care of comatose head injured patients, phenobarbitone was the only pharmacological means employed for the purpose of brain protection. In our opinion, mannitol has dangerous side effects, and troubles in controlling increased ICP after its overuse have been reported (4).

There are no indications available in our data about the possible beneficial mechanisms of phenobarbitone treatment. However it is known, that most of the reduction in cerebral metabolism caused by barbiturates occurs at low doses and prior to marked suppression of spontaneous cortical electrical activity (11). The early administration of phenobarbitone could therefore protect the brain by a reduction of its energy requirements. Anyhow, it would be difficult to explain why treated patients had a lower mortality rate than non-treated patients with similar characteristics, without postulating some brain protecting properties of sedative doses of phenobarbitone.

The distribution of final outcome groups is in accordance with results published from other centers (2, 4, 6, 8). This finding suggests, that similar results can be obtained without the use of steroids and mannitol.

Conclusion

This study reveals that moderate doses of phenobarbitone were positively related to survival and recovery of brain function in the early management of comatose head-injured patients. It also suggests that steroids and mannitol can be done without, as the total outcome distribution in this series was not different from other recent studies of comparable patients. Considering this, future randomized clinical trials dealing with barbiturate therapy of comatose brain-injured patients, could then be studies of pharmacological monotherapy and therefore less confusing.

Summary

Influencing the outcome after severe head injury by means of medical treatment is still controversial. During the years 1977 and 1978 143 patients were admitted, all of whom on arrival fulfilled the criteria of traumatic coma. Median age was 18 years (range: 1 to 85 years). Forty per cent had focal, surgical lesions which were promptly treated.

In June 1981 the patients were re-examined and final outcome grouping assessed. Principles of general treatment were:

1. Endotracheal intubation and moderate hyperventilation;
2. Sedative doses of a slowly eliminated barbiturate (phenobarbitone = phenemal);
3. External drainage of ventricular cerebrospinal fluid; and
4. No steroids or osmotic diuretics.

The effect of treatment upon outcome was examined according to distributions of prognostic features - coma score on admission, age, and traumatic pathology - in two different periods, i.e. within 24 hours and within 7 days after the injury. Our findings show, that treatment with sedative doses of phenobarbitone was significantly related to both survival, and to recovery of brain function. Furthermore, the strength of relationship regarding survival increased during the first week of injury. Significantly fewer deaths occurred in patients receiving 1000 to 3000 mg phenobarbitone within the seven days after injury.

References

1. Armitage, P.: Statistical methods in medical research. Oxford: Blackwell Scientific Publications 1971
2. Becker, D.P., Miller, J.D., Ward, J.D., Greenberg, R.P., Young, H.F., Salakas, R.: The outcome from severe head injury with early diagnosis and intensive management. J. Neurosurg. 47, 491-502 (1977)
3. Blalock, H.M.: Social statistics. Tokyo: McGraw-Hill Kogakusha Ltd. 1972
4. Bowers, S.A., Marshall, L.F.: Outcome in 200 consecutive cases of severe head injury treated in San Diego County: A prospective analysis. Neurosurgery 6, 237-242 (1980)
5. Collice, M., Rossanda, M., Beduschi, A. et al.: Management of head injury by means of ventricular fluid pressure monitoring: In: Intracranial Pressure III. Beks, J.W.F., Bosch, D.A., Brock, M. (eds.), pp. 101-109. Berlin, Heidelberg, New York: Springer 1976
6. Gennarelli, Th.A., Spielman, G.M., Langfitt, Th.W., Gildenberg, P.L., Harrington, T., Jane, J.A., Marshall, L.F., Miller, J.D., Pitts, L.H.: Influence of the type of intracranial lesion on outcome from severe head injury. A multicenter study using a new classification system. J. Neurosurg. 56, 26-32 (1982)
7. Jennett, W.B., Bond, M.: Assessment of outcome after severe brain damage: A practical scale. Lancet 1, 480-484 (1975)
8. Jennett, W.B., Teasdale, G., Galbraith, S., Pickard, J., Grant, H., Braakman, R., Avezaat, C., Maas, A., Minderhout, J., Vecht, C.J., Heiden, J., Small, R., Caton, W., Kurze, T.: Severe head injuries in three countries. J. Neurol. Neurosurg. Psychiat. 40, 291-298 (1977)
9. Marsh, M.L., Marshall, L.F., Shapiro, H.M.: Neurosurgical intensive care. Anesthesiology 47, 149-163 (1977)
10. Marshall, L.F., Smith, R.W., Shapiro, H.M.: The outcome with aggressive treatment in severe head injuries. Part II: Acute and chronic barbiturate administration in the management of head injury. J. Neurosurg. 50, 26-30 (1979)

11. Michenfelder, J.D.: The interdependence of cerebral function and metabolic effects following massive doses of thiopental in the dog. Anesthesiology 41, 231-236 (1974)

12. Overgaard, J., Haase, J., Hein, O., Laursen, B., Rønde, F., Barnkop, F., Hinke, H.: Trafiktraumatiske hjernelæsioner. Ugeskr. Læger 132, 695-705 (1970)

13. Overgaard, J., Hvid-Hansen, O., Land, A.M., Pedersen, K.K., Christensen, S., Haase, J., Hein, O., Tweed, W.A.: Prognosis after head injury based on early clinical examination. Lancet 2, 631-635 (1973)

14. Shapiro, H.M., Galindo, A., Wyte, S.R. et al.: Rapid intraoperative reduction of intracranial hypertension with thiopentone. Br. J. Anaesth. 45, 1057-1062 (1973)

15. Shapiro, H.M., Wyte, S.R., Loeser, J.: Barbiturate-augmented hypothermia for reduction of persistent intracranial hypertension. J. Neurosurg. 40, 90-100 (1974)

16. Teasdale, G., Jennet, W.B.: Assessment of coma and impaired consciousness: A practical scale. Lancet 2, 81-84 (1974)

17. Teasdale, G., Parker, L., Murray, G., Knill-Jones, R., Jennett, G.: Predicting outcome of individual patients in the first week after severe head injury. Acta Neurochir. Suppl. 28, 161-164 (1979)

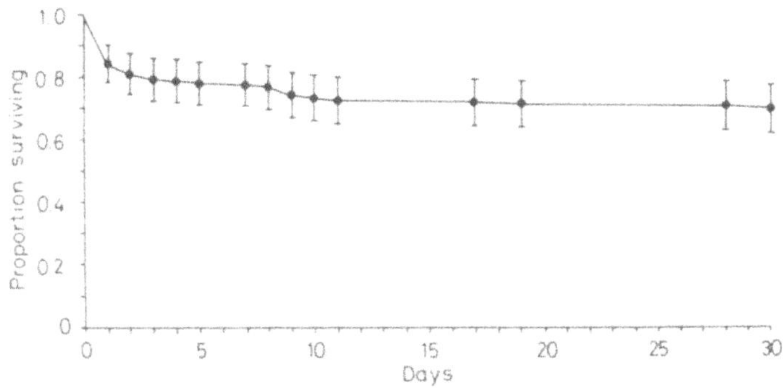

Fig. 1. Survival rates within 30 days of injury in 143 patients with severe head injury. Vertical bars are 95% confidence limits

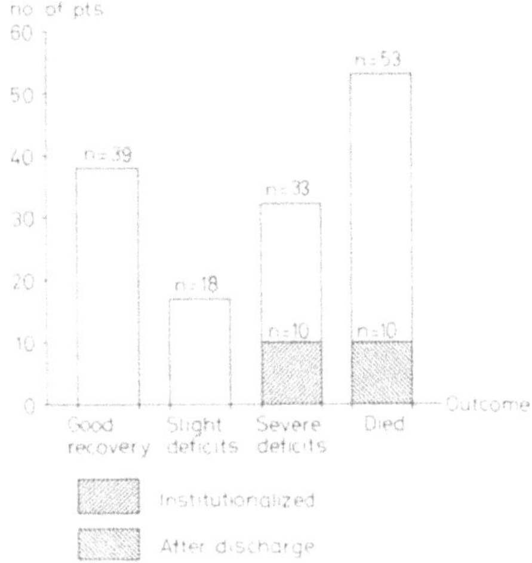

Fig. 2. Final outcome grouping of 143 patients with severe head injury based on re-examination approximately 3 1/2 years after the injury

Caloric Stimulation Pattern in Traumatic Coma and Some rCBF Changed Induced by Caloric Vestibular Stimulation

U. Schønsted-Madsen, G. Fogstrup, J. Overgaard, and T. Rosendal
Department of Neurological Surgery, Odense University Hospital, DK-5000 Odense C

Introduction

Oculo-vestibular reflexes can be elicited in the comatose by caloric stimulation with ice-cold water. As early as 1910 ROSENFELDT (11) concluded that the type of reaction was a measure of the depth of coma. In 1952 KLINGON (5) investigated brain stem vascular disorders, in 1958 ETHELBERG and VERNET (13) studied supratentorial intracranial tumors and in 1962 BLEGVAD (1) wrote on barbiturate intoxication and oculo-vestibular reflexes. Head injured patients were studied in 1972 by ZILSTORFF and POULSEN (10). BLEGVAD as well as the two last mentioned authors did their studies with water at 30° C.

In this communication oculo-vestibular reactions elicited by caloric stimulation (HALLPIKE) are compared with reactions elicited by ice-cold water.

Methods

The patients we have studied were comatose after head injury. The management procedures included nasal intubation plus mechanical ventilation to moderate hypocapnia. Sedation was by slowly eliminated barbiturate (phenobarbitone). Blood levels were analysed daily.

Irrigation of the external auditory meatus was performed on both sides. Three separate levels of water temperature were used, 0° C, 30° C and 44° C. Irrigation lasted forty seconds and the amount of water used was 250 ml. Eyeball reactions and pupillary reaction were observed through Frenzel's glasses. There was an interval of five minutes between stimulations.

In two patients regional cerebral blood flow was measured by the intracarotid bolus injection of Xenon 133.

Results

A total of 402 independent caloric stimulations was performed as described above. A reaction was elicited in 238 instances whereas no reactions were observed after 164 stimulations.

These reactions were followed during the entire period of coma which lasted from 8 to 23 days. Reactions to caloric stimulation changed

Advances in Neurosurgery, Vol. 11
Edited by H.-P. Jensen, M. Brock, and M. Klinger

from one day to another but generally the reaction returned to normal with time.

Despite the many differences between patients all six patients studied showed nystagmus in response to the caloric stimulation, just before any awakening could be recorded. However, the general pattern of development was: No observeable reaction, internuclear paresis, conjugate eye movements, nystagmus. There were other types of reaction which occurred in an unsystematic way, and which we have called: atypical reactions.

The types of reaction were unrelated to serum levels of phenobarbitone.

The reactions elicited by right and left sided caloric stimulation are compared in Table 2. The type of reaction we have called "atypical reactions" contained in all cases an ipsilateral pupillary dilation associated with motor unrest when irrigation was done with ice-cold water and shortly before recovery of consciousness. In these patients the reaction to irrigation with water at 30° C and 44° C elicited horizontal eyeball undulations or vertical nystagmus. In the two patients who had CBF studied during caloric stimulation, regional CBF was unchanged after ipsilateral stimulation with water at 30° C and 44° C. Stimulation performed with ice-cold water elicited a short lasting increase of cerebral blood flow whereas intracranial pressure was unchanged. Only the external auditory canal ipsilateral to the cerebral hemisphere studied had the caloric stimulation done.

Table 1. Types of vestibulo-ocular reaction elicited by caloric stimulation in traumatic coma

No reaction
Internuclear palsy
Conjugate eye movements
Nystagmus
Atypical reactions

Table 2. Types of vestibulo-ocular reaction elicited by caloric stimulation in traumatic coma

Water temperature	0° C	30° C	44° C
No reaction	35	51	85
Internuclear palsy	13	5	4
Conjugate eye movements	48	54	21
Nystagmus	29	19	20
Atypical reactions	9	5	4
Total	134	134	134

Table 3. Comparison of vestibulo-ocular reactions elicited by caloric stimulation at three different temperatures in traumatic coma

Water temperatures	Reactions Identical	Different	None	Total
0° C / 30° C	24	40	70	134
0° C / 44° C	10	62	62	134
30° C / 44° C	17	44	73	134

Discussion

The reactions elicited by external caloric stimulation associated with pathological levels of consciousness are usually explained by various degrees of damage to the brain stem. SZENTAGOTHAI (12) argues that normal nystagmus can only be elicited when three interconnected neurons are intact and he also adds multineuronal connections. This circuit arises in the crista ampullaris from where it passes to the oculomotor nucleus and consists of three different neurons of which the second probably involves the medial longitudinal fasciculus. LORENTO de NO (6-8) demonstrated that alternative pathways were the median sectors of the reticular formation in the brain stem.

According to this an increasing disturbance of function in the reticular formation should result in an abolition of the fast components of the nystagmus and only slow conjungate reactions can be elicited.

It is further argued that a lesion of the medial longitudinal fasciculus is responsible for the internuclear paralysis under which only an ipsilateral movement is observed (2).

The six patients had all these reactions elicited for a shorter or longer period after head injury. The pronounced changing of reaction pattern is not completely understood. Serum barbiturate levels have been pointed out as being responsible for such changes in reactivity (BLEGVAD 1962) but we are not sure about this explanation, because all serum levels of barbiturate in these patients were high, but not toxic. The elicited pupillary dilation produced by irrigation with ice-cold water is hardly explained by the above-mentioned reflex mechanism. We do not regard the pupillary dilation as a specific vestibular reaction but are more inclined to relate this reaction to nociceptive mechanisms (9). Irrigation of the external auditory meatus with ice-cold water may therefore not be a specific stimulus for testing vestibular reactivity.

References

1. Blegvad, B.: Caloric vestibular reaction in unconscious patients. Arch. Otolanyngol. 75, 504-514 (1962)
2. Cogan, D.G.: Neurology of the ocular muscles. Springfield/Ill.: Thomas 1948
3. Ethelberg, S., Vernet, K.: Vestibulo-ocular reflex disorders of the anterior internuclear paralysis type in supratentorial space-taking lesions. Acta Psychiat. Scand. 33, 268-282 (1958)
4. Jørgensen, P.B.: Clinical deterioration prior to brain death. Acta Neurochirurgica 28, 29 (1973)
5. Klingon, G.H.: Caloric stimulation in localization of brain stem lesions in a comatose patient. A.M.A. Arch. Neurol. Psychiat. 68, 233-235 (1952
6. Lorente de Nó, R.: Untersuchungen über die Anatomie und der Physiologie des Ohrlabyrinthes und des nervus octavus. Mschr. Ohrenheilk. 61, 1066, 1152 and 1300 (1927)
7. Lorento de Nó, R.: Vestibulo-ocular reflex arc. Arch. Neurol. Psychiat. 30, 245 (1933)
8. Lorento de Nó, R.: Analysis of the activity of the chains of internuncial neurons. J. Neurophysiol. 1, 207 (1938)

9. Plum, F., Posner, J.B.: The diagnosis of stupor and coma. F.A. Davis Comp., Philadelphia, USA. 2nd ed. 33-53, 19..

10. Poulsen, J., Zilstorff, K.: Prognostic value of the caloric-vestibular test in the unconscious patient with cranial trauma. Acta Neurol. Scand. 48, 282-292 (1972)

11. Rosenfeld, M.: Untersuchungen über den calorischen Nystagmus bei Gehirnkranken mit Störungen des Bewußtseins. 2. Ges. Neurol. Psychiat. 3, 271 (1910)

12. Szentagothai, J.: The elementary vestibulo-ocular reflex arc. J. Neurophysiol. 13, 395 (1950)

Early Management and Observations in Traumatic Coma. The Significance of Moderate Hyperventilation and the Occurrence of Sudden Blood Pressure Falls

O. Fedders, M. Tange, P. Sehested, and J. Overgaard

Department of Neurological Surgery, Odense University Hospital, DK-5000 Odense C

Introduction

Controversy still exists regarding the early management of comatose head injured patients and the significance of intensive observations. This presentation deals with the effect on outcome of mechanical ventilation to moderate hypocapnic levels and the occurrence of sudden falls in blood pressure.

Material and Method

A total of 143 patients were admitted during a two year period. All patients fulfilled the criteria of traumatic coma (2).

A X^2 test was used to test for differences with a probability level less than 5% ($p < 0.05$) taken as significant.

Figure 1 shows the neurological observations ranked as the sum score of the Glasgow Responsiveness scale. There were no differences ($p>0.1$) in the sizes of the groups. One third of the patients scored from 3 to 4 and approximately 50% scored from 5 to 7. Two thirds were male and the mean age was 28 years. Half of the patients were younger than 20 years of age. Forty per cent had promptly treated focal, surgical lesions and the others had diffuse brain lesions.

Apart from operation, management procedures followed simple principles for the intensive care of comatose brain injured patients. This included controlled mechanical ventilation to moderate hypocapnic levels, i.e. a stabilised $PaCO_2$ of 30 to 35 mm Hg within one hour, daily water and electrolyte balance etc.

Steroids and osmotic diuretics were not used.

Among standard procedures, moderate doses of a slowly eliminated barbiturate (phenobarbitone) were used, primarily for sedation and elimination of abnormal motor activity, especially extensor and spastic flexor responses.

Results

The duration of respirator treatment was inversely related to the coma score on admission (Fig. 2), and ranged from 12 days in coma score 4 to four days in coma score 8.

Advances in Neurosurgery, Vol. 11
Edited by H.-P. Jensen, M. Brock, and M. Klinger

Figure 3 shows the re-awakening characteristics according to traumatic pathology. The figure shows the mean duration of the coma period and the re-awakening phases. In addition, the mean duration of respirator treatment within the groups has been added. The respirator treatment was usually stopped, when the patient first opened his eyes. Focal surgical lesions had the shortest average coma period and re-awakening phase in the subgroups of traumatic pathology.

Analysis has shown that three factors influence outcome: Coma score on admission, age and the type of cranio-cerebral injury. Appropriate subgroups are:

1. Coma score 3 to 4, 5 to 7 and 8,
2. Age < 20 years, age > 20 years, and
3. Contusions and focal, surgical lesions.

Early primary brain injury related to deaths occurring within 30 days of the injury, was 30%. Final outcome grouping was based on re-examination approximately 3 1/2 years after injury. There was a good recovery in 27%, 13% had slight deficits, 23% had severe deficits and the mortality rate was increased to 37% (1).

Table 1 shows that 104 (73%) patients were treated with controlled ventilation on the respirator, whereas 39 (27%) patients were not treated. The mortality rate among treated patients were significantly lower than among patients not treated ($p < 0.02$).

Subgroup-analysis, patients with the same prognostic features being tested for differences of mortality rates according to the treatment, showed a significant higher survival rate regarding coma score 3 to 4 ($p < 0.05$) and contusions ($p < 0.01$).

Final outcome groupings showed a significant difference ($p < 0.05$) in favour of respirator treatment, but mainly because of a reduction in mortality.

Table 1. Respirator treatment. Subgroups of prognostic factors and mortality rates within subgroups distributed according to treatment. Final outcome subgroups distributed between treatments

Prognostic groups, subgroups, and final outcome	Treatment No. of cases	Dead No.	%	No treatment No. of cases	Dead No.	%
Coma scores 3 - 4	28	15	54	19	17	89
5 - 7	66	10	15	15	1	7
8	10	0	0	5	0	0
Age groups < 20 years	57	8	14	20	7	35
> 20 years	47	17	36	19	11	58
Traumatic pathology						
Contusions	61	9	15	25	11	44
Focal lesions	43	16	37	14	7	50
Final outcome						
Recovery	44	-	-	60	-	-
Non-recovery	13	-	-	26	-	-

The observation during the early phase after the injury, i.e. within 24 hours, that sudden falls in blood pressure (BP falls) occurred, the hypothesis developed, that this event might influence outcome.

Recognizing, that the blood pressure varied considerably between patients, the effect of the following criteria for sudden falls in BP on outcome was investigated:

1. A sudden fall in systolic BP, i.e. within 15 minutes, of 80 mm Hg or more;
2. A mean arterial fall in BP within 15 minutes, from more than 80 to 60 mm Hg or less.

The falls in BP could not be related to cerebral perfusion pressure, deviations in $PaCO_2$ level or extracerebral lesions.

Fifty-three (37%) patients fulfilled the criteria of sudden falls in BP and the mortality rate was 50%. By contrast, the mortality rate was 19% in patients who did not experience falls in BP (Table 2).

Table 2. Subgroups of prognostic factors and mortality rates distributed according to whether sudden falls in BP occur or not. Final outcome-subgroups distributed according to the same criteria

Prognostic groups, subgroups, and final outcomes	No sudden fall in BP			Sudden fall in BP		
	No. of cases	Dead No.	%	No. of cases	Dead No.	%
Coma scores 3 - 4	23	13	57	24	19	79
5 - 7	53	4	8	28	7	25
8	14	0	0	1	0	0
Coma score 5 - 7						
Recovery	33	-	-	8	-	-
Non-recovery	20	-	-	20	-	-
Age groups < 20 years	57	8	14	20	7	35
> 20 years	33	9	27	33	19	58
Age > 20 years						
Recovery	8	-	-	0	-	-
Non-recovery	25	-	-	33	-	-
Traumatic pathology						
Contusions	58	7	12	28	13	46
Focal lesions	32	10	31	25	13	52
Contusions						
Recovery	39	-	-	8	-	-
Non-recovery	19	-	-	20	-	-
Focal lesions						
Recovery	9	-	-	1	-	-
Non-recovery	23	-	-	24	-	-

These figures are significantly different ($p < 0.001$). Further it is shown that falls in BP were more frequent in those aged more than 20 years ($p < 0.01$).

The occurrence of falls in BP in the following subgroups, were significantly associated with a higher mortality rate: GCS 5 to 7 ($p < 0.05$), age > 20 years ($p < 0.05$), and contusions ($p < 0.01$).

The effect of falls in BP on recovery of brain function = good recovery plus slight deficits were likewise associated with a significantly lower recovery rate: all patients ($p < 0.001$), GCS 5 to 7 ($p < 0.01$), age > 20 years ($p < 0.05$), contusions ($p < 0.01$), and focal, surgical lesions ($p < 0.05$).

Conclusion

This study shows that endotracheal intubation and controlled mechanical ventilation was positively related to survival, but had no effect upon recovery of brain function i.e. good recovery + slight deficits. On the other hand the proportion of severely disabled patients was not increased.

In addition the occurrence of sudden falls in BP as arbitrarily defined here, strongly suggests a profound adverse effect upon survival as well as recovery of brain function. However, the observations need further investigations concerning the pathogenesis and pathophysiological consequences, since the phenomenon predominantly seems to occur in the very early phase after injury.

The advantage of its prevention is obvious.

References

1. Overgaard, J., Hvid-Hansen, O., Land, A.M.: Prognosis after head injury based on early clinical examination. Lancet 2, 631-635 (1973)
2. Teasdale, G., Jennett, W.B.: Assessment of coma and impaired consciousness: A practical scale. Lancet 2, 81-84 (1974)

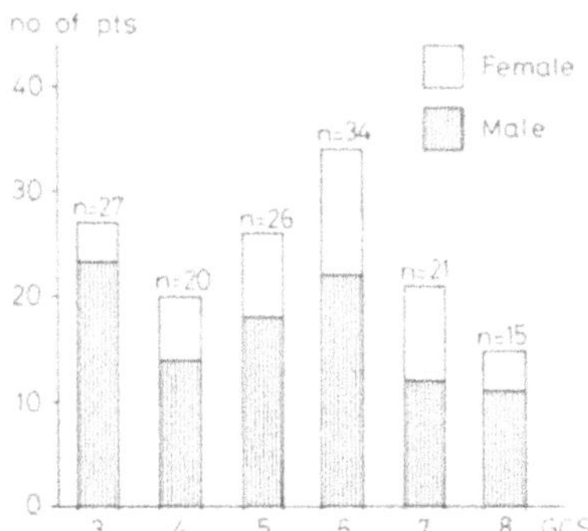

Fig. 1. GCS / sex distribution

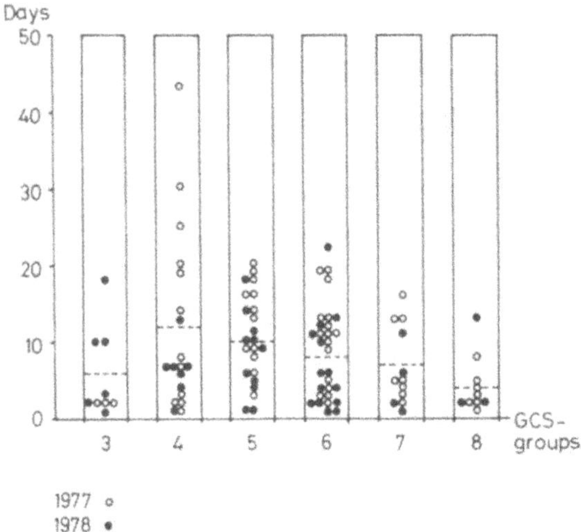

Fig. 2. Duration of respirator treatment in GCS-groups

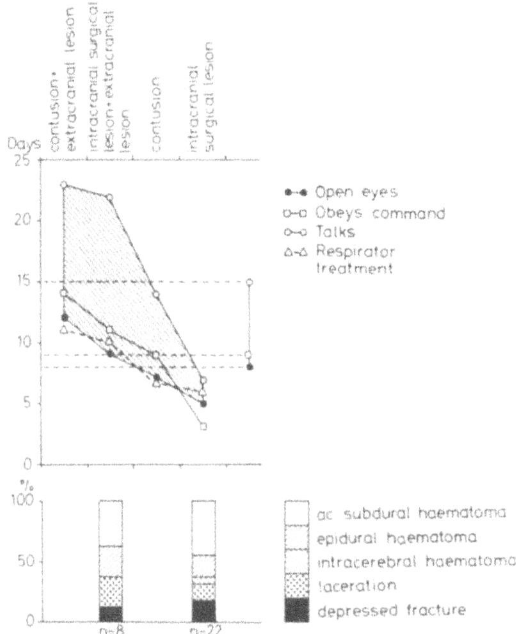

Fig. 3. Factors influencing re-awaking after traumatic coma/diagnoses

Effects of Bradykinin on Blood-Brain-Barrier Function and Pial Microcirculation*

A. Unterberg,** M. Wahl, and A. Baethmann***

Institut für Chirurgische Forschung der Universität München, Klinikum Grosshardern, Marchioninistrasse 15, D-8000 München 70

Introduction

Development of vasogenic brain edema is frequently observed in several cerebral disorders, as e.g. ischemia or severe head injury. Uptake of vasogenic edema fluid into brain parenchyma is caused by an opening of the blood-brain-barrier. It has recently been demonstrated that components of the plasma kallikrein-kininsystem enter cerebral tissue under these circumstances (1).

Moreover, formation of kinins was noted in areas of focal necrosis; in perifocal edematous brain, if additional ischemia developed. Kinins were reported to raise markedly the permeability of peripheral vessels for plasma proteins and to induce brain swelling (2, 3). Hence, formation of kinins in a vasogenic insult leading to brain edema could enhance the dysfunction of the blood-brain-barrier and affect the cerebral microcirculation. This would aggravate the process of secondary brain damage, as e.g. edema. The current study was performed to elucidate the effect of bradykinin on the permeability of the blood-brain-barrier and on the pial microcirculation.

Methods

Mongrel cats of both sexes (2.4 $\pm$ 0.3 kg b.w.) were anesthetised with chloralose (50 mg/kg) and mechanically ventilated.

After opening the skull in the parieto-occipital region, a fluid reservoir was made around a cranial window using rapid polymerizing dental cement. After opening of the dura mater under paraffin oil, the exposed cerebral cortex was irrigated with different solutions via plastic tubes perforating the wall of the reservoir.

Fluorescein-labelled dextran (FITC-dextran) of different molecular weight or Na^+-fluorescein, respectively, were injected i.e. as blood-brain-barrier indicators. Penetration of a marker from the intravas-

*Supported by Deutsche Forschungsgemeinschaft: Ba 452/5

**Dedicated to Prof. Dr. Dr. h.c. W. Brendel on his 60th birthday

***We appreciate the gift of FITC-dextran from Pharmacia AB, Uppsala/ Sweden. We thank Dr. B. Endrich and Prof. Dr. K. Meßmer who made available the fluorescence microscope. The technical and secreterial help of Frl. U. Goerke, A. Müller and I. Moll is gratefully appreciated

cular space into the parenchyma indicative of an increased barrier-permeability was observed intravitally by fluorescence-microscopy.

As a control, the exposed cerebral cortex was irrigated with mock CSF for 30 minutes. In a normal preparation extravasion of the indicators was not observed. Subsequently, mock CSF containing bradykinin in rising concentrations was substituted for the control solution. The duration of exposure to each bradykinin concentration was 30 minutes.

The response of the diameter of pial arteries, arterioles, veins and venules, as well as extravasion of the barrier-indicators into the tissue were documented by microphotography.

Results and Discussion

The results on the response of the blood-brain-barrier to bradykinin are summarized in Table 1. In contrast to the peripheral vessels, bradykinin (10^{-9} - 10^{-2} M) did not increase the permeability of cerebral vessels for FITC-dextran of different molecular weight (MW: 20.000 - 70.000). However, leakage of the blood-brain-barrier was induced by bradykinin in concentrations of only 10^{-6} M, or higher, if Na^+-fluorescein was employed as barrier-indicator. Bradykinin concentrations of 10^{-9} - 10^{-7} M had no effect. Extravasion of Na^+-fluorescein became visible initially in few areas around small venules, stained by the fluorescence-marker which intensified and became confluent at higher concentrations.

In Table 2, the changes in diameter of pial arteries and arterioles (ø 60 - 250 µm) are given, one minute after irrigation with different bradykinin solutions. As seen, pial arteries dilate upon exposure to the peptide in a dose-dependent manner. At 4 x 10^{-6} M of bradykinin, the diameter increased to 110% ($p < 0.01$), at 4 x 10^{-5} M to 137% of normal. These findings are in perfect agreement with the comparable data of WAHL (4). Dilation of arterioles and small arteries was found to decrease within 30 minutes of continuous irrigation. Such a response may be explained by tachyphylaxis.

In contrast to the arteries, a moderate constriction of veins and venules (ø 50 - 270 µm) was detected. Constriction was at a maximum after 30 minutes of irrigation, starting at bradykinin-concentrations of 4 x 10^{-7} M (s. Table 3). With a bradykinin-concentration of 4 x 10^{-3} M constriction reduced the diameter to 86% of normal ($p < 0.01$).

Table 1. Effects of bradykinin on permeability of the blood-brain-barrier

Bradykinin (10^{-9} - 10^{-2} M) after 30 min	ø (D20, D40, D70)
Bradykinin (10^{-9} - 10^{-7} M) after 30 min	ø (Na^+-fluorescein)
Bradykinin (10^{-6} M and higher) after ca. 15 - 20 min	Opening for Na^+-fluorescein

ø, no increase of permeability; D, FITC-dextran (MW: 20.000; 40.000; 70.000)

Table 2. Change of vascular diameter (in % of control value) of pial arterioles and arteries (60 - 250 µm) after one minute of exposure to bradykinin in rising concentrations

Mock CSF		Bradykinin (M) 4×10^{-7}	4×10^{-6}	4×10^{-5}	4×10^{-4}
$\bar{x}$	100%	102.9%	109.0%	137.2%	125.6%
SD	-	16.1	11.7	19.2	10.5
n	21	8	17	19	8
		n.s.	$p < 0.01$	$p < 0.001$	$p < 0.001$

Table 3. Change of vascular diameter (in % of control value) of pial venules and veins (50 - 270 µm) after 30 minutes of exposure to bradykinin in rising concentrations

Mock CSF		Bradykinin (M) 4×10^{-7}	4×10^{-6}	4×10^{-5}	4×10^{-4}	4×10^{-3}
$\bar{x}$	100%	95.3%	91.6%	93.2%	86.1%	86.3%
SD	-	17.9	15.7	10.3	9.8	7.6
n	24	10	13	6	6	
		n.s.	$p < 0.05$	$p < 0.05$	$p < 0.02$	$p < 0.01$

Conclusions

The findings demonstrate that bradykinin induces selective opening of the blood-brain-barrier. Calculation of the molecular radius (Stokes' radius) of fluorescein (i.e. 5.5 Å) suggests that functional pores of at least 11 Å in diameter were opened by bradykinin.

Pores through the blood-brain interface of this size would suffice to allow penetration of electrolytes together with water from the intravascular compartment into the cerebral parenchyma. Moreover, filtration of fluid would be intensified by arteriolar dilation and venous constriction.

The findings support our concept that formation of kinins in vasogenic brain edema enhance a primary insult to the blood-brain-barrier and induce dysfunction of the cerebral microcirculation. This would result in an increased formation and spread of vasogenic brain edema. Therapeutic inhibition of the kinin mechanisms in cerebral injury should be expected to limit secondary brain damage.

References

1. Baethmann, A., Maier-Hauff, K., Lange, M., Kempski, O.: Activation of the kallikrein-kinin-system in brain tissue secondary to cerebral injury and ischemia. J. Cerebr. Blood Flow Metabol. 1, Suppl. 1, 219-219 (1981)

2. Svensjö, E., Arfors, K.-E., Raymond, R.M., Grega, G.J.: Morphological and physiologically correlation of bradykinin-induced macromolecular efflux. J. Amer. Physiol. 236, 600-606 (1979)

3. Unterberg, A., Baethmann, A., Geiger, R.: The kallikrein-kinin-system as mediator in cerebral edema. Agents and Actions, Suppl. 9, 402-407 (1982)

4. Wahl, M.: The effect of morphine, encephalin and bradykinin on cerebral vascular resistance in cats. J. Cerebr. Blood Flow Metabol. 1, Suppl. 1, 323-324 (1981

Comparative Study of the Blood Flow Velocities in the Basilar and the Carotid Artery

W.-D. Möller[1] and K. Wolschendorf[2]

[1] Zentrum Nervenheilkunde im Klinikum der Universität Kiel, Röntgendiagnostik, Niemannsweg 147, D-2300 Kiel

[2] Institut für Angewandte Physik der Universität Kiel, Olshausenstrasse 40-60, D-2300 Kiel

Introduction

The recently developed computer tomography has proved to have advantages over angiographic methods for certain diagnostic purposes, but in the domain of hemodynamic measurements, rapid serial angiography is the only technique so far which allows the determination of blood flow velocities with sufficient accuracy.

In combination with a special photometric evaluation technique which is called quotient densitometry this method has proved very useful for the study of hemodynamics as has been shown by MÖLLER and WOLSCHENDORF 1978.

From several clinical aspects which are described below it is presumed that the blood flow velocity in the basilar artery might be different from that of the carotid artery and the present investigations were performed to obrain quantitative information about the relation of the flow velocities in these two vessels.

Materials and Methods

The present experiments were carried out with a group of 18 patients of various ages and both sexes. All persons were normal cases with a sound circulatory system which showed no other pathological findings.

The contrast medium was injected by the retrograde brachial technique (30 ml in 2 sec) and under halothane anesthesia. It has been shown by MÖLLER and WOLSCHENDORF 1978 that halothane anesthesia accelerates the cerebral blood flow by about 25% and for the present investigations this acceleration factor was assumed to be the same for the carotid and the basilar system.

For the radiographic records ten normal angiograms (two frames/sec), and seven rapid serial angiographies (4 - 6 frames/sec) of a cut-film changer were utilized. From the latter the decisive data were obtained since in spite of the low time-resolution, the large-field radiography of the cut-film changer proved to be advantageous over the cinefilm because of the considerably lower influence of quantum noise, as was shown by WOLSCHENDORF 1982. Lateral angiograms were taken, but in addition an AP view was used to calculate the true length of the vessels from the two radiographic projections.

Advances in Neurosurgery, Vol. 11
Edited by H.-P. Jensen, M. Brock, and M. Klinger

Digital quotient densitometry was applied for the photometric evaluation of the serial angiogram - a technique which was developed in the Institute of Applied Physics of the University of Kiel and the principle of which is shown in Fig. 1. This technique utilizes a reference system of defined composition which is also depicted on the X-ray film. A pair of photodiode-equipped detectors samples the optical density both of the vessel under investigation and the reference system. After preamplification the data acquisition system performs the division of the measurement and the reference signal thus yielding a density signal of significantly improved signal-to-noise-ratio. The densitometric bolus curves obtained in this way can be stored on a floppy disk and can be reproduced by either a printer or a plotter.

Results and Discussion

From the densitometric bolus curves of the serial angiograms the mean transit time between two corresponding points along the vessel is calculated. These points of measurement were the bony carotid canal and the carotid siphon for the carotid system with an average distance of 10.2 cm and the atlas kinking and the peak of the basilar artery for the basilar system with an average distance of 11.1 cm.

The blood flow velocities which were determined from the mean transit time and the travelling distance along the vessel ranged from about 22 cm/sec to 29 cm/sec for the carotid artery with an average value of 25.3 cm/sec and from about 11 cm/sec to 18 cm/sec for the basilar artery with an average value of 14.5 cm/sec.

From these results which are shown in Table 1 we obtain a factor of 1.77 for the ratio of the two velocities. Moreover the average volume flow was calculated and found to be 5.73 cm^3/sec for the carotid and 1.79 cm^3/sec for the basilar system which corresponds to a ratio of 3.25.

These results appear to be in good agreement with the observations of VOLK 1982 who found that the relation between the radiological evidence of arteriosclerosis in the carotid and the basilar region is about 2 : 1. VOLK supposed that a lower blood flow velocity in the basilar artery might account for the different degree of arteriosclerosis in these vessels.

Moreover REUTHER and DORNDORF 1977 and FIELDS 1981 reported that thrombocyte aggregation inhibitors showed little or no significant effect in the vertebro-basilar system. If we assume that the hemodynamic factor is decisive for the development of insults,these findings are also supported by our results.

Table 1. Average values of blood flow velocities and volume flows in the carotid and the basilar system

Average flow velocity in cm/sec		Average volume flow in cm^3/sec	
In the carotid artery	25.3	In the carotid artery	5.73
In the basilar artery	14.5	In the basilar artery	1.79
Velocity ratio	1.77	Volume flow ratio	3.25

References

1. Fields, W.S.: Primär- und Sekundärprophylaxe von Hirn- und Herzinfarkt mit Aggregationshemmern. In: Prophylaxe venöser, peripherer, kardialer und zerebraler Gefäßkrankheit mit Acetylsalicylsäure. Breddin, K., Loew, D., Überla, K., Dorndorf, W., Marx, R. (eds.). Stuttgart, New York: Schattauer 1981

2. Möller, W.D., Wolschendorf, K.: The dependance of cerebral blood flow on age. Eur. Neurol. 17, 276-279 (1978)

3. Reuther, R., Dorndorf, W.: Zur Wirkung von Aspirin bei Patienten mit zerebraler Ischämie - Ergebnisse einer Doppelblindstudie. In: Acetylsalicylsäure bei zerebrovaskulären und kardiovaskulären Erkrankungen. Breddin, K., Dorndorf, W., Loew, D., Marx, R. (eds.), pp. 89-96. Berlin: Bayer 1977

4. Volk, H.P.: Zum Zusammenhang zwischen angiographisch faßbaren Gefäßveränderungen im Carotis- und Basilarisstromgebiet bei Patienten mit zerebralem Gefäßprozeß. Inauguraldissertation, Kiel 1982

5. Wolschendorf, K.: Information content of cinedensitometric blood flow measurements. Symposium "Röntgenkontrastmittel in der angiologischen Funktionsdiagnostik". Konstanz 1982

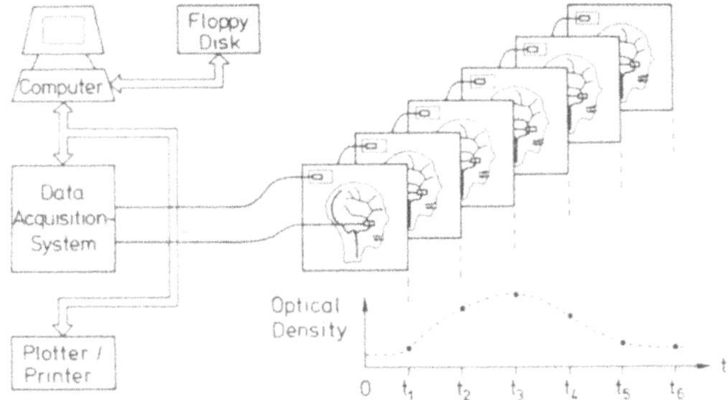

Fig. 1. Functional diagram of an angiodensitometric measurement system

Intraoperative Doppler Flow Measurements in Cerebrovascular Surgery

G. Fischdick, H. Wassmann, and L. Diaz
Neurochirurgische Klinik der Universität Bonn, Sigmund-Freud-Strasse 25, D-5300 Bonn-Venusberg

Introduction

Intraoperative evaluation of blood flow volume in cerebrovascular surgery like carotid endarterectomy and extracranial-intracranial arterial bypass (EIAB) surgery is of great significance as far as the flow, the change of blood flow volume and the control of the functioning of an inline shunt are concerned. Since 1979 we use Doppler vessel clips, bearing in mind the necessity to test simultaneously their validity and intraoperative practicability.

Methods

We use a directional ultrasonic Doppler flow meter system Delalande with transmitted frequency of 4 MHz. Intraoperative measurements were performed by Doppler vessel clips with a diameter of 2, 5 and 10 mm. Validity of these Doppler clips was tested in comparison to electromagnetic flow measurements. In dissected arteries of three dogs ten measurements at a time were performed under controlled constant blood pressure after calibrating the electromagnetic flow meter.

Furthermore the Doppler probes with a diameter of 2, 5 and 10 mm were tested in dissected isolated arteries perfused by a pump with heparinized blood, varying the flow velocity. In a sample of 25 patients with circumscribed stenosis of the internal carotid artery near the bifurcation, flow was determined pre-, intra- and postoperatively. The function and flow volume of an inline shunt used in eight patients was also tested. In 45 patients undergoing EIAB surgery the 2 mm Doppler clip was used to control bypass function and to determine blood flow volume passing through the anastomosis immediately after the shunt construction.

Results

Determination of Accuracy of Doppler Vessel Clips

In 10 different positions of a dissected dog artery perfusion was determined by the Doppler technique using 2, 5 and 10 mm Doppler clips. These values were compared with electromagnetic flow measurements during a constant systolic blood pressure of 120 mm Hg. The deviation from these values was $\pm$ 12.8 to 15.2% (Table 1). During *in vitro* examinations the perfusion of the arteries was changed and the

 Advances in Neurosurgery, Vol. 11
Edited by H.-P. Jensen, M. Brock, and M. Klinger

Table 1. Divergences of Doppler values from electromagnetic and in vitro measurements

Doppler clip (mm)	Blood flow volume (ml)	Deviation (%)
Electromagnetic flow measurements		
2	52	± 12.8
5	188	± 14.3
10	753	± 15.2
In vitro measurements:		
2	9.4 - 97.8	± 17.0
5	59 - 330	± 17.2
10	188 - 753	± 11.0

deviation of the Doppler values from the real flow volume was determined (Fig. 1), resulting in ± 11 to ± 17% (Table 1).

Intraoperative Use of Doppler Vessel Clips

In a 45-year-old patient with right-sided transient ischemic attacks a reduced flow over the left supratrochlear artery and the common and internal carotid artery was found, and also a turbulence over the carotid bifurcation (Fig. 2). Angiography showed a circumscribed severe stenosis of the left internal carotid artery. Intraoperatively we measured a flow in the left common carotid artery of 282 ml/min and in the left internal carotid artery of 94 ml/min. After carotid endarterectomy Doppler examinations showed an increase of blood flow volume in the internal carotid artery of 125%. Postoperative angiography showed a patent carotid artery (Fig. 3).

Altogether changes of carotid perfusion after endarterectomy in 25 patients with circumscribed stenosis of internal carotid artery were determined, and showed an average increase of 56 ml/min.

In eight cases the use of an inline shunt during endarterectomy was necessary. Thereby the function and blood flow through the shunt ranging from 52 to 97 ml/min was determined. Furthermore a Doppler vessel clip was used in EIAB surgery. In a 46-year-old patient with right internal carotid occlusion a mean velocity over the parietal branch of the right superficial temporal artery of 16 cm/sec was measured. After performing the EIAB a perfusion of the anastomosis of 52 ml/min was determined intraoperatively by Doppler probe. The increase of diastolic Doppler values was observed as a sign of cerebral perfusion resistance. Ten days after operation an increase of velocity of 85% was registered over the bypass artery. Furthermore, the temporal artery now showed Doppler-sonographically the signs of an internal carotid artery with raised end diastolic pressure (1, 2) (Fig. 4).

Altogether we examined 45 patients intraoperatively during EIAB operations. Blood flow values through the bypass from 3.7 to 65 ml/min with a mean value of 36 ml/min were measured. In spite of well pulsating donor vessel in two cases the Doppler revealed a perfusion of only 1 to 2 ml/min, so that a revision of the bypass was possible.

Summarizing Remarks

The investigations of others and our first experiments concerning the accuracy of determining blood flow volume by Doppler vessel clip showed that only minor fluctuations in vessel diameter can lead to deviations of 50% from the real flow volume due to the square function between velocity and volume (3, 4).

Although we paid attention to that fact in these measurements, we found a deviation of Doppler-probe-values *in vitro* and *in vivo* up to 17%. Nevertheless taking these divergences into consideration, the intraoperative use of this method seems to be of value, since the relative change of perfusion in carotid endarterectomy and critical examination of sufficient inline shunt function is well demonstrated by this simple method.

In EIAB surgery intraoperative flow measurements enable us not only to control bypass function, but also to gauge the function of the anastomosis, so that in case of dysfunction an immediate revision of the bypass is possible.

Our experience indicates intraoperative Doppler flow volume measurement is a suitable method for evaluating cerebrovascular changes, particularly in patients undergoing carotid endarterectomy or bypass surgery.

References

1. Gottwald, K.: Doppler-Sonographie der extrakraniellen hirnversorgenden Gefäße. EEG-Labor 3, 121-137 (1981)
2. Hopman, H., Gratzl, O., Schmiedek, P., Schneider, I.: Doppler-Sonographie bei mikrovaskulärem Bypass. Neurochirurgia 19, 190-196 (1976)
3. Messmer, K., Meisner, H., Weishaar, E.F.: Flow tracings of three types of flow meters in acute dog experiments. In: New findings in blood flowmetry. Cappelen, I.R. (ed.), pp. 136-145. Oslo 1968
4. Thal, H.U., Leem, W.: Animal experiments with Doppler flow transducers. Acta Neurochirurgica 28, 275-277 (1979)

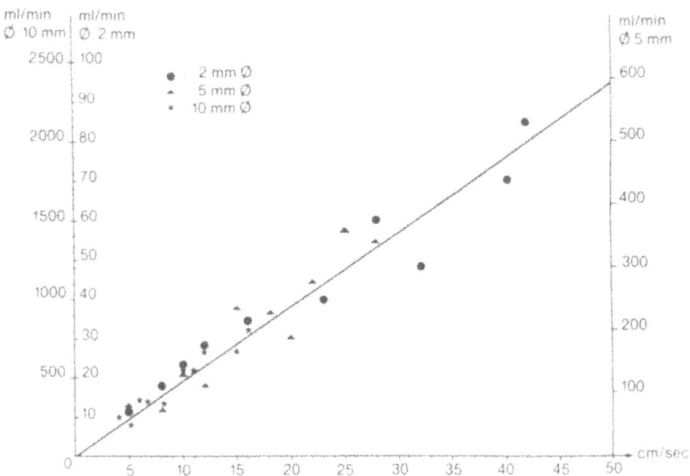

Fig. 1. *In vitro* Doppler flow examinations performed by 2, 5 and 10 mm Doppler vessel clip in comparison to real flow

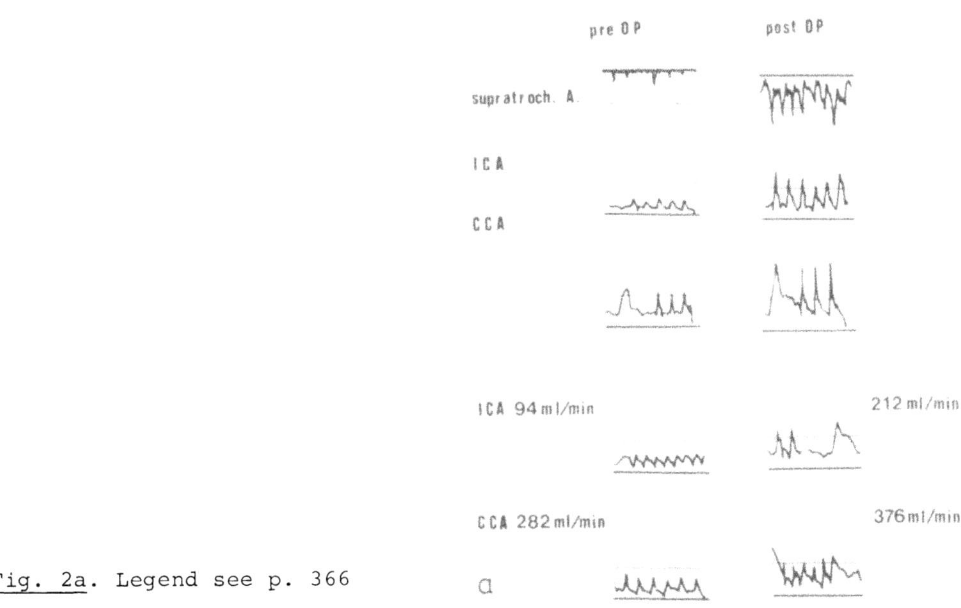

Fig. 2a. Legend see p. 366

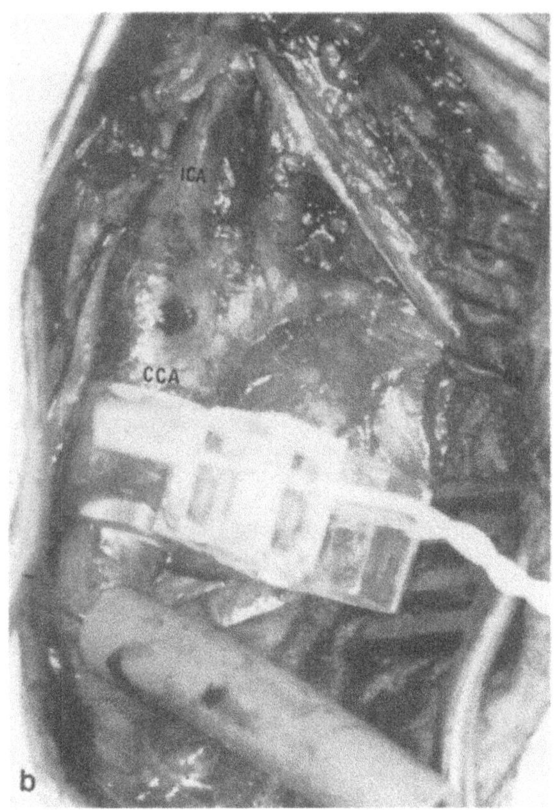

Fig. 2. a Pre-, intra- and postoperative Doppler examinations in a patient with stenosis of the left internal carotid artery (ICA). (CCA = common carotid artery); b Intraoperative blood flow measurement by Doppler vessel clip

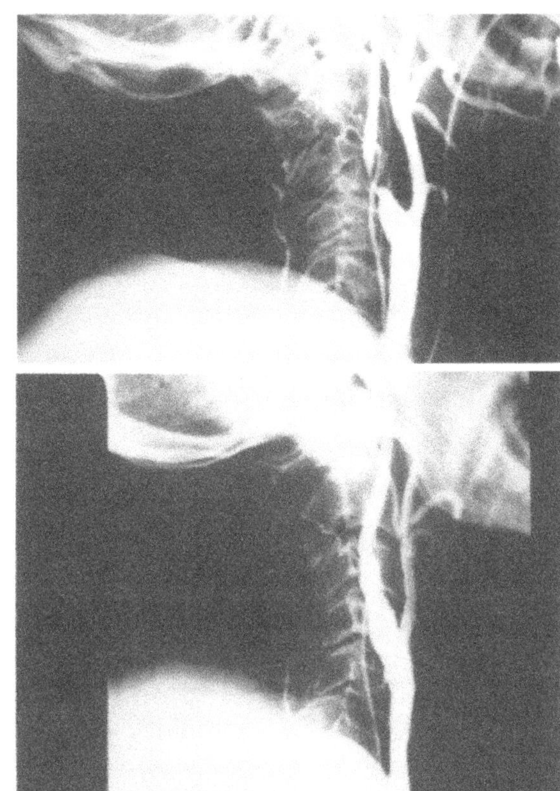

Fig. 3. Angiography before (above) and after endarterectomy of a stenosis of the left internal carotid artery

pre O P
S T A
percut.

intraop. flow
Ø 2 mm
52ml/min

post O P
S T A
percut.

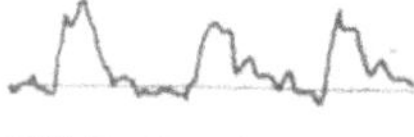

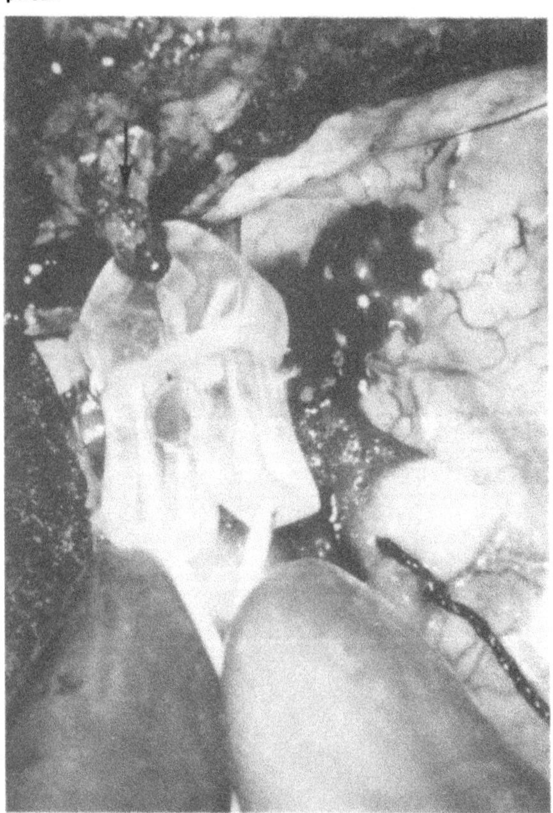

Fig. 4. *(Above)* Doppler examinations before, during and after operation in a patient with occlusion of the right internal carotid artery, undergoing bypass surgery. *(Below)* Intra-operative blood flow measurements through the bypass artery (→) by Doppler vessel clip.

Intraoperative Doppler Sonography

J. Gilsbach and A. Harders
Neurochirurgische Klinik, Albert-Ludwig-Universität Freiburg, Hugstetter Strasse 55, D-7800 Freiburg

Introduction

Hemodynamic information available during the operation can make neurovascular procedures safer and simpler.

In this context possible information required from a method of hemodynamic examination is for direction and velocity of flow, normal and abnormal flow pattern, and patency. This information can theoretically be obtained by the atraumatic, repeatable and simple procedure of Doppler-Sonography.

Direct intraoperative Doppler sonographic measurements, however, have rarely been published up to now (FRIEDRICH 1980, HITCHON 1979, NORNES 1979, 1980, MORITAKE 1981, MÜLLER 1980, THAL 1981, WASSMANN 1980), because technique and probe dimensions of commercially produced instruments for Doppler-Sonography are poorly adapted to small vessels.

Material and Methods (Table 1)

For one and a half years we have been using a pulsed 20 MHz prototype[1]. This instrument is able to detect small and slow flow sample volumes in vessels less than 1 mm in diameter.

The pulsed technique, owing to its unique crystal for emission and reception, together with its high frequency, allows the development of very small probes.

Table 1. Intraoperative direct Doppler sonography (20 MHz PUDVM) Case material

Diagnosis	No. of cases
Tumor	31
Aneurysm	31
Bypass	20
Angioma	12
Stenosis/anastomosis (gerbil, rat)	50

1 INSERM, Lyon

Advances in Neurosurgery, Vol. 11
Edited by H.-P. Jensen, M. Brock, and M. Klinger

Small and well centred samples within the vessel lumen can be measured thanks to a gate circuit, thus avoiding interference from adjacent vessels.

Most measurements are performed in the central stream using the smallest possible gate, because the built in zero-cross frequency counter works precisely and because results from different vessels are easier to compare.

Lately we have been using a real time audiospectrum analyser[1] which gives better results particularly at high frequencies and with a poor signal to noise ratio (Figs. 1 and 2).

Results

Animals

In order to test the validity of Doppler-Sonography for EIAB we produced stenosis of varying degrees as well as end-to-end and end-to-side anastomosis in 50 laboratory animals (rats and gerbils). During measurements the Doppler probe was applied to the vessels by means of a micromanipulator in order to obtain stable conditions. Poststenotic pressure via the external carotid artery was monitored in 20 animals with common carotid stenosis in order to control the hemodynamic efficiency. Retrograde *in vivo* angiography via the external carotid or the brachial artery was performed in 30 animals.

Eighty-eight per cent of the intended and all of the unintentional narrowings could be detected by measuring increased velocities in the area of stenosis, even in vessels with unchanged total flow.

Undetectable stenoses were either too insignificant or too severe and too abrupt so that they did not allow increased velocities.

EIAB

In 18 out of 20 patients with EIAB no flow impairment could be detected intraoperatively in the area of anastomosis or in the afferent and efferent vessels. One anastomosis appeared normal externally, but the STA was thrombosed and had to be recanalized. Another anastomosis resulted in kinking and narrowing of the receiving vessel.

All of the 20 anastomoses were patent at the end of the procedure and remained open during the subsequent course after angiographic and Doppler sonographic controls (Fig. 3).

Aneurysms

We made complete recordings of all adjacent vessels in 31 patients with saccular aneurysms, before and after clipping of the aneurysm. In addition the flow in the aneursym was measured if at all possible. Complete occlusion by means of a clip was ascertained with Doppler-sonography and confirmed by puncture or resection of the aneurysm. Invisible narrowing of efferent arteries was detected and corrected by repositioning of the clip in two cases. In one poorly accessible PICA aneurysm exclusion of the sac as well as patency of the vertebral

1 Carolina Medical Electronics

artery and PICA were checked solely by Doppler-Sonography, since only the proximal portion of the clip was visible (Fig. 4).

Angiomas

Distinction between arteries and veins was sometimes difficult at the surface of the angioma, but could be accomplished in most cases at the border. Identification of arteries was facilitated by the unequivocal detection of flow direction. Knowledge of the direction of flow was particularly helpful in cases where intraoperative embolization was attempted, thus avoiding faulty embolizations in the direction of the heart.

Conclusions

Provided that the above-mentioned technical performances are available, direct intra-operative Doppler-sonography represents a valuable aid in neurovascular surgery, as it can identify hemodynamic changes in any exposed vessels after operative measures. In addition, this method can be used in laboratory animals in order to investigate mechanical and pharmacological influences on small vessels.

References

1. Friedrich, H., Hänsel-Friedrich, G., Seeger, W.: Intraoperative Doppler-sonographie an Hirngefäßen. Neurochirurgia 23, 89-98 (1980)
2. Hitchon, P.W., Kassel, N.F., Carlstrom, T.A., McDonnell, D.E.: The Doppler ultrasonic flowmeter as an adjunct to operative management of cerebral arteriovenous malformations. Surg. Neurol. 11, 345-347 (1979)
3. Moritake, K., Handa, H., Yonekawa, Y., Nagata, I.: Ultrasonic Doppler assessment of hemodynamics: - superficial temporal artery - middle cerebral anastomosis. Surg. Neurol. 13, 249-257 (1980)
4. Müller, H.R., Gratzl, O.: Assessment of STA/MCA bypass function by Doppler ultrasonic techniques (velocity and flow measurements). Fifth International Symposium on microvascular anastomoses for cerebral ischemia, p. 83. Wienna 1980
5. Nornes, H., Grip, A.: Hemodynamic aspects of cerebral arteriovenous malformations. J. Neurosurg. 53, 456-464 (1980)
6. Nornes, H., Grip, A., Wikeby, P.: Intraoperative evaluation of cerebral hemodynamics using directional Doppler technique; Part 1: Arteriovenous malformations. J. Neurosurg. 50, 145-151 (1979)
7. Nornes, H., Grip, A., Wikeby, P.: Intraoperative evaluation of cerebral hemodynamics using directional Doppler technique. Part 2: Saccular aneurysms. J. Neurosurg. 50, 570-577 (1979)
8. Thal, H.-U.: Evaluation of patency during and after extracranial-intracranial bypass operations using Doppler sonography. Neurological Surgery, p. 384. Bastracts of the 7th International Congress of Neurological Surgery 1981
9. Wassmann, H., Fischdick, G., Holbach, K.-H.: Ultrasonic Doppler assessment of hemodynamics in extra-intracranial arterial bypass (EIAB) surgery. Fifth International Symposium on microvascular anastomoses for cerebral ischemia, p. 85. Vienna 1980

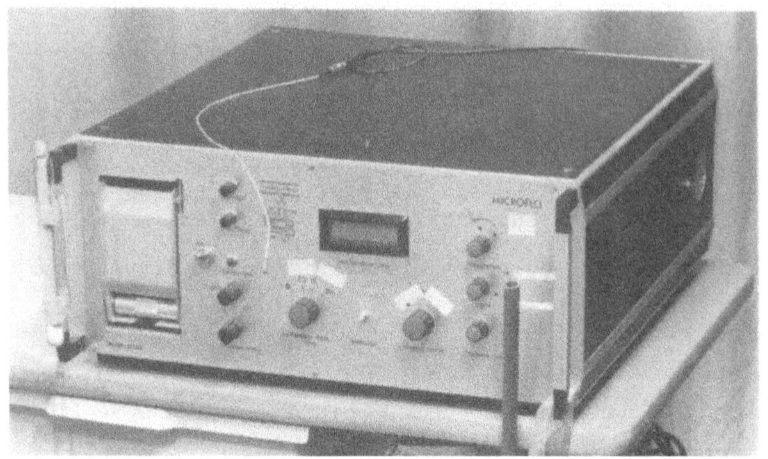

Fig. 1. 20 MHz pulsed ultrasonic velocity meter (Microflo) with probe

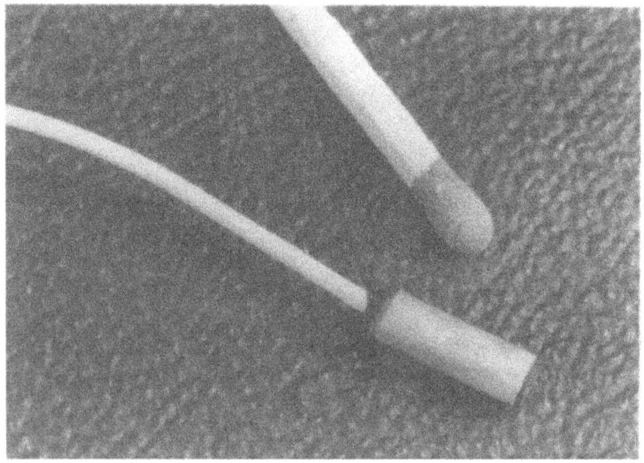

Fig. 2. Doppler probe

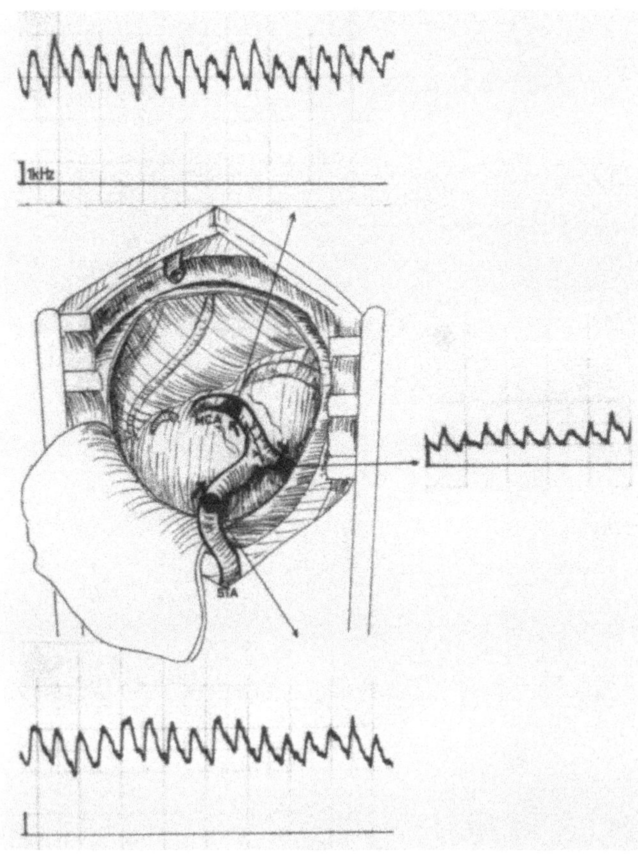

Fig. 3. Intraoperative recordings in EIAB

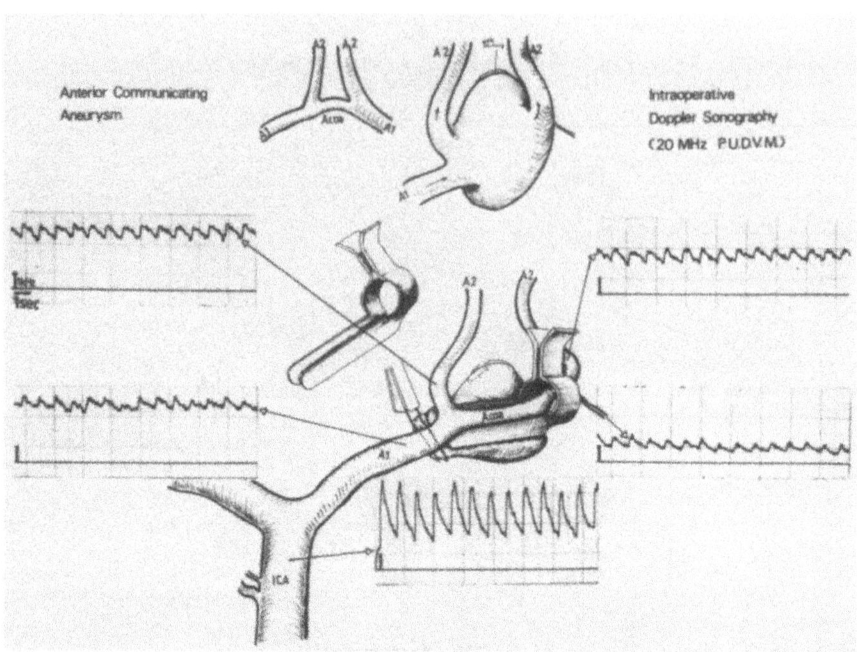

Fig. 4. Intraoperative recordings of adjacent vessels after clipping of an anterior communicating aneurysm

Comparative Study of Doppler Sonography and Angiography in Extracranial-Intracranial Anastomosis

Z. A. Jamjoon, H. M. Mehdorn, and H. C. Nahser

Neurochirurgische Klinik des Universitätsklinikum der Gesamthochschule Essen, Hufelandstrasse 55, D-4300 Essen

Introduction

Although cerebral angiography has nowadays become a safe procedure, its potential risks must be always considered, especially when patients with cerebrovascular disease are involved (4, 5, 9). In contrast to angiography, Doppler sonography is noninvasive, easily repeatable and has already been recognized as being reliable in detecting significant (> 50%) stenoses and occlusions of extracranial cerebral arteries (2, 3, 8). Its accuracy in demonstrating the function of extracranial-intracranial anastomosis has been questioned, however by several authors, including the Cooperative EC-IC-Bypass Study (10).

In this paper we describe the results of Doppler sonography in patients, who underwent EC-IC anastomosis, and discuss the value of this method in the postoperative follow-up examination of the function of the bypass, as compared to angiography.

Patients and Methods

In a series of 60 patients who underwent EC-IC anastomoses (Table 1) pre- and postoperative Doppler sonography was performed using a directional Doppler velocimeter DUD 400 (Debimetre Ultrasonic Delalande) with an emitting frequency of 4 MHz.

Table 1. Preoperative angiographic diagnosis on the side of operation in 60 patients who underwent EC-IC anastomoses

Artery	Type of lesion	No. of cases
Middle cerebral artery	Stenosis	7
	Occlusion	11
	Giant aneurysm	1
Internal carotid artery	Stenosis	7
	Occlusion	31
Combined	Stenosis / occlusion	4

Advances in Neurosurgery, Vol. 11
Edited by H.-P. Jensen, M. Brock, and M. Klinger

For the evaluation of the postoperative Doppler sonography a grading system that consisted of four criteria was established (Table 2). A total score of 7 - 8 was used to describe "good", 5 - 6 "fair", 2 - 4 "poor" and 0 - 1 no bypass function.

Pre- and postoperative angiography was performed, in most patients, as a transfemoral, selective angiography, although occasionally, retrograde brachial angiography was performed.

Using three criteria and a similar scoring system (Table 3) the bypass function in angiography was independently determined by two other investigators.

Results

Pre-operatively the pulse curve of the superficial temporal artery (STA) and its branches is similar to that of the external carotid artery which is characterised by a rapid systolic ascent as well as a rapid diastolic descent with only little diastolic flow above zero. This pulsatile flow pattern is thought to be due to the relatively high peripheral vascular resistance. After successful anastomosis to a cortical vessel, which offers comparatively little resistance, the diastolic descent is slowed and the diastolic flow is markedly raised above zero. Consequently the pulse curve resembles that of the internal carotid artery (Fig. 1).

Table 2. Doppler sonographic criteria

		Score
Extracranial donor artery pulse can be detected and followed up to the bone window	Yes	1
	No	0
Change of flow characteristics from external carotid type preoperatively to internal carotid type postoperatively	No change	0
	Diastolic fall of flow is slowed	1
	Slight increase of diastolic flow	2
	Prominent increase of diastolic flow	3
Interruption of pulse signals at the bone window by digital compression of the donor artery proximal to the Doppler probe	Yes	1
	No	0
Effect of compression of the non-anastomising branch on the flow in the main artery	Slight or no decrease	3
	Moderate decrease	2
	Prominent decrease	1
	Diastolic flow falls to zero, weak pulse signals	0

Table 3. Angiographic criteria

		Score
Intracranial filling through the bypass	No intracranial filling	0
	Only recipient cortical artery is visualized	1
	2 cortical vessels are filled	2
	3 or more cortical vessels are filled	3
Postoperative change of diameter of the donor artery	Diameter decreased	0
	No change	1
	Diameter increased up to twice	2
	Diameter increased more than twice	3
Flow through the bypass	Slower than in the other extracranial vessels	0
	Same as in the other extracranial vessels	1
	Earlier than in the other extracranial vessels	2

A good function of the anastomosis is recognized angiographically by the filling of more than three cortical vessels and the increased flow through the dilated extracranial donor artery; in Doppler sonography, it is characterised by a prominent postoperative elevation of the diastolic flow above zero (Fig. 2). On digital compression of the nonanastomosed branch the flow in the main trunk of the superficial temporal artery will be only slightly depressed. The extracranial portion of the donor artery can be easily detected and followed up to the bone opening, where the pulse signals can be interrupted by digital compression proximal to the position to the Doppler probe.

In contrast there is no significant increase of the diastolic flow in the postoperative Doppler sonograms of cases with poor bypass function. The only hemodynamic change is a slowing of the diastolic descent of the pulse curve. Since it may sometimes be difficult to follow the donor artery along its whole extracranial extent, it should be possible to interrupt pulse signals at the craniectomy somehow by digital compression proximal to the position of the probe before the patency of the bypass can be assumed (Fig. 3).

When the bypass is occluded no hemodynamic changes in the main trunk of the STA could be observed postoperatively (Fig. 4). Pulse signals at the bone opening originate from the cortical vessels, since they are not influenced by digital compression proximal to the Doppler probe.

Among the 60 anastomoses which were examined Doppler sonography indicated a good function in 31 cases, a fair function in 18 cases, a poor function in 9 cases and no bypass function in 2 cases. A comparison of these results with those of cerebral angiography is shown in Fig. 5. Although Doppler sonography estimation of the bypass function was identical with cerebral angiography in only 70% of all cases no discrepancies between both methods were observed as far as the patency of the bypass was concerned. The differences between Doppler sonography and angiography are apparent only in the observations of the estimated function of patent anastomoses.

In order to determine whether the postoperative bypass function might be influenced by the nature of the preoperative vascular lesion the postoperative results of bypass function in Doppler sonography and angiography were related to the underlying vascular lesion (Fig. 6). While in 60% of the cases with internal carotid artery occlusion a good postoperative bypass function was noticed, this was true in only 17% of the cases with middle cerebral artery stenosis which achieved comparative results. These findings suggest that the postoperative bypass function may depend on the pressure gradient between donor and recipient artery, since the higher this gradient, the greater is the volume of blood which will flow through the anastomosis. These hemodynamic factors seem to contribute to the postoperative dilatation of the donor artery. In Fig. 7 the postoperative change in flow as detected by Doppler sonography is related to the postoperative increase of diameter of the donor artery.

In 86% of the cases in which the postoperative diameter of the donor artery increased more than twofold there was a marked increase of the diastolic flow in Doppler sonography, compared to 14% of the cases in which no diameter change was observed.

Discussion

Previous studies have already indicated that Doppler sonography is highly reliable in detecting bypass patency (1, 6).

Its accuracy is estimated as high as 92 - 96%, but false positive results were found to be the major error of the method. However in these studies no attention was paid to the hemodynamic changes in the donor artery.

MORITAKE et al. were the first who gave a detailed description of the hemodynamics in the vessels involved in extracranial-intracranial anastomoses. They showed that it is possible on the basis of these changes to differentiate between good and poor bypass function.

In this study a grading system with four criteria including hemodynamic changes was employed to evaluate the postoperative bypass function. Our data suggest that Doppler sonography is as reliable as angiography when the patency of the bypass is concerned. Furthermore there was a good correlation between the Doppler sonography and angiography in estimating the relative function of a patent anastomosis.

These results allow the following conclusions:

1. Doppler sonography is a valid method for rapid and reliable follow-up study of extracranial-intracranial anastomoses.
2. Angiography is necessary for pre-operative investigations and, in the postoperative period, whenever new symptoms of cerebral

ischemia occur and the indication for another surgical procedure is discussed.

References

1. Ausman, J.I., Diaz, F.G.: Correlation of noninvasive Doppler and angiographic evaluation of extracranial-intracranial anastomoses. In: Microsurgery for cerebral ischemia. Peerless, S.J., McCormick, C.W. (eds.). Berlin, Heidelberg, New York: Springer 1980
2. Barnes, R.W., Russel, H.E., Bone, G.E., Slaymaker, E.E.: Doppler cerebrovascular examination: Improved results with refinement in technique. Stroke 84, 468-471 (1977)
3. Büdingen, H.J., v. Reutern, G.M., Freund, H.J.: Doppler-Sonographie der extrakraniellen Hirnarterien. Stuttgart: Thieme 1982
4. Eisenberg, R.L., Bank, W.O., Hedgcock, M.W.: Neurologic complications of angiography for cerebrovascular disease. Neurology 30, 895-897 (1980)
5. Faught, E., Trader, S.D., Homma, G.R.: Cerebral complications of angiography for transient ischemia and stroke: Prediction of risk. Neurology 29, 4-15 (1979)
6. Hopman, H., Gratzl, O., Schmiedeck, P., Schneider, I.: Doppler-Sonographie bei mikrovaskulärem Bypass. Neurochirurgia 19, 190-196 (1976)
7. Keller, H.M., Meier, W.E., Zumstein, B.: Nichtinvasive Doppler-Ultraschall-Abklärung zerebrovaskulärer Patienten: Karotis- und Vertebralis Doppleruntersuchung. In: Ultraschall-Dopplerdiagnostik, Kriessmann, A., Bollinger, A. (eds.). Stuttgart: Thieme 1978
8. Lindner, D.W., Hardy, W.G., Thomas, L.M. et al.: Angiographic complications in patients with cerebrovascular disease. J. Neurosurgery 19, 179-185 (1962)
9. Moritake, K., Handa, H., Yonekawa, Y., Izumia, N.: Ultrasonic Doppler assessment of hemodynamics in superficial temporal artery middle cerebral artery anastomosis. Surgical Neurology 13, 249-257 (1980)
10. Protocol, Cooperative Study on Extracranial-Intracranial anastomoses. London/Canada 1978

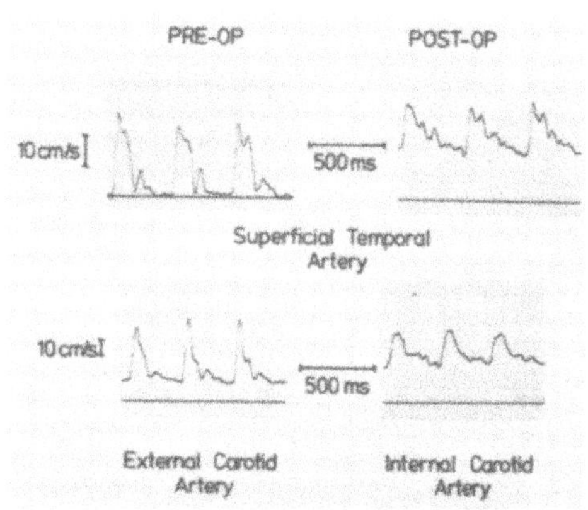

Fig. 1. Flow characteristics as detected by Doppler sonography

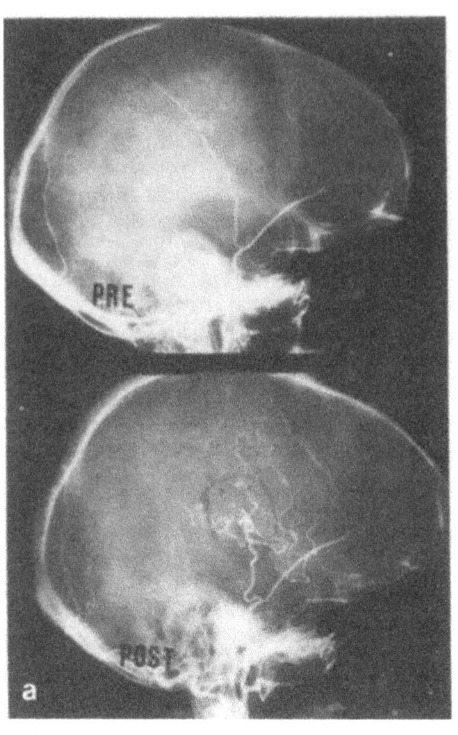

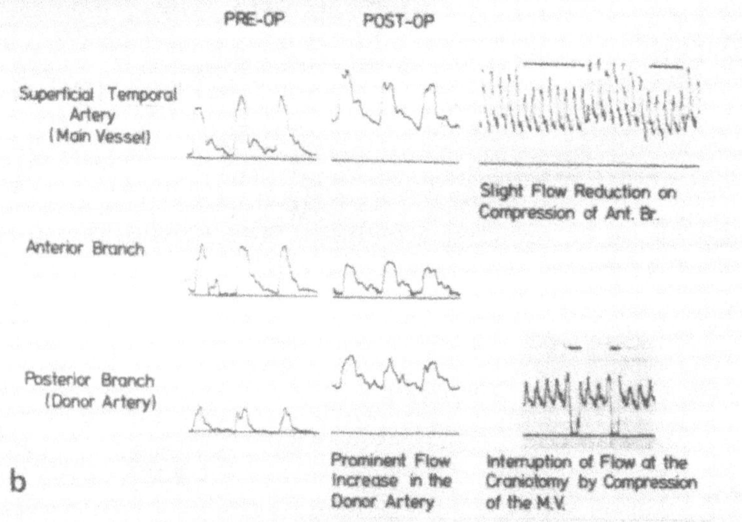

Fig. 2a, b. Good function of an EC-IC anastomosis: angiography showing filling of the entire territory of the MCA through the bypass and also the corresponding Doppler sonography

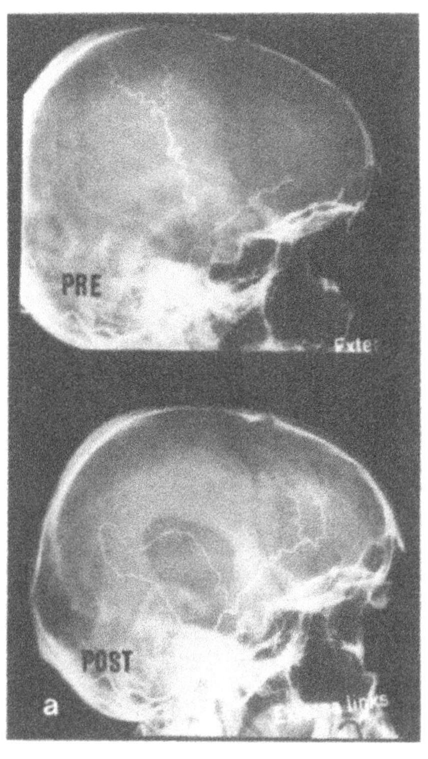

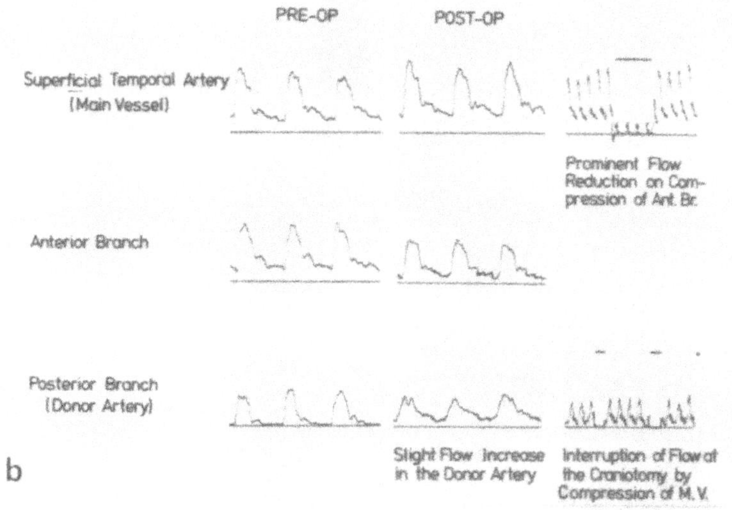

Fig. 3a, b. Poor function of an EC-IC anastomosis: angiography showing filling of only the distal branch of a MCA-vessel

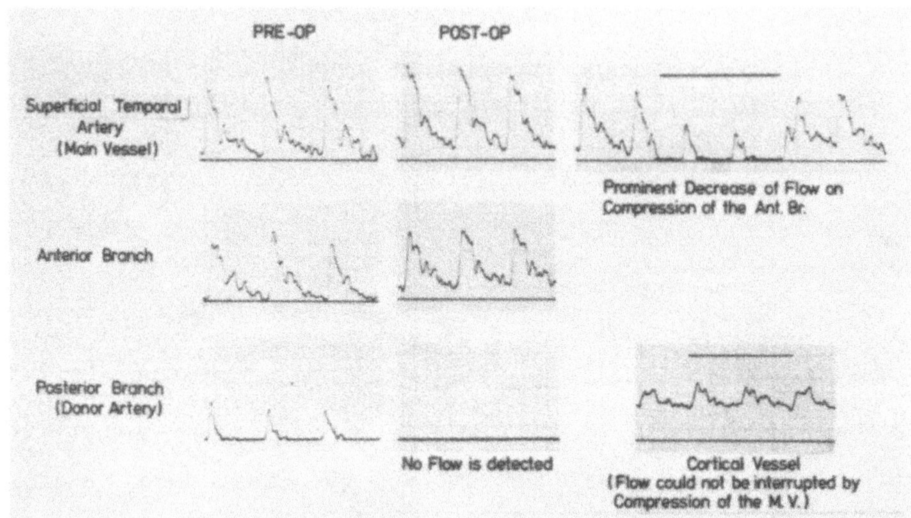

Fig. 4. Doppler sonography of an occluded EC-IC bypass

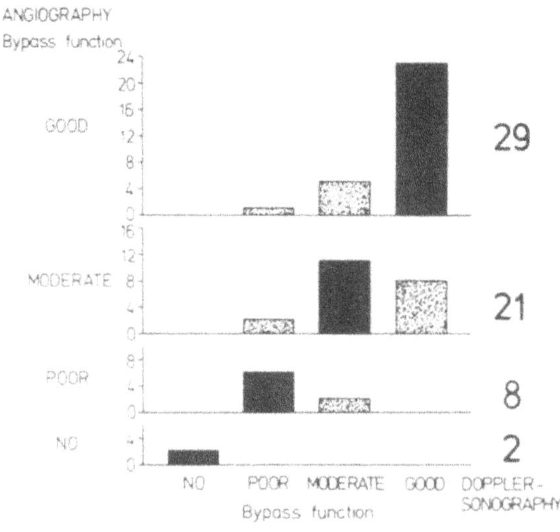

Fig. 5. Post-operative results of Doppler-sonography in 60 EC-IC anastomoses as compared to angiography

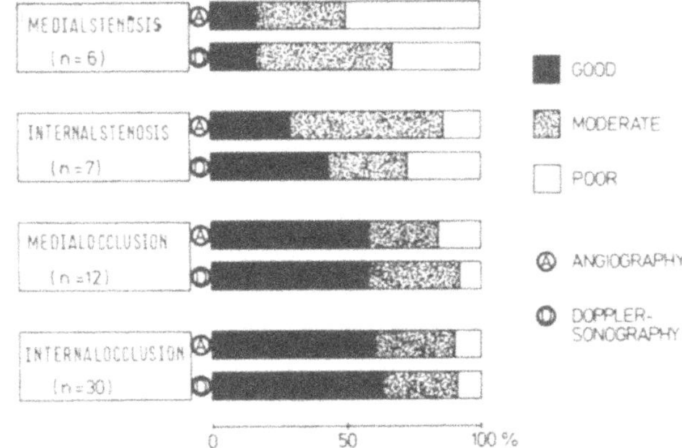

Fig. 6. Correlation of bypass function to the type of vascular lesion

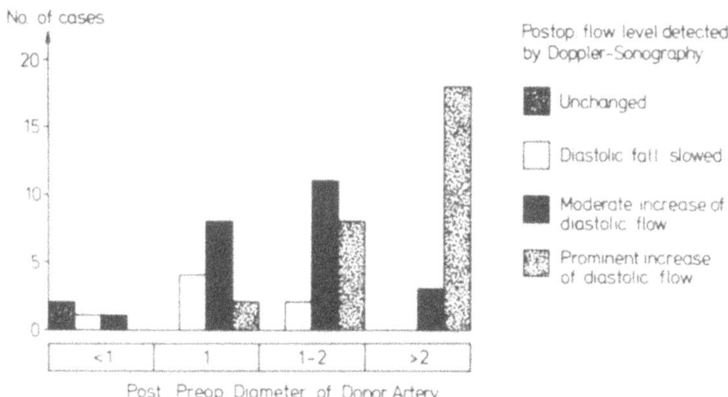

Fig. 7. Correlation of flow increase observed in Doppler-sonography with diameter increase of the donor artery

Super-Selective External Carotid Artery Embolization

D.A. Batchelor[1], E.B. Bongartz[2], J.P. von Berkel[1], B. Matricali[2], D.M.A. Agenant[1], and Z.D. Goedhart[2]

[1] Radiological Department of the Slotervaart Municipal Hospital Amsterdam, NL-1066 EC Amsterdam
[2] Neurosurgical Department of the Slotervaart Municipal Hospital Amsterdam, NL-1066 EC Amsterdam

The value of selective catheterization and embolisation of branches of the external carotid artery are emphasized in relation to the pre-operative treatment of meningiomas and palliative or sometimes curative treatment of other vascular tumors or malformations.

Pre-operative embolisation of meningiomas (6, 7) has been shown to reduce operative blood loss, operating time is shortened and there appear to be fewer post-operative complications. The long-term effect on growth rate in other inoperable lesions has yet to be evaluated. In certain forms of arteriovenous malformation a cure may be effected (8, 2).

The technique is relatively simple and safe. In the case of meningiomas its effectiveness of course depends on the degree to which branches of the external carotid artery contribute to the vascularization.

The importance of selective angiography with detailed subtraction prior to embolisation is emphasized. The degree of success depends a great deal on identifying all the feeding arteries.

In most cases embolisation can be accomplished using the diagnostic catheter. We have found a number I headhunter catheter ideal in the majority of patients. The feeding artery is selectively catheterised and single pieces of embolic material in diluted contrast medium are slowly injected under fluoroscopic control. The branches of the artery feeding the lesion have an increased blood flow. This "steal" effect causes lodgement of the emboli in the feeding artery. The aim is to occlude the feeding branches as peripherally as possible. This reduces the possibility of other vessels taking over the supply of the lesion and avoids occlusion of other branches that may cause unwanted complications. Emboli are introduced until the feeding branches are occluded or until the flow in the main artery is seen to stagnate. This is important in reducing the risk of reflux of embolic material. A control angiographic series is then made to assess the results.

We use simple embolic materials (3-5). Pre-operative embolisation is performed with gelfoam particles. The patient is then operated upon within the next 3 - 5 days (7). In palliative embolisation, or when a more permanent occlusion is desired, lyophilized dura mater is used (5).

Up to now we have not encountered any serious complications. Minor degrees of pain and warmth are experienced by some patients but these usually subside within 24 - 48 hours.

Advances in Neurosurgery, Vol. 11
Edited by H.-P. Jensen, M. Brock, and M. Klinger

When carefully performed, after thorough assessment of the arterial anatomy this simplified method of embolisation has proved to be of considerable help to the neurosurgeon and has benefited the patient.

Four illustrative cases are described. Three meningiomas as examples of our technique and a thoracic spinal arterio-venous malformation to illustrate that the same technique can be used with benefit in other extra-cerebral arterial territories of interest to the neurosurgeon.

References

1. Katzen, B.T. et al.: Transcatheter therapeutic arterial embolisation. Rad. 120, 523-531 (1976)
2. Brismar et al.: Therapeutic embolisation in the external carotid artery region. Ada Rad. Diag. 19 (1978)
3. Bank, W.O. et al.: Gelfoam embolisation. A simplified technique. AJR 132, 299-301 (1979)
4. Berenstein et al.: Gelatin sponge in therapeutic neuroradiology. A subject review. Rad. 141, 105-112 (1981)
5. Berenstein et al.: Catheter and material selection for transarterial embolisation. I. Catheters. II. Materials. Radiology 132, 619-639 (1979)
6. Hekster, R.E.M. et al.: Presurgical transfemoral catheter embolisation to reduce operative blood loss - technical note. J. Neurosurg. 14 (1980)
7. Hieshima, G.B. et al.: Pre-operative embolisation of meningiomas. J. Surg. Neurol. 4 (1980)
8. Hekster, R.E.M. et al.: Transfemoral catheter embolisation. A method of treatment of glomus jugulare tumors. Neuroradiology 5, 208-214 (1973)

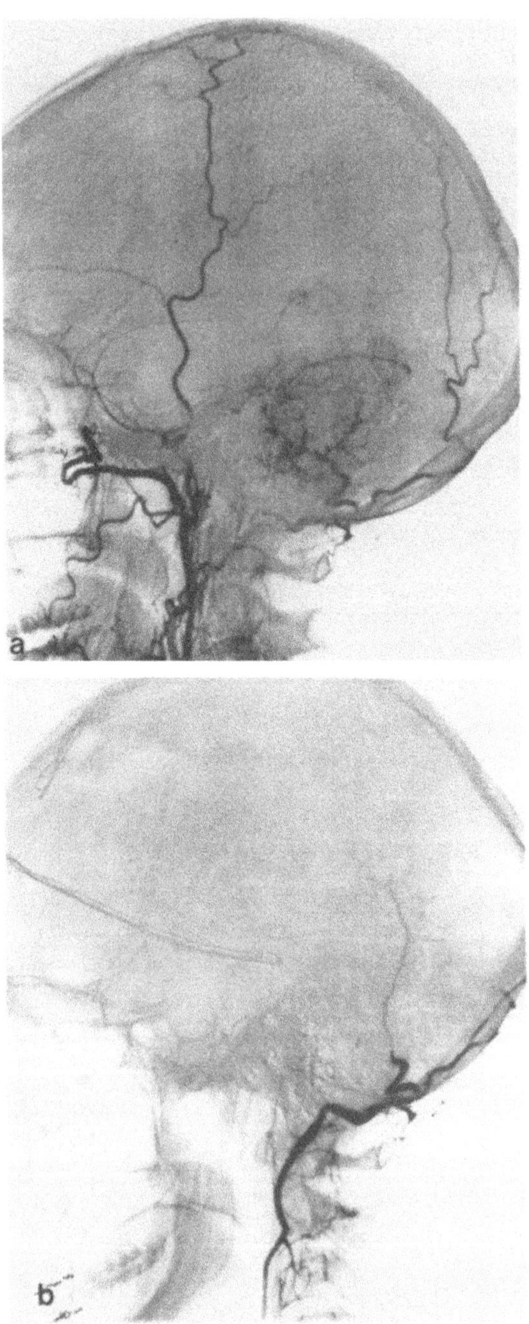

Fig. 1. a External carotid injection showing branches of the occipital artery supplying an infra-tentorial meningioma. b Occipital artery injection after embolisation showing peripheral occlusion of the feeding branches

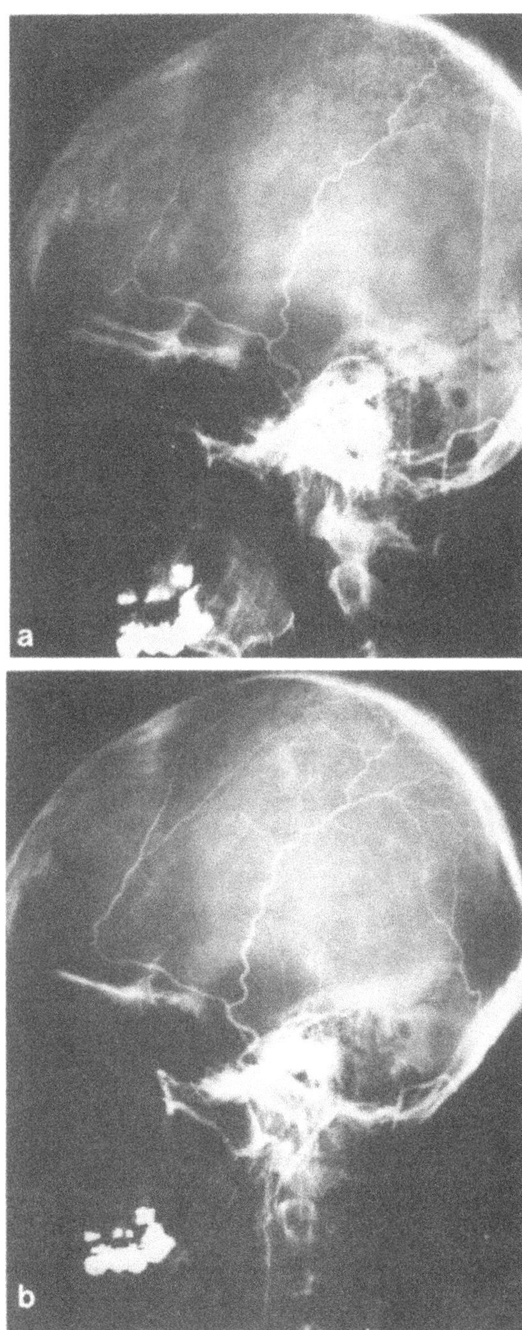

Fig. 2. a External carotid injection showing the anterior branch of the superficial temporal artery supplying a falx meningioma that was invading bone. b The same series after embolisation. Again note the peripheral occlusion

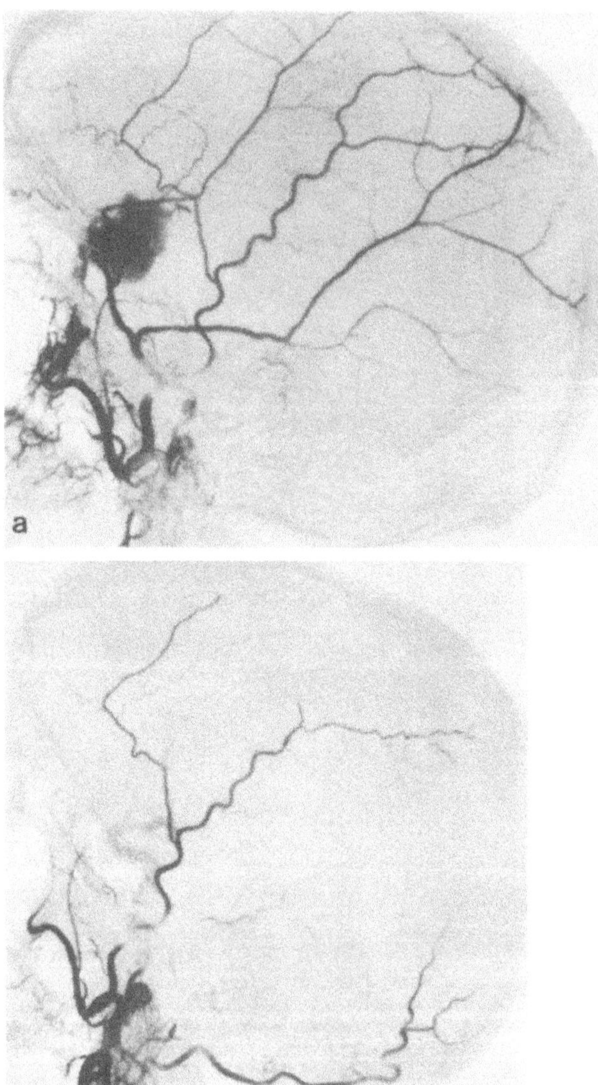

Fig. 3. a External carotid injection showing vascularization by the anterior branch of the middle meningeal artery. b After embolization in the maxillary artery the emboli have selectively occluded the feeding artery and the vascular blush is no longer visible

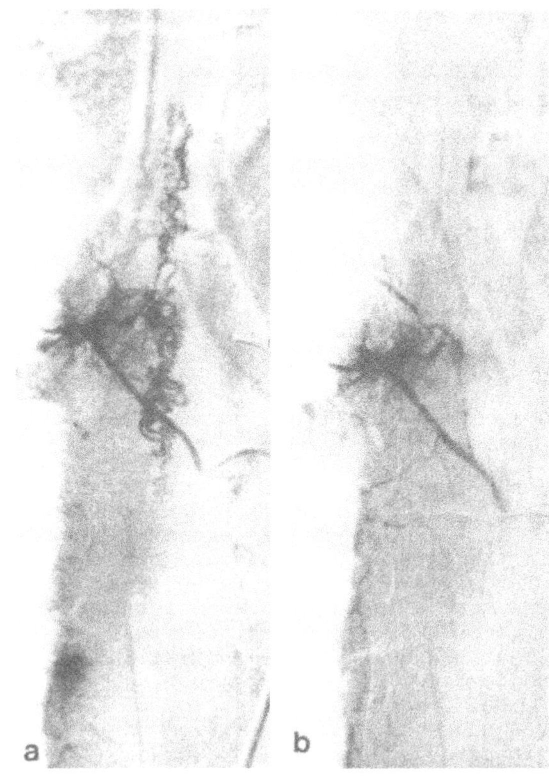

Fig. 4. a Sixth right intercostal artery injection showing an arteriovenous malformation. This was the only feeding artery found. The artery of Adamkiewitz was normal. b Same series after embolisation. The feeding is selectively occluded. The other branches of the intercostal artery remain open

Neurosurgical Aspects of Ectatic Intracranial Vessels

E. B. Bongartz[1], D. Moffie[2], J. P. van Berkel[3], B. Matricali[1], D. A. Batchelor[3], and Z. D. Goedhart[1]

[1] Neurosurgical Department, Municipal Hospital Slotervaart, NL-1066 EC Amsterdam
[2] Neurological Department, Municipal Hospital Slotervaart, NL-1066 EC Amsterdam
[3] Radiological Department, Municipal Hospital Slotervaart, NL-1066 EC Amsterdam

Introduction

Although ectatic intracranial vessels are not so common, they should be included in the differential diagnosis of space-occupying lesions in the cranium. We consider it better to talk about ectatic vessels of fusiform aneurysms rather than megadolicho arteries. This being due to the large discrepancies concerning the length of intracranial arteries (3). These ectatic lesions were described long ago by pathologists and anatomists. Since the introduction and development of angiography, we are now more aware of this disease process. An ectatic vessel can produce a variety of symptom complexes. Until now, only a few neurosurgeons have tried to tackle this problem surgically and with varying amounts of success (1, 13). For most of us it is more a question of trying to alleviate the symptoms.

Patients

1. A 49-year-old hypertensive female patient suddenly developed a right-sided hemiparesis with complete aphasia. There was no loss of consciousness. Slight right sided hyper-reflexia and sensory loss were noted. Pupillary reactions were normal. The cerebrospinal fluid was normal, but electroencephalography showed diffuse abnormalities. Left carotid angiography demonstrated dilatation and elongation of the anterior cerebral artery, both intracranially and in the first branch. The right anterior cerebral artery was also filled and this was also dilated. Right carotid angiography demonstrated a dilated middle cerebral artery with dilatation and elongation in some branches of the Sylvian complex (Figs. 1 and 2). The patient gradually improved, without treatment, and was discharged from hospital two months later.

2. A 71-year-old female with rheumatoid arthritis and hypertension complained of headache and vomiting. Over the preceding months she had become ataxic and had experienced a few "drop" attacks. Neurological examination revealed a right sided Babinski and further minor deficits. CT scan demonstrated a large tumor in the posterior fossa with accompanying hydrocephalus (Figs. 3 and 4). Before insertion of a ventriculo-atrial drain, angiography was performed. This revealed elongation and dilatation of the basilar artery (Figs. 5 and 6). Postoperatively the patient gradually improved, but later died of an aspiration pneumonia caused by dysphagia of central origin.

Advances in Neurosurgery, Vol. 11
Edited by H.-P. Jensen, M. Brock, and M. Klinger

Discussion

The first patient showed a picture of an acute brain infarct. Although this could not be proved by angiography it had to be treated as a circulatory abnormality.

Concerning the second patient, the CT scan revealed a space-occupying lesion producing obstruction of the aqueduct of Sylvius.

It is obvious that these degenerative vessel abnormalities can, in certain circumstances, mimic more common vascular lesions, tumors or other diseases (Table 1). In most cases all one can do is try to alleviate the symptoms or give supportive therapy, as in most cases operative correction or extirpation is impossible.

Concerning the symptoms, obstructive hydrocephalus is present in one case and in the other there is a picture of normal pressure hydrocephalus. This fairly typical symptom complex is thought to be caused by the continual pulsations of the top of the elongated basilar artery against the floor of the third ventricle (6, 7). An infarct-like picture has also been described (10). Also the fusiform aneurysmal dilatation can present as a space-occupying lesion (12, 15). The most common presentation is that of a cranial nerve palsy (2, 5, 11).

Pathological examination of fusiform aneurysms in children has demonstrated extensive defects with degeneration of the tunica muscularis and the elastic lamina. In the context of the basilar artery, defective union of primitive arteries on the ventral side of the neural tube could be a factor (9). However these ectatic vascular abnormalities are seen mostly with arteriosclerosis and hypertension. It has been noted (4, 14) that in a population with a high incidence of arteriosclerosis and hypertension, these vascular abnormalities are relatively uncommon. A congenital or developmental factor is therefore plausible. This so-called "underdevelopment" or "medial necrosis" described by KRAULAND (8) may be the expression of a disease process, first attacking a *locus minoris resistentiae*.

We have come to the conclusion, contrary to that recently published in the literature (5, 11) that these ectatic lesions cannot be differentiated from other lesions by CT alone. This is perhaps best illustrated by our second patient. We now believe that in any case of doubt angiography should be performed in order to decide about possible treatment.

Table 1. Possible symptoms in case of ectatic intracranial vessels

Vertebro-basilar insufficiency / TIA
Space-occupying lesions
Hydrocephalus - obstructive - normal pressure
Cranial nerve palsies

References

1. Ammerman, B.J., Smith, D.R.: Giant fusiform middle cerebral aneurysm: Successful treatment utilizing microvascular bypass. Surg. Neurol. 7, 255-257 (1977)
2. Boeri, R., Passerini, A.: The megadolichobasilar anomaly. J. Neurol. Sci. 1, 475-484 (1964)
3. Busch, W.: Beitrag zur Morphologie und Pathologie der Arteria basilaris (Untersuchungsergebnisse bei 1000 Gehirnen). Arch. Psych. z. d. ges. Neur. 208, 326-344 (1966)
4. Dandy, W.E.: Intracranial arterial aneurysms. New York: Comstock 1947
5. Deeb, Z.L., Jannetta, P.J., Rosenbaum, A.E., Kerber, C.W., Drayer, B.P.: Tortuous vertebrobasilaris arteries causing cranial nerve syndromes: Screening by computed tomography. J. Comput. Assist. Tomogr. 3, 774-778 (1979)
6. Ekbom, K., Greitz, T., Kugelberg, E.: Hydrocephalus due to ectasia of the basilar artery. J. Neurol. Sci. 8, 465-477 (1969)
7. Hirsh, L.F., Gonzalez, C.F.: Fusiform basilar aneurysm simulating carotid transient ischemic attacks. Stroke 10, 598-601 (1979)
8. Krauland, W.: Die Aneurysmen der Schlagadern am Hirn- und Schädelgrund. In: Handbuch der speziellen pathologischen Anatomie und Histologie, Vol. 13, Part 1B (Erkrankungen des zentralen Nervensystems I), Lubarsch, O., Henke, F. (eds.), p. 1511. Berlin, Heidelberg, New York: Springer 1957
9. Matricali, B., van Dulken, H.: Aneurysm of fenestrated basilar artery. Surg. Neurol. 15, 189-191 (1981)
10. Moffie, D.: Fusiform aneurysm of intracranial arteries. Psychiat. Neurol. Neurochir. 71, 85-91 (1968)
11. Moseley, I.F., Holland, I.M.: Ectasia of the basilar artery: The breadth of the clinical spectrum and the diagnostic value of computed tomography. Neuroradiology 18, 83-91 (1979)
12. Probram, H.F.W., Hudson, J.D., Joynt, R.J.: Posterior fossa aneurysms presenting as mass lesions. Radiology 105, 334-340 (1969)
13. Segal, H.D., McLaurin, R.L.: Giant serpentine aneurysm. Report of 2 cases. J. Neurosurg. 46, 115-120 (1977)
14. Taptas, J.N.: Les dilatations et allongements de l'artère carotide interne. Rev. Neurol. 80, 338 (1948)
15. Tulleken, C.A.F.: Giant aneurysms of the posterior fossa presenting as space-occupying lesions. Clin. Neurol. Neurosurg. 79, 161-186 (1976)

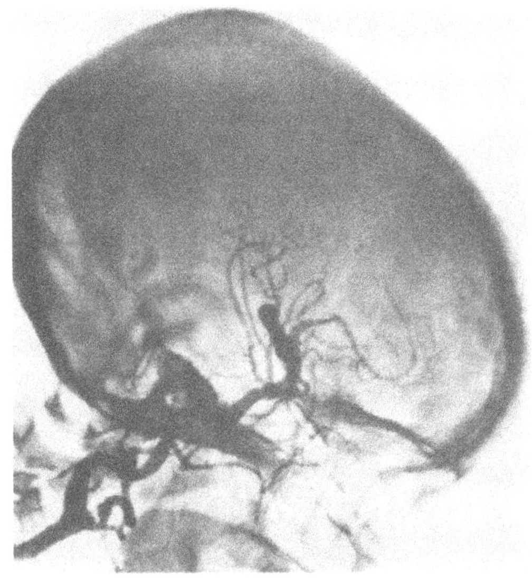

Fig. 1. Left carotid arteriography; the internal carotid artery is too wide and too long. There is dilatation of the anterior and middle cerebral arteries

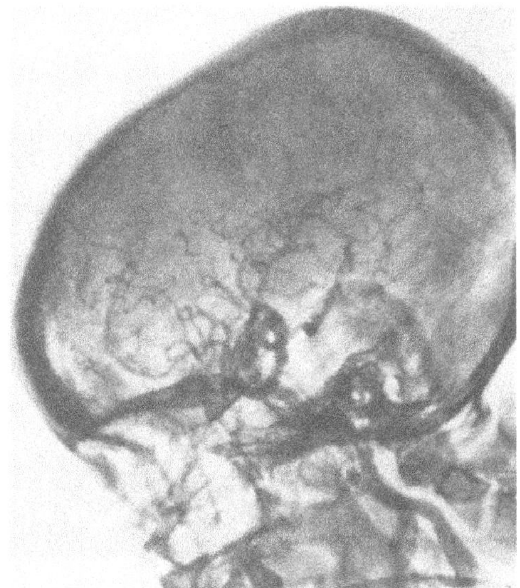

Fig. 2. Right carotid arteriography; the first part of the Sylvian artery forms a loop and is very dilated

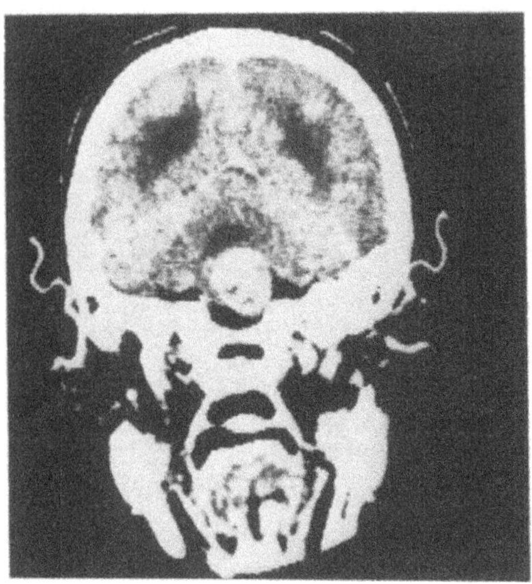

Fig. 3. Postcontrast scan showing a large posterior fossa lesion. There is little evidence to suggest the possibility of an ectatic basilar artery

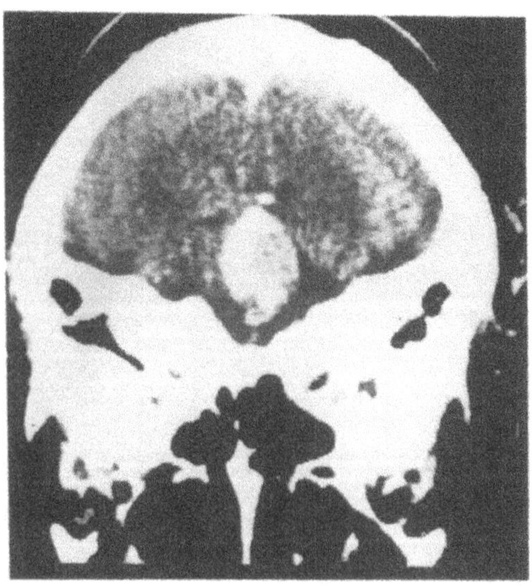

Fig. 4. This postcontrast scan demonstrates the area of increased density adjacent to the clivus. The posterior horns of the ventricles are dilated

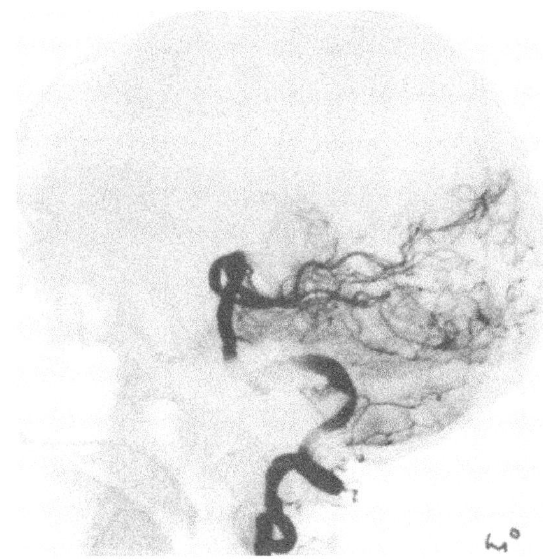

Fig. 5. Left vertebral arteriogram showing elongation and irregular dilatation of the first part of the basilar artery

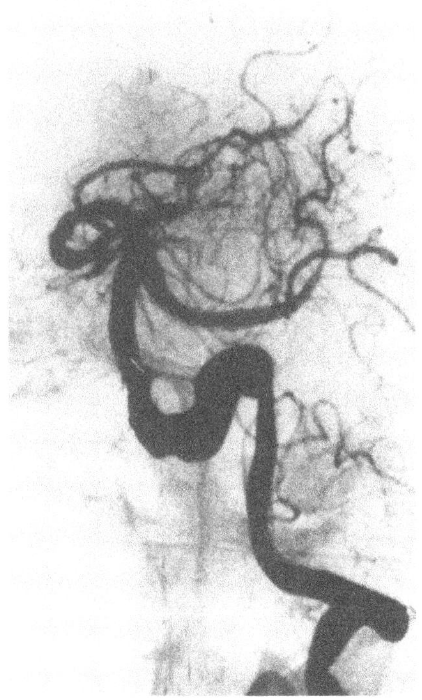

Fig. 6. Oblique projection of vertebral arteriogram again showing the irregular dilatation and elongation

Subject Index

Advances in Neurosurgery

Volume 10

Computerized Tomography - Brain Metabolism - Spinal Injuries

Editors: **W. Driesen, M. Brock, M. Klinger**

1982. 186 figures, 75 tables. XXVIII, 408 pages
Distributions rights for Japan:
Nankodo Co. Ltd. Tokyo
ISBN 3-540-11115-8

Contents: Computerized Tomography. - Brain Metabolism. - Spinal Injuries. Free Topics. - Subject Index.

Volume 9

Brain Abscess and Meningitis Subarachnoid Hemorrhage: Timing Problems

Editors: **W. Schiefer, M. Klinger, M. Brock**

1981. 219 figures, 134 tables. XIX, 519 pages
Distribution rights for Japan:
Nankodo Co. Ltd. Tokyo
ISBN 3-540-10539-5

Contents: Brain Abscess and Meningitis. - Subarachnoid Hemorrhage: Timing Problems. - Free Topics. - Subject Index.

Volume 8

Surgery of Cervical Myelopathy Infantile Hydrocephalus: Long-Term Results

Editors: **W. Grote, M. Brock, H.-E. Clar, M. Klinger, H.-E. Nau**

1980. 178 figures in 215 separate illustrations, 138 tables. XVII, 456 pages
Distributions rights for Japan:
Nankodo Co. Ltd. Tokyo

Contents: Cervical Myelopathy. - Hydrocephalus in Childhood. - Free Topics. - Subject Index.

Springer-Verlag
Berlin
Heidelberg
New York
Tokyo

www.ingramcontent.com/pod-product-compliance
Ingram Content Group UK Ltd.
Pitfield, Milton Keynes, MK11 3LW, UK
UKHW061656190726
13853UKWH00008B/2228
* 9 7 8 3 6 6 2 0 5 5 9 0 8 *